Physical Properties of Polymers

Second Edition

By

James E. Mark
University of Cincinnati

Adi Eisenberg
McGill University

William W. Graessley
Princeton University

Leo Mandelkern
Florida State University

Edward T. Samulski
University of North Carolina
Chapel Hill

Jack L. Koenig
Case Western Reserve University

George D. Wignall
Oak Ridge National Laboratory

ACS Professional Reference Book

American Chemical Society, Washington, DC   1993

Library of Congress Cataloging-in-Publication Data

Physical properties of polymers, second edition / by James E. Mark . . . [et al.]—2nd ed.
    p. cm.
    Includes bibliographical references and indexes.

    ISBN 0–8412–2505–2 (cloth)—ISBN 0–8412–2506–0 (paper).

      1. Polymers. 2. Chemistry, Physical and theoretical.
I. Mark, James E., 1934–   . II. American Chemical Society.
TA455.P58P474 1992
620.1′92—dc20   92–35330
                                                                                    CIP

Copy editing and indexing: A. Maureen Rouhi
Production: Donna Lucas
Acquisition: Robin Giroux and Cheryl Shanks

Printed and bound by United Book Press, Baltimore, MD

The paper used in this publication meets the minimum requirements of American National Standard for Information Sciences—Permanence of Paper for Printed Library Materials, ANSI Z39.48–1984.   ∞

1993 ACS Books
Advisory Board

About the Authors

James E. Mark received his B.S. degree in 1957 in chemistry from Wilkes College and his Ph.D. from the University of Pennsylvania. After serving as a postdoctoral fellow from 1962 to 1964 at Stanford University under Paul J. Flory, he served as assistant professor of chemistry at the Polytechnic Institute of Brooklyn from 1964 to 1967. He then moved to the University of Michigan, where he became a full professor in 1972. In 1977, he assumed the position of professor of chemistry, and in 1987, he was named the first Distinguished Research Professor at the University of Cincinnati.

Dr. Mark's research interests pertain to the physical chemistry of polymers, including configuration-dependent properties, conformational energies of chain molecules, the elasticity of polymer networks, and polymer-coated electrodes. He has lectured extensively in polymer chemistry and has published approximately 400 papers. He is a fellow of both the New York Academy of Sciences and the American Physical Society.

James E. Mark

Adi Eisenberg received his B.S. degree from Worcester Polytechnic Institute and his M.S. and Ph.D. degrees from Princeton University. From 1962 to 1967, he was an assistant professor at the University of California at Los Angeles. From 1967 to 1974, he was an associate professor at McGill University. He was a visiting professor in 1973 and 1974 at the Weizmann Institute in Rehovot, Israel, and at Kyoto University in Japan. Since 1975, he has been a full professor at McGill University, and he is now director of Polymer at McGill, a research center devoted to the study of polymers. He has been a consultant for such companies as the Jet Propulsion Laboratory in California, Owens Illinois in Ohio, and GTE in Massachusetts. He has worked on more than 200 articles and patents in the polymer field and has written sixbooks on ion-containing polymers. He is a member of the editorial advisory boards of *Macromolecular Reviews; Journal of Polymer Science, Polymer Physics Edition;* and *Applied Physics Communications* and is a member of the advisory committee for the Institute for Amorphous Studies in Michigan.

Adi Eisenberg

William W. Graessley received B.S. degrees in both chemistry and chemical engineering from the University of Michigan, stayed on there for graduate work, and received his Ph.D. in 1960. After four years with Air Reduction Company, he joined the Chemical Engineering and Materials Science departments at Northwestern University. In 1982 he returned to industry as a senior scientific adviser at Exxon Corporate Laboratories and moved in 1987 to his present position, professor of chemical engineering at Princeton University. He has published extensively on radiation cross-linking of polymers; polymerization reactor engineering; molecular aspects of polymer rheology; rubber network elasticity; and, most recently, the thermodynamics of polymer blends. In 1979–80 he was a senior visiting fellow at Cambridge University. His honors and awards include an NSF Predoctoral Fellowship, the Bingham Medal (Society of Rheology), the Whitby Lectureship (University of Akron), the High Polymer Physics Prize (American Physical Society), and membership in the National Academy of Engineering.

William W. Graessley

Leo Mandelkern received his undergraduate degree from Cornell University in 1942. After serving with the armed forces, he returned to Cornell and received his Ph.D. in 1949. He remained at Cornell in a postdoctoral capacity until 1952 and then joined the National Bureau of Standards, where he was a member of the staff from 1952 to 1962. From 1962 to the present, he has been a professor of chemistry and biophysics at The Florida State University. In 1984, Florida State recognized him with its highest faculty honor, the Robert O. Lawton Distinguished Professor Award. Among other awards he has received are the Arthur S. Fleming Award (1958), the American Chemical Society (ACS) Award in Polymer Chemistry (1975), the ACS Award in Applied Polymer Science (1989), the Florida Award of the ACS (1984), the George Stafford Whitby Award (1988) and the Charles Goodyear Medal (1993) from the Rubber Division of the ACS, and the Mettler Award of the North American Thermal Analysis Society (1984). The Society of Polymer Science, Japan, has given him the award for Distinguished Service in Advancement of PolyScience (1993). He has been, or is, a member of the editorial boards of the *Journal of the American Chemical Society*, *Journal of Polymer Science*, *Macromolecules*, *Journal of Mechanochemistry and Cell Motility*, and *ChemTracts*.

Leo Mandelkern

Edward T. Samulski

Edward T. Samulski attended Clemson University and Princeton University prior to two years as an NIH Postdoctoral Fellow at the University of Groningen and the University of Texas, Austin. In 1972, he joined the faculty at the University of Connecticut. In 1988 he moved to the University of North Carolina at Chapel Hill as professor of chemistry. During the same period, he held visiting professor appointments at the University of Paris, the Weizmann Institute of Science, and the IBM Research Laboratories in San Jose, CA, and in 1985–86, he was a Science & Engineering Research Council senior visiting fellow at the Cavendish Laboratory.

Dr. Samulski is a founding editor of the journal *Liquid Crystals*, and is a fellow of the American Physical Society and the American Association for the Advancement of Science. His research interests center on the physical chemistry of macromolecules; molecular structure and dynamics in low-molar-mass and polymer liquid crystals; new high-performance polymers; and novel, oriented, macromolecular assemblies for nonlinear optics.

Jack L. Koenig

Jack L. Koenig received his Ph.D. in 1960 from the University of Nebraska after doing his research in theoretical spectroscopy. In 1973 he joined the National Science Foundation. He is presently the J. Donnell professor of macromolecular science and physical chemistry at Case Western Reserve University. He is active in spectroscopic research and is the director of the molecular spectroscopy laboratory. His interests include Raman spectroscopy, Fourier transform infrared spectroscopy, solid-state NMR spectrometry, and NMR imaging. He is well-known for his basic work in spectroscopic characterization of polymeric materials and has more than 400 publications to his credit, including the ACS Professional Reference Book *Spectroscopy of Polymers*.

George D. Wignall received his Ph.D. in physics from Sheffield University (England) in 1966. After postdoctoral fellowships at the Atomic Energy Research Establishment (Harwell, England) and the California Institute of Technology, he joined Imperial Chemical Industries from 1969 to 1979, where he applied X-ray and neutron scattering techniques to the study of polymer structure. During this time, he initiated some of the first small-angle neutron scattering (SANS) experiments on deuterium-labeled polymers, which gave for the first time direct information on chain configurations in the bulk state. In 1979 he joined the Oak Ridge National Laboratory and helped construct the 30-meter SANS facility, which was one of the first instruments available to the U.S. scientific community. He has collaborated with many visiting scientists in studies of polymer structure, thermodynamics, and phase behavior and has more than 120 publications to his credit. He has lectured widely on these topics and has organized several symposia on small-angle scattering from polymers. In 1988 he received the Martin Marietta Significant Event Award for his contribution to the elucidation of isotope effects in polymers. He is a member of the American Chemical Society and a fellow of the American Physical Society.

George D. Wignall

Contents

Preface

When we noticed that the first edition of this book (published in 1984) was being used either as a supplementary text or as the sole textbook in introductory polymer courses, we decided it was time to bring out an expanded second edition. All of the chapters contain general introductory material and comprehensive literature citations designed to give newcomers to the field an appreciation of the subject and how it fits into the general context of polymer science. For pedagogical purposes, the contents have been subdivided into two parts, "Physical States of Polymers" and "Some Characterization Techniques". A new chapter has been added to each part: "The Mesomorphic State" (Samulski) covers the rapidly developing subject of liquid-crystalline polymers; "Scattering Techniques" (Wignall) emphasizes the potential of small-angle neutron scattering in contemporary characterization of bulk polymers. The original five chapters: "The Rubber Elastic State" (Mark), "The Glassy State and the Glass Transition" (Eisenberg), "Viscoelasticity and Flow in Polymer Melts and Concentrated Solutions" (Graessley), "The Crystalline State" (Mandelkern), and "Molecular Spectroscopy" (Koenig), have been revised and updated. This expanded edition should provide ample core material for a one-term survey course at the graduate or advanced undergraduate level. Although the chapters have been arranged in a sequence that may be readily adapted to the classroom, each chapter is self-contained and may be used as an introductory source for these seven topics.

JAMES E. MARK
University of Cincinnati
Cincinnati, OH 45221–0172

Physical States
of Polymers

Chapter 1

The Rubber Elastic State

James E. Mark

Department of Chemistry and the Polymer Research Center, University of Cincinnati, Cincinnati, OH 45221-0172

Basic Concepts

The most useful way to begin an article on rubberlike elasticity is to define it and then to discuss what types of materials can exhibit this very unusual behavior. Accordingly, rubber elasticity may be defined operationally as very large deformability with essentially complete recoverability. For a material to exhibit this type of elasticity, three molecular requirements must be met: (1) the material must consist of polymeric chains, (2) the chains must have a high degree of flexibility, and (3) the chains must be joined into a network structure (*1–3*).

The first requirement arises from the fact that the molecules in a rubber or elastomeric material must be able to alter dramatically their arrangements and extensions in space in response to an imposed stress, and only a long-chain molecule has the required very large number of spatial arrangements of very different extensions. This versatility is illustrated in Figure 1 (*3*), which depicts a two-dimensional projection of a random spatial arrangement of a relatively short polyethylene chain in the amorphous state. The spatial configuration shown was computer-generated, in as realistic a manner as possible. The correct bond lengths and bond angles were used, as was the known preference for *trans* rotational states about the skeletal bonds in any *n*-alkane molecule. A final feature taken into account is the fact that rotational states are interdependent; what one rotational skeletal bond does, depends on what the adjoining skeletal bonds are doing (*4*). One important feature of this typical configuration is the relatively high spatial extension of some parts of the chain. This feature is due to the preference for the *trans* rotational states, already mentioned, which are essentially planar zigzag and thus of high extension. The second important feature is the fact that, despite these

2505–2/93/0003$15.25/1

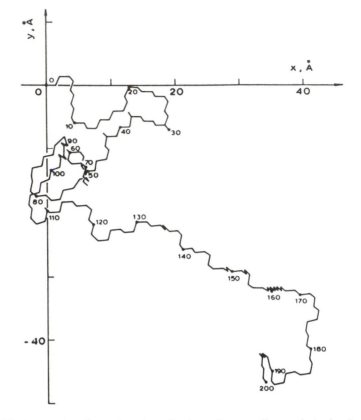

FIGURE 1. A two-dimensional projection of an *n*-alkane chain having 200 skeletal bonds (*3*). The end-to-end vector starts at the origin of the coordinate system and ends at carbon atom number 200.

preferences, many sections of the chain are quite compact. Thus, the overall chain extension (as measured by the end-to-end separation) is quite small. For even such a short chain, the extension could be increased approximately fourfold by simple rotations about skeletal bonds, without any need for distortions of bond angles or increases in bond lengths.

The second requirement for rubberlike elasticity specifies that the different spatial arrangements be *accessible*; that is, changes in these arrangements should not be hindered by constraints that might result from inherent chain rigidity, extensive chain crystallization, or the very highly viscous nature of the glassy state (*1, 2, 5*).

The third requirement allows elastomeric recoverability. A network structure is obtained by joining together or cross-linking pairs of segments, approximately one out of a hundred; the cross-linked segments prevent stretched polymer chains from irreversibly sliding by one another. In a network structure (Figure 2; *5*), the cross-links may be either chemical bonds (as would occur in sulfur-vulcanized natural rubber) or physical aggregates (for example, the small crystallites in a partially crystalline polymer or the glassy domains in a multiphase block copolymer; *3*). Additional information on the cross-linking of chains is given later in *Preparation of Networks*.

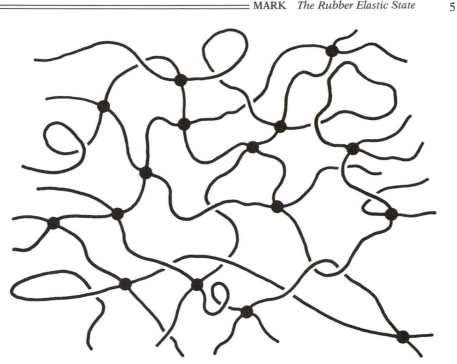

FIGURE 2. Schematic sketch of a typical elastomeric network. (Reproduced with permission from reference 5. Copyright 1988 John Wiley and Sons, Inc.)

Origin of Elastic Retractive Force

The molecular origin of the elastic force (f) exhibited by a deformed elastomeric network can be elucidated through thermoelastic experiments, which examine the temperature dependence of either the force at constant length (L) or the length at constant force (*1, 3*). Consider a thin metal strip stretched with a weight W to a point short of that giving permanent deformation, as shown in Figure 3 (*3*). An increase in temperature (at constant force) would increase the length of the stretched strip in what would be considered the "usual" behavior. Exactly the opposite, a *shrinkage*, is observed in the case of a stretched elastomer! For purpose of comparison, the result observed for a gas at constant pressure is included in the figure. Raising the gas temperature would, of course, cause an increase in volume (V), as required by the ideal gas law.

The explanation for these observations is given in Figure 4 (*3*). The primary effect of stretching the metal is the increase in energy (ΔE) caused by changing the distance (d) of separation between the metal atoms. The stretched strip retracts to its original dimension upon removal of the force, because this retraction is associated with a decrease in energy. Similarly, heating the strip at constant force causes the usual expansion arising from increased oscillations about the minimum in the asymmetric potential energy curve. For the elastomer, however, the major effect of the deformation is the stretching out of the network chains, which substantially reduces their entropy (*1–3*). Thus, the retractive force arises primarily from the tendency of the system to increase its entropy toward

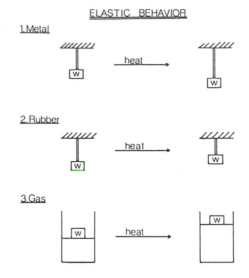

FIGURE 3. Results of thermoelastic experiments carried out on a typical metal, rubber, and gas (3).

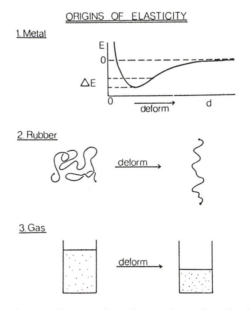

FIGURE 4. Sketches explaining the observations described in Figure 3 in terms of the molecular origin of the elastic force or pressure (3).

the (maximum) value it had in the undeformed state. An increase in temperature increases the chaotic molecular motions of the chains and thus increases the tendency toward this more-random state. As a result, the length decreases at constant force, or the force increases at constant length. This behavior is strikingly similar to that of a compressed gas, in which the extent of deformation is given by the reciprocal volume ($1/V$). The pressure of the gas is also largely entropically derived, with an increase in deformation (i.e., increase in $1/V$) also corresponding to a decrease in entropy. Heating the gas increases the driving force toward the state of maximum entropy (infinite volume or zero deformation). Thus, increasing the temperature increases the volume at constant pressure or increases the pressure at constant volume.

This surprising analogy between a gas and an elastomer (which is a condensed phase) carries over into the expressions for the work of deformation (dw). For a gas, dw is, of course, $-pdV$, in which p is pressure. For an elastomer, however, this pressure–volume term is generally essentially negligible. For example, network elongation is known to take place at very nearly constant volume (*1, 3*). The corresponding work term (dw) now becomes $+fdL$; the difference in sign is due to the fact that a positive w corresponds to a decrease in the volume of a gas but to an increase in the length of an elastomer. Adiabatically stretching an elastomer increases its temperature in the same way that adiabatically compressing a gas (for example, in a diesel engine) will increase its temperature. Similarly, an elastomer cools on adiabatic retraction, just as a compressed gas cools in the corresponding expansion. The basic point here is the fact that the retractive force of an elastomer and the pressure of a gas are both primarily entropically derived, and as a result, the thermodynamic and molecular descriptions of these otherwise dissimilar systems are very closely related.

Some Historical High Points

Experimental Approaches

The simplest of the thermoelastic experiments described earlier were first carried out many years ago, by J. Gough, back in 1805 (*1, 2, 5, 6*). The discovery of vulcanization (i.e., curing of rubber into network structures) by C. Goodyear and N. Hayward in 1839 was important in this regard, because it permitted the preparation of samples that could be investigated in this regard with much greater reliability. Such more-quantitative experiments were carried out by J. P. Joule, in 1859, in fact, only a few years after entropy was introduced as a concept in thermodynamics in general! Another important experimental finding relevant to the development of these molecular ideas was the fact that deformations of rubberlike materials, other than swelling, occurred essentially at constant volume as long as crystallization was not induced (*1*). (In this sense, the deformation of an elastomer differs from that of a gas.)

Theoretical Approaches

A molecular interpretation of the fact that rubberlike elasticity is primarily entropic in origin had to await H. Staudinger's demonstration in the 1920s that polymers were covalently bonded molecules and not some type of association complex best studied by the colloid chemists (*1*). In 1932, W. Kuhn used this observed constancy in volume to point out that the changes in entropy must therefore involve changes in orientation or configuration of the network chains. These basic qualitative ideas are shown in Figure 5 (5), where the arrows represent some typical end-to-end vectors of the network chains.

Later in the 1930s, W. Kuhn, E. Guth, and H. Mark first began to develop quantitative theories based on the idea that the network chains undergo configurational changes, by skeletal bond rotations, in response to an imposed stress (1, 2). More-rigorous theories began with the development of the "phantom network" theory by H. M. James and E. Guth in 1941 and the "affine model" theory by F. T. Wall and by P. J. Flory and J. Rehner, Jr., in 1942 and 1943.

These theories, and some of their modern-day refinements, are described in the following sections.

Basic Postulates

Several important postulates are used in the development of the molecular theories of rubberlike elasticity (5).

The first postulate specifies that, although intermolecular interactions are certainly present in elastomeric materials, they are independent of chain configuration and are therefore also independent of deformation. In effect, the assumption is that rubberlike elasticity is entirely of *intramolecular* origin.

The second postulate states that the free energy of the network is separable into two parts, a liquidlike part and an elastic part, with the liquidlike part not depending on deformation. This postulate permits elasticity to be treated independently of other properties characteristic of solids and liquids in general.

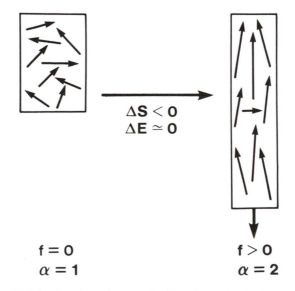

FIGURE 5. Sketch showing changes in length and orientation of network end-to-end vectors upon elongation of a network. Vectors lying approximately perpendicular to the stretching direction (i.e., horizontally) get *compressed*. (Reproduced with permission from reference 5. Copyright 1988 John Wiley and Sons, Inc.)

Some theories assume further that the deformation is affine, that is, that the network chains move in a simple linear fashion with the macroscopic deformation. Most, but not all, theories assume additionally that the network chains have end-to-end distances that have a Gaussian distribution. Non-Gaussian theories, however, have been developed for network chains that are unusually short or stretched close to the limits of their extensibility (*2*).

Some Rubberlike Materials

Because high flexibility and mobility are required for rubberlike elasticity, elastomers generally do not contain stiffening groups such as ring structures and bulky side chains (*2*, *5*). These characteristics are evidenced by the low glass transition temperatures (T_g) exhibited by these materials. (The structural features of a polymeric chain conducive to low values of T_g are discussed by A. Eisenberg in Chapter 2.) These polymers also tend to have low melting points, if any, but some do undergo crystallization upon sufficiently large deformations. Examples of typical elastomers include natural rubber and butyl rubber (which do undergo strain-induced crystallization) and poly(dimethylsiloxane), poly(ethyl acrylate), styrene–butadiene copolymer, and ethylene–propylene copolymer (which usually do not). The crystallization of polymers in general is discussed by L. Mandelkern in Chapter 4.

Some polymers are not elastomeric under normal conditions but can be made so by raising the temperature or adding a diluent ("plasticizer"). Polyethylene is in this category because of its high degree of crystallinity. Polystyrene, poly(vinyl chloride), and the biopolymer elastin are also of this type, but because of their relatively high glass transition temperatures (*5*).

A final class of polymers is inherently nonelastomeric. Examples are polymeric sulfur because its chains are too unstable, poly(*p*-phenylene) because its chains are too rigid, and thermosetting resins because their chains are too short (*5*).

Preparation of Networks

One of the simplest ways to introduce the cross-links required for rubberlike elasticity is to carry out a copolymerization in which one of the comonomers has a functionality (ϕ) of three or higher (*5*, *7*). This method, however, has been used primarily to prepare materials so heavily cross-linked that they are in the category of thermosets rather than elastomeric networks (*8*).

A sufficiently stable network structure also can be obtained by physical aggregation of some of the chain segments onto filler particles, by formation of microcrystallites, by condensation of ionic side chains onto metal ions, by chelation of ligand side chains to metal ions, and by microphase separation of glassy or crystalline end blocks in a triblock copolymer (*5*). The main advantage of these materials is the fact that the cross-links are generally only temporary, which means that such materials frequently exhibit reprocessibility. This temporary nature of the cross-linking can, of course, be a disadvantage also, because the materials are rubberlike only as long as the aggregates are not broken up by high temperatures, presence of diluents or plasticizers, etc.

Structure of Networks

Before commenting further on elasticity experiments, however, it is useful to digress briefly to establish the relationship between the three most widely used measures of the cross-link density. The first measure of cross-link density involves the number (or number of moles) of network chains (v), where a network chain is defined as one that extends from one cross-link to another. This quantity is usually expressed as the chain density, v/V, where V is the volume of the (unswollen) network (1). The second measure, directly proportional to v/V, is the density of cross-links, μ/V. The relationship between the number of cross-links (μ) and the number of chains (v) must obviously depend on the cross-link functionality (ϕ). The two most important types of networks in this regard are the tetrafunctional ($\phi = 4$) networks, almost invariably obtained upon joining two segments from different chains, and the trifunctional networks, obtained for example when forming a polyurethane network by end linking hydroxyl-terminated chains with a triisocyanate. The relationship between μ and v is illustrated in Figure 6 (9), which shows sketches of simple, perfect tetrafunctional (Figure 6a) and trifunctional (Figure 6b) network structures. They are simple in the sense of having small enough values of μ and v for them to be easily counted, and perfect in the sense of not having any dangling ends or elastically ineffective loops (chains with both ends attached to the same cross-link). The tetrafunctional network yields a μ/v of 4/8 or 1/2, and the trifunctional network yields a μ/v of 4/6 or

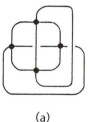

(a)

(b)

FIGURE 6.   Sketches of some simple, perfect networks having (a) tetrafunctional and (b) trifunctional cross-links (indicated by the heavy dots). (Reproduced with permission from reference 9. Copyright 1982 *Rubber Chemistry and Technology*.)

2/3. In general, for a perfect φ-functional network, the number of cross-link attachment points (φμ) equals the number of chain ends (2ν): μ = (2/φ)ν (1).

Another (inverse) measure of cross-link density is the molecular weight between cross-links M_c. M_c is simply the density (ρ; g · cm⁻³) divided by the number of moles of chains (ν/V; mol · cm⁻³): $M_c = \rho/(\nu/V)$ (1). A related structural quantity that is important in the more-modern theories is the cycle rank (ξ), which is the number of chains that have to be cut to reduce the network to a tree with no closed cycles at all: ξ = (1 − 2/φ)ν (5).

Theory

Phenomenological Approach

The phenomenological approach to rubberlike elasticity is based on continuum mechanics and symmetry arguments rather than on molecular concepts (2, 7, 10). It attempts to fit stress–strain data with a minimum number of parameters, which are then used to predict other mechanical properties of the same material. Its best known result is the Mooney–Rivlin equation, which states that the modulus of an elastomer should vary linearly with reciprocal elongation (2).

The Affine Model

This theory, and any other molecular theory of rubberlike elasticity, is based on a chain distribution function, which gives the probability of any end-to-end separation, r. The characteristics of this type of distribution function are given in Figure 7 (1). What is required is a function that answers the question, "If a chain starts at the origin of the coordinate system shown, what is the probability that the other end will be in an infinitesimal volume $dV = dx\,dy\,dz$ around some specified values of x, y, and z?"

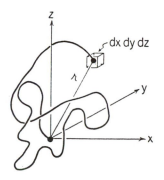

FIGURE 7. A spatial configuration of a polymer chain, with some quantities used in the distribution function for the end-to-end distance, r. (Reproduced with permission from reference 1. Copyright 1953 Cornell University Press.)

The simplest molecular theories of rubberlike elasticity are based on the Gaussian distribution function for the end-to-end separations of the network chains (i.e., chain sequences extending from one cross-link to another; 1–3):

$$w(r) = \left(\frac{3}{2\pi\langle r^2\rangle_0}\right)^{3/2} \exp\left(-\frac{3r^2}{2\langle r^2\rangle_0}\right) \tag{1}$$

In this equation, $\langle r^2\rangle_0$ represents the dimensions of the free chains as unperturbed by excluded-volume effects (1). These excluded-volume interactions arise from the spatial requirements of the atoms making up the polymeric chain and are thus similar to those occurring in gases. They are more complex for networks, however, in that they have an intramolecular as well as an intermolecular origin. If present, excluded-volume interactions increase the dimensions of a polymer chain in the same way that they can increase the pressure of a gas. The Gaussian distribution function in which $\langle r^2\rangle_0$ resides is applied to the network chains in both the stretched and unstretched states. The Helmholtz free energy (F) of such a chain is given by the simple variant of the Boltzmann relationship shown in the first part of the following equation:

$$F(T) = -kT\ln w(r) = C(T) + \left(\frac{3kT}{2\langle r^2\rangle_0}\right)r^2 \tag{2}$$

In equation 2, $C(T)$ is a constant at a specified absolute temperature T, and k is the Boltzmann constant.

Consider now the process of stretching a network chain from its random undeformed state with r components of x, y, and z, to the deformed state with r components of $\alpha_x x$, $\alpha_y y$, and $\alpha_z z$ (where the α_x, α_y, and α_z are molecular deformation ratios). The free energy change (ΔF) for a single network chain is then simply given by equation 3:

$$\Delta F = \left(\frac{3kT}{2\langle r^2\rangle_0}\right)\left[\left(\alpha_x^2 x^2 + \alpha_y^2 y^2 + \alpha_z^2 z^2\right) - \left(x^2 + y^2 + z^2\right)\right] \tag{3}$$

Because the elastic response is essentially entirely intramolecular (1–3), the free energy change for v network chains is just v times the result from equation 3:

$$\Delta F = \left(\frac{3vkT}{2\langle r^2\rangle_0}\right)\left[\left(\alpha_x^2 - 1\right)\langle x^2\rangle + \left(\alpha_y^2 - 1\right)\langle y^2\rangle + \left(\alpha_z^2 - 1\right)\langle z^2\rangle\right] \tag{4}$$

In equation 4, the brackets around x^2, y^2, and z^2 specify their values when averaged over the v chains. In this model, the strain-induced displacements of the cross-links or junction points are assumed to be *affine* (i.e., linear) in the macroscopic strain. In this case, the deformation ratios (α_x, α_y, and α_z) are obtained directly from the dimensions of the sample in the strained state (L_x, L_y, and L_z) and in the initial, unstrained state (L_{xi}, L_{yi}, and L_{zi}):

$$\alpha_x = L_x/L_{xi}, \alpha_y = L_y/L_{yi}, \alpha_z = L_z/L_{zi} \tag{5}$$

The dimensions of the cross-linked chains in the undeformed state (indicated by subscript i) are given by the Pythagorean theorem:

$$\langle r^2 \rangle_i = \langle x^2 \rangle + \langle y^2 \rangle + \langle z^2 \rangle \tag{6}$$

Also, the isotropy of the undeformed state requires that the average values of x^2, y^2, and z^2, be the same:

$$\langle x^2 \rangle = \langle y^2 \rangle = \langle z^2 \rangle \tag{7}$$

Thus, the chain dimensions are given by equation 8:

$$\langle r^2 \rangle_i = 3\langle x^2 \rangle = 3\langle y^2 \rangle = 3\langle z^2 \rangle \tag{8}$$

and the elastic free energy of deformation is given by equation 9:

$$\Delta F = \left(\frac{vkT}{2} \right) \left(\frac{\langle r^2 \rangle_i}{\langle r^2 \rangle_0} \right) (\alpha_x^2 + \alpha_y^2 + \alpha_z^2 - 3) \tag{9}$$

In the simplest theories (*1–3*), $\langle r^2 \rangle_i$ is assumed to be identical to $\langle r^2 \rangle_0$; that is, it is assumed that the cross-links do not significantly change the chain dimensions from their unperturbed values. Equation 9 may then be approximated by equation 10:

$$\Delta F \cong \left(\frac{vkT}{2} \right) (\alpha_x^2 + \alpha_y^2 + \alpha_z^2 - 3) \tag{10}$$

Equations 9 and 10 are basic to the molecular theories of rubberlike elasticity and can be used to obtain the elastic equations of state for any type of deformation (*1–3*), that is, the equations interrelating stress, strain, temperature, and number or number density of network chains. Their application is best illustrated for the case of elongation, which is the type of deformation used in majority of experimental studies (*1–3*). This deformation occurs at essentially constant volume, and thus, a network stretched by the amount $\alpha_x = \alpha > 1$ would have its perpendicular dimensions compressed by the following amounts:

$$\alpha_y = \alpha_z = \alpha^{-1/2} < 1 \tag{11}$$

Accordingly, for elongation, the first part of the equation is as follows:

$$\Delta F = \left(\frac{vkT}{2} \right) (\alpha^2 + 2\alpha^{-1} - 3) = fdL \tag{12}$$

Because the Helmholtz free energy is the "work function" and the work of deformation is fdL (where $L = \alpha L_i$), as shown in the second equality, the elastic force may be obtained by differentiating equation 12 to obtain equation 13:

$$f = (\partial \Delta F / \partial L)_{T,V} = \left(\frac{\nu kT}{L_i}\right)(\alpha - \alpha^{-2}) \tag{13}$$

In equation 13, the expression $(\partial \Delta F / \partial L)_{T,V}$ indicates that differentiation is at constant T and V.

The nominal stress $f^* \equiv f/A^*$, where A^* is the undeformed cross-sectional area) is then given by equation 14:

$$f^* \equiv f/A^* = (\nu kT/V)(\alpha - \alpha^{-2}) \tag{14}$$

In equation 14, ν/V is the density of network chains (the number per unit volume V, which is equal to $L_i A^*$).

The elastic equation of state in the form given in equation 14 is strikingly similar to the molecular form of the equation of state for an ideal gas:

$$p = NkT(1/V) \tag{15}$$

In equation 15, the stress was replaced by the pressure, and the number density of network chains was replaced by the number of gas molecules (N). Similarly, because the stress was assumed to be entirely entropic in origin, f^* is predicted to be directly proportional to T at constant α (and V), as is predicted for the pressure of the ideal gas at constant $1/V$. The strain function $(\alpha - \alpha^{-2})$ is somewhat more complicated than $1/V$, because the near incompressibility of the elastomeric network superposes compressive effects (given by the term $-\alpha^{-2}$) on the simple elongation (α) being applied to the system. This situation is illustrated by the approximately horizontal end-to-end vector shown schematically in Figure 5.

Also frequently used in elasticity studies is the "reduced stress" or modulus, $[f^*]$, defined in the first part of equation 16:

$$[f^*] \equiv f^* v_2^{1/3}/(\alpha - \alpha^{-2}) = \nu kT/V \tag{16}$$

The reduced stress, $[f^*]$, includes a factor that makes it applicable to networks that have been swelled with a low-molecular-weight diluent, which is frequently done to facilitate the approach to elastic equilibrium. This factor, which is the cube root of the volume fraction of polymer in the network ($v_2^{1/3}$), takes into account the fact that a swollen network has fewer chains passing through unit cross-sectional area and that the chains are stretched because of the presence of the diluent (1).

The Phantom Model

In this model, the chains are viewed as having zero cross-sectional area and as being able to pass through one another as "phantoms" (2, 5, 11, 12). The cross-links undergo considerable fluctuations in space, and in the deformed state, these fluctuations occur in an asymmetric manner to reduce the strain below that imposed macroscopically. The deformation thus viewed is very non-affine. Because of this reduction in strain sensed by the network chains, the modulus predicted by equation 16 would be diminished by the factor $A_\phi < 1$:

$$[f^*] = A_\phi \nu kT/V \tag{17}$$

In the limit of the very non-affine deformation that would be exhibited by a phantom network, A_ϕ is given by equation 18:

$$A_\phi = 1 - 2/\phi \tag{18}$$

For a trifunctional network ($\phi = 3$), A_ϕ is $1/3$, and for a tetrafunctional one, A_ϕ is $1/2$; A_ϕ approaches unity in the limit of very high cross-link functionality (5).

The Constrained-Junction Model

Recent work suggests that the elongation of a network generally falls between the affine and phantom limits (*13–16*). At low deformations, chain–junction entangling suppresses the fluctuations of the junctions, and the deformation is relatively close to the affine limit. Figure 8 shows schematically some of the results of the "constrained-junction" theory based on this qualitative idea (*14–16*). For the two limits, the affine deformation and the non-affine deformation in the phantom network limit, the reduced stress should be independent of α. Because of junction fluctuations, the value for the phantom limit should be reduced, however, by the factor $(1 - 2/\phi)$ for a ϕ-functional network, as illustrated in Figure 8 for the case $\phi = 4$. The experimentally observed decreases in reduced

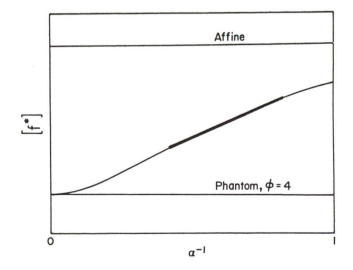

FIGURE 8. A schematic diagram qualitatively showing theoretical predictions (*14–16*) for the reduced stress as a function of a reciprocal elongation, α^{-1}.

stress with increasing α are shown as the heavier portion of the theoretical curve. An increase in elongation disentangles the chains somewhat from the junctions, and the fluctuations increase in magnitude, most markedly in the direction of the deformation. This change causes the chains to sense a deformation smaller than that imposed macroscopically, and consequently, the deformation becomes more non-affine. The modulus thus decreases until phantomlike behavior is reached in the limit of very high elongations. The extent to which the fluctuations are constrained is described by a constraint parameter κ, which is essentially infinite in the affine limit and zero in the phantom limit. One great success of this type of theory is the explanation (14–16) it provides for the previously puzzling decrease in modulus almost always observed with increasing elongation (for low and moderate elongations) and represented by the Mooney–Rivlin equation (2). The increases in modulus frequently observed at very high deformations are discussed in the section *Networks at Very High Deformations*.

The Constrained-Chain Model

This refinement of the constrained-junction model is based on reexamination of the constraint problem and evaluation of some neutron-scattering estimates of actual junction fluctuations (17, 18). It was concluded that the suppression of fluctuations was overestimated in the theory, presumably because the entire effect of the interchain interactions was arbitrarily placed on the junctions. The theory was therefore revised to make it more realistic by spreading the effects of the constraints along the entire network chain contours (19). This revision also improved the agreement between theory and experiment.

Some Other General Models

One of the most interesting alternative approaches is the "slip-link" model, which incorporates the effects of entanglements (20, 21) along the network chains directly in the elastic free energy (22). Still other approaches are the "tube" model (23) and the van der Waals model (24).

Rotational Isomeric State Representation of the Network Chains

An approach (25–28) that takes direct account of the structural differences between chemically different elastomers is based on the rotational isomeric state representation of the chains (4). In this approach, all the structural features that distinguish one type of elastomeric chain from another are taken into account, as was done in the generation of the spatial configuration shown in Figure 1. The required bond lengths, skeletal bond angles, locations of rotational states, and rotational state energies are obtained from data on small molecules and then used in a Monte Carlo method to generate a large number of spatial configurations that are representative of the specified chain structure at the specified chain length and temperature. The values of the end-to-end separation (r) for these various configurations are then calculated and, in effect, put into boxes corresponding to different ranges of r. Representing the number of chains in a given range by the height of a bar and displaying these bars as a function of r then give the usual type of bar graph. A smooth curve put through the levels of this bar graph then represents the distribution of r, which can be used to replace the approximate Gaussian distribution. Such distributions are particularly useful for chains that are known to be non-Gaussian (for example, because of their shortness or because of their being stretched close to the limits of their extensibility).

Going from the usual structureless molecular theories of rubberlike elasticity to ones taking into account the structural features that distinguish one type of polymer from another (7) parallels going from the theory of ideal gases to the van der Waals theory of nonideal gases. The advantage of incorporating structural features in both cases is a more realistic portrayal of the system, but this advantage is gained at the loss of universality (in that additional information specific to the chosen system is required). Useful theories for liquid-crystalline polymers (29) may be particularly important in this regard.

Some of the elastic equations of state resulting from these various approaches are discussed further in later sections.

Some Experimental Details

Mechanical Properties

Most studies of mechanical properties of elastomers have been carried out in elongation because of the simplicity of this type of deformation (5). The apparatus typically used to measure the force required to give a specified elongation in a rubberlike material is, indeed, very simple (Figure 9; 3). The elastomeric strip is mounted between two clamps, with the lower clamp fixed and the upper clamp attached to a movable force gauge. A recorder is used to monitor the output of the gauge as a function of time to obtain equilibrium values of the force suitable for comparisons with theory. The sample is generally protected with an inert atmosphere, such as nitrogen, to prevent degradation, particularly for measurements carried out at elevated temperatures. Both the sample cell and surrounding constant-temperature bath are made of glass to permit use of a cathetometer or a traveling microscope to obtain values of the strain, by measurements of the distance between two lines marked on the central portion of the test sample.

Some typical studies with other types of deformation, namely, biaxial extension or compression, shear, and torsion, are described in the section *Other Types of Deformation*.

Swelling

This nonmechanical property is also much used to characterize elastomeric materials (1, 2, 5, 7). It is an unusual deformation in that volume changes are of central importance, rather than being negligible. Swelling

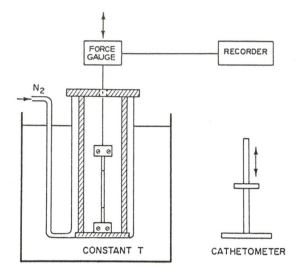

FIGURE 9. Apparatus for carrying out stress–strain measurements on an elastomer in elongation (*3*).

is a three-dimensional dilation in which the network absorbs solvent and reaches an equilibrium degree of swelling at which the free energy decrease due to the mixing of the solvent with the network chains is balanced by the free energy increase due to the stretching of the chains. In this type of experiment, the network is typically placed in an excess of solvent, which it imbibes until the dilational stretching of the chains prevents further absorption. This equilibrium extent of swelling can be interpreted to yield the degree of cross-linking of the network, provided that the polymer–solvent interaction parameter χ_1 is known. Conversely, if the degree of cross-linking is known from an independent experiment, then the interaction parameter can be determined. The equilibrium degree of swelling and its dependence on various parameters and conditions provide, of course, additional tests of theory.

The classic swelling theory developed by Flory and Rehner gives the relationship (*1*):

$$v/V = -\left[\ln(1 - v_{2m}) + v_{2m} + \chi_1 v_{2m}^2\right]/A'_\phi V_1 v_{2S}^{2/3}\left(v_{2m}^{1/3} - \omega v_{2m}\right)] \quad (19)$$

In equation 19, v/V is the cross-link density, v_{2m} is the volume fraction of polymer at swelling equilibrium, χ_1 is the polymer–solvent interaction parameter (*1*), A'_ϕ is a structure factor equal to unity in the affine limit, V_1 the molar volume of the solvent, v_{2S} is the volume fraction of polymer present during cross-linking, and ω is an entropic volume factor equal to $2/\phi$.

In a refined theory developed by Flory (*30*), the extent to which the swelling deformation is non-affine depends on the looseness with which the cross-links are embedded in the network structure. This looseness depends in turn on both the structure of the network and its degree of equilibrium swelling. In one version of this theory, the resulting equation is as follows:

$$v/V = -\left[\ln(1 - v_{2m}) + v_{2m} + \chi_1 v_{2m}^2\right]/F_\phi V_1 v_{2S}^{2/3} v_{2m}^{1/3} \tag{20}$$

The factor F_ϕ characterizes the extent to which the deformation in swelling approaches the affine limit and is given by equation 21:

$$F_\phi = (1 - 2/\phi)\left[1 + (\mu/\xi)K\right] \tag{21}$$

In equation 21, ξ is the cycle rank of the network and K is a function of v_{2m}, κ, and p [$K = f(v_{2m}, \kappa, p)$] (*30*), where κ is a parameter specifying constraints on cross-links, and p is a parameter specifying dependence of cross-link fluctuations on the strain (*30*). This theory is somewhat more difficult to apply, because it contains parameters that are not present in the simpler theory and that are not always available, even for some relatively common and important elastomers.

Optical and Spectroscopic Properties

An example of a relevant optical property is the birefringence of deformed polymer networks (*7*). Strain-induced birefringence can be used to characterize segmental orientation, both Gaussian and non-Gaussian elasticity, crystallization and other types of chain ordering, and short-range correlations (*2, 5*). Other optical and spectroscopic techniques are also important, particularly with regard to segmental orientation. Some examples are fluorescence polarization, deuterium NMR, and polarized infrared spectroscopy (*5, 7, 31*). The application of spectroscopy to the characterization of polymers in general is covered by J. L. Koenig in Chapter 6.

Scattering

The most useful technique of this type in the study of elastomers is small-angle neutron scattering, for example, from deuterated chains in a nondeuterated host (*5, 7*). One application has been the determination of the degree of randomness of the chain configurations in the undeformed state, an issue of great importance with regard to the basic postulates of elasticity theory. Of even greater importance is determination of the manner in which the dimensions of the chains follow the macroscopic dimensions of the sample, that is, the degree of affineness of the deformation. This relationship between the microscopic and macroscopic levels in an elastomer is one of the central problems in rubberlike elasticity. The use of neutron-scattering measurements in the characterization of polymers in general is discussed by G. D. Wignall in Chapter 7.

Some small-angle X-ray scattering techniques have also been applied to elastomers. Examples are the characterization of fillers precipitated into elastomers and the corresponding incorporation of elastomers into ceramic matrices, in both cases to improve mechanical properties (*5, 32*).

Comparisons between Theory and Experiment

Dependence of Stress on Deformation

Most experimental results used to evaluate theory have been carried out in elongation. Correspondingly, these results will be emphasized in this section, with some results on other deformations discussed briefly in the later section *Other Types of Deformation*.

A typical stress–strain isotherm obtained on a strip of cross-linked natural rubber as described earlier is shown in Figure 10 (*2, 3*). The units for the force are generally newtons, and the curves obtained are usually checked for reversibility. In this type of representation, the area under the curve is frequently of considerable interest, because it is proportional to the work of deformation, $w = \int f dL$. Its value up to the rupture point is, thus, a measure of the toughness of the material.

The initial part of the stress–strain isotherm shown in Figure 10 is of the expected form in that the nominal stress, f^*, approaches linearity with the deformation ratio, α, as α becomes sufficiently large to make the α^{-2} term in equation 14 negligibly small. The large increase in f^* at high deformation in the case of natural rubber is due largely, if not entirely, to strain-induced crystallization, as is described in a later section on non-Gaussian effects (*Networks at Very High Deformations*). The melting point of the polymer is inversely proportional to the entropy of fusion, which is significantly diminished when the chains in the amorphous network remain stretched out because of the applied deformation. The melting point is thereby increased, and it is in this sense that the stretching "induces" the crystallization of some of the network chains, as shown schematically in Figure 11 (*33*). Removal of the force generally reduces the elevated melting point back to its original reference value. The effect is qualitatively similar to the increase in melting point generally observed upon an increase in pressure on a low-molecular-weight substance in the

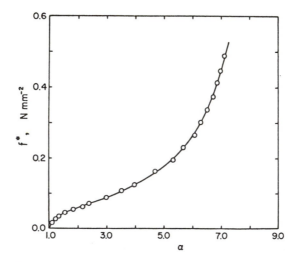

FIGURE 10. Stress–elongation curve for natural rubber in the vicinity of room temperature (*2, 3*).

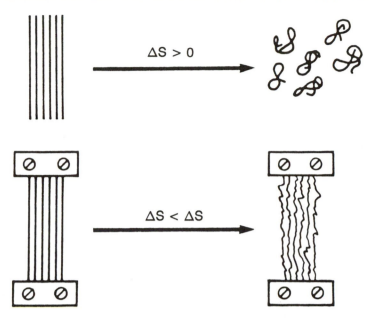

FIGURE 11.   Sketch explaining the increase in melting point with elongation in the case of a crystallizable elastomer. (Reproduced from reference 33. Copyright 1984 American Chemical Society.)

crystalline state. The crystallites thus formed act as physical cross-links, increasing the modulus of the network. The properties of both crystallizable and noncrystallizable networks at high elongations are discussed further in the later section *Networks at Very High Deformations*.

Additional deviations from theory are found in the region of moderate deformation upon examination of the usual plots of modulus against reciprocal elongation (*2, 34*). Although equation 16 predicts the modulus to be independent of elongation, the modulus generally decreases significantly upon an increase in α, as already mentioned. Typical results, obtained on swollen and unswollen networks of natural rubber, are shown in Figure 12 (*34*). The intercepts and slopes of such linear plots are generally called the Mooney–Rivlin constants $2C_1$ and $2C_2$, respectively, in the semiempirical relationship $[f^*] = 2C_1 + 2C_2\alpha^{-1}$. The slope, $2C_2$, a measure of the discrepancy from the predicted behavior, decreases to an essentially negligible value as the degree of swelling of the network increases. As described earlier, the more-refined molecular theories of rubberlike elasticity (*13–16*) explain this decrease by the gradual increase in the non-affineness of the deformation as the elongation increases toward the phantom limit, as shown schematically in Figure 8.

In these theories, the degree of entangling around the cross-links is of primary importance, because this will determine the firmness with which the cross-links are embedded in the network structure. This type of chain–cross-link entangling is illustrated in Figure 13 (*35*). For a typical degree of cross-linking, 50–100 cross-links are closer to a given cross-link than those directly joined to it through a single network chain. The configurational domains, thus, generally overlap severely. The degree of

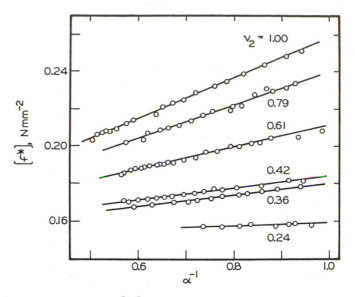

FIGURE 12. The modulus, $[f^*]$, shown as a function of reciprocal elongation (α^{-1}), as suggested by the semiempirical Mooney–Rivlin equation $[f^*] = 2C_1 + 2C_2\alpha^{-1}$ (*2, 34*). The elastomer is natural rubber, both unswollen and swollen with *n*-decane (*34*). Each isotherm is labeled with the volume fraction of polymer in the network.

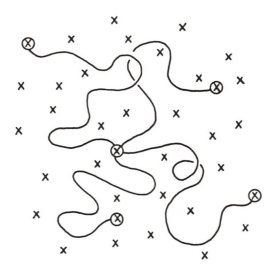

FIGURE 13. Typical configurations of four chains emanating from a tetra-functional cross-link in a polymer network prepared in the undiluted state (*35*).

overlapping is a measure of the firmness with which the cross-links are embedded and, thus, of the extent to which the idealized, affine deformation is approached. As already mentioned, stretching out the network chains decreases this degree of entangling and thereby permits increased cross-link fluctuations, which are then asymmetric. The modulus thus decreases and approaches the value predicted for a phantom network, in which entangling is impossible and cross-link fluctuations are unimpeded. This concept also explains the essentially constant modulus at high degrees of swelling illustrated in Figure 12. Large amounts of diluent "loosen" the cross-links so that the deformation is highly non-affine even at low deformations, and thus, the modulus changes relatively little upon an increase in elongation.

Dependence of Stress on Temperature

The assumption of a purely entropic elasticity leads to the prediction (equation 14) that the stress should be directly proportional to the absolute temperature at constant α (and V). The extent to which there are deviations from this direct proportionality may therefore be used as a measure of the thermodynamic nonideality of an elastomer (5, 36–41). In fact, the definition of ideality for an elastomer is that the energetic contribution (f_e) to the elastic force (f) be zero. This quantity is defined as follows:

$$f_e \equiv (\partial E/\partial L)_{V,T} \tag{22}$$

In equation 22, E is energy, L is length, and the differentiation is at constant volume (V) and temperature (T). The definition closely parallels the requirement of ideality for a gas, that $(\partial E/\partial V)_T$ be zero.

Force–temperature ("thermoelastic") measurements may therefore be used to obtain experimental values of the fraction of the force that is energetic in origin (f_e/f). Such experiments carried out at constant volume are the most direct and can be interpreted through use of the following, purely thermodynamic relationship:

$$f_e/f = -T[\partial \ln(f/T)/\partial T]_{V,L} \tag{23}$$

Because, however, it is very difficult to maintain constant volume in these experiments, they are usually carried out at constant pressure instead and are then interpreted by using the following equation:

$$f_e/f = -T[\partial \ln(f/T)/\partial T]_{p,L} - \beta T/(\alpha^3 - 1) \tag{24}$$

In equation 24, β is the thermal expansion coefficient of the network. This relationship was obtained by using the Gaussian elastic equation of state to correct the data to constant pressure (36, 37, 39, 40).

These energy changes are intramolecular (36, 37, 39, 40) and arise from transitions of the chains from one spatial configuration to another (because different configurations generally correspond to different intramolecular energies; 4). They are thus obviously related to the temperature coefficient of the unperturbed dimensions:

$$f_e/f = Td \ln \langle r^2 \rangle_0/dT \tag{25}$$

This quantitative relationship is obtained by keeping the $\langle r^2 \rangle_i$ factor in equation 9 distinct from $\langle r^2 \rangle_0$. Because this type of nonideality is intramolecular, it is not removed by diluting the chains (swelling the network) nor by increasing the lengths of the network chains (decreasing the degree of cross-linking). In this respect, elastomers are rather different from gases, which can be made to behave ideally by decreasing the pressure to a sufficiently low value.

Some typical thermoelastic data, obtained on amorphous polyethylene, are shown in Figure 14 (*37, 40*). Their interpretation with equation 24 indicates that the energetic contribution to the elastic force (f_e) is large and negative. These results on polyethylene (*37*) may be understood by using the information given in Figure 15. The preferred (lowest energy) conformation of the chain is the all-*trans* form, because *gauche* states (at rotational angles of $\pm 120°$) cause steric repulsions between CH_2 groups (*4*). Because this conformation has the highest possible spatial extension, stretching a polyethylene chain requires switching some of the *gauche* states (which are of course present in the randomly coiled form) to the alternative *trans* states (*4, 37, 39, 40*). These changes decrease the conformational energy and are the origin of the negative type of ideality represented in the experimental value of f_e/f. This physical picture also explains the decrease in unperturbed dimensions upon an increase in temperature. The additional thermal energy increases the number of the higher energy *gauche* states, which are more compact than the *trans* states.

The opposite behavior is observed with poly(dimethylsiloxane) (Figure 16). The all-*trans* form is again the preferred conformation; the relatively long Si–O bonds and the unusually large Si–O–Si bond angles reduce steric repulsions in general, and the *trans* conformation places CH_3 side groups at distances of separation where they are strongly attractive (*4, 39, 40*). Because of the inequality of Si–O–Si and O–Si–O bond angles,

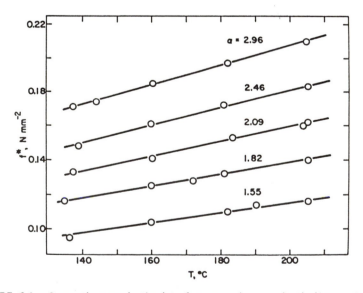

FIGURE 14. Some thermoelastic data for amorphous polyethylene networks at constant length (*37*). Each curve is labeled with the value of the elongation (α) at the highest temperature of measurement. (Reproduced with permission from reference 40. Copyright 1976 John Wiley and Sons, Inc.)

Results

$$\frac{f_e}{f} = T\frac{d \ln <r^2>_o}{dT} = -0.45$$

Interpretation

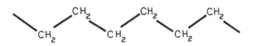

FIGURE 15.   Thermoelastic results on amorphous polyethylene networks and their interpretation in terms of the preferred, all-*trans* conformation of the chain (*3, 4*).

Results

$$\frac{f_e}{f} = T\frac{d \ln <r^2>_o}{dT} = 0.25$$

Interpretation

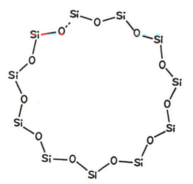

FIGURE 16.   Thermoelastic results on poly(dimethylsiloxane) networks and their interpretation in terms of the preferred, all-*trans* conformation of the chain (*3, 4*). For purposes of clarity, the two methyl groups on each silicon atom have been deleted.

however, this conformation is of very low spatial extension, approximating a closed polygon. Stretching a poly(dimethylsiloxane) chain, therefore, requires an increase in the number of *gauche* states. Because *gauche* states are of higher energy, the deviations from ideality for these networks are positive (*4, 39, 40*).

Thermoelastic results also are used to test some of the assumptions used in the development of the molecular theories. Specifically, the results given in Table I (*40*) indicate that the ratio f_e/f is essentially independent of the degree of swelling of the network, and this conclusion supports the postulate made earlier that intermolecular interactions do not contribute significantly to the elastic force. The assumption is further supported by the results given in Table II (*40*), which show that the values of the temperature coefficient of the unperturbed dimensions obtained from thermoelastic experiments are in good agreement with those obtained from viscosity–temperature measurements on the isolated chains in dilute solution.

Also, because intermolecular interactions do not affect the force, they must be independent of the extent of the deformation and, thus, independent of the spatial configurations of the chains. Therefore, the amorphous

Table I. Effect of Dilution on f_e/f

Polymer	Diluent	$v_2{}^a$	f_e/f
Polyethylene	None	1.00	$-0.42\ (\pm0.04)$
	Diethylhexyl azelate	0.80–0.30	$-0.44\ (\pm0.10)$
	$n\text{-}C_{30}H_{62}$	0.50	-0.64
	$n\text{-}C_{32}H_{66}$	~ 0.30	$-0.50\ (\pm0.06)$
Natural rubber	None	1.00	$0.17\ (\pm0.03)$
	$n\text{-}C_{16}H_{34}$	0.98–0.34	$0.18\ (\pm0.04)$
	$n\text{-}C_{10}H_{22}$	0.65–0.36	$0.13\ (\pm0.01)$
	Paraffin oil	0.40	$0.19\ (\pm0.02)$
	Decalin^b	~ 0.20	$0.14\ (\pm0.02)$
trans-1,4-Polyisoprene	None	1.00	$-0.10\ (\pm0.05)$
	Paraffin oil	0.40	$-0.13\ (\pm0.02)$
	Decalin^b	~ 0.18	$-0.20\ (\pm0.04)$

^{a} Volume fraction of polymer in the network.
^{b} Swelling equilibrium.

Table II. Comparison of $d\ln\langle r^2\rangle_0/dT$ Deduced from Thermoelastic Measurements on Networks with Values from Viscometric Measurements on Isolated Chains

Polymer	$d\ln\langle r^2\rangle_0/dT \times 10^3$	
	$f - T^a$	$[n] - T^b$
Polyethylene	$-1.05\ (\pm0.10)$	$-1.10\ (\pm0.07)$
		$-1.09\ (\pm0.04)$
		$-0.8\ (\pm0.1)$
Poly(*n*-pentene-1), isotactic	$0.34\ (\pm0.04)$	$0.52\ (\pm0.05)$
Polystyrene, atactic	$0.37\ (\pm0.06)$	$0.56\ (\pm0.10)$
Polyisobutylene	$-0.19\ (\pm0.11)$	$-0.28\ (\pm0.05)$
		$-0.4\ (\pm0.1)$
Polyoxyethylene	$0.23\ (\pm0.02)$	$0.2\ (\pm0.2)$
Poly(dimethylsiloxane)	$0.59\ (\pm0.14)$	$0.52\ (\pm0.20)$

^{a} Thermoelastic results.
^{b} Viscometric results.

chains must be in random, unordered configurations, the dimensions of which should be the unperturbed values (*1*). This conclusion has now been amply verified, in particular by neutron-scattering studies on undiluted amorphous polymers by numerous research groups (*40*).

Dependence of Stress on Network Structure

Until recently, relatively little reliable quantitative information was available on the relationship of stress to structure, primarily because of the uncontrolled manner in which elastomeric networks were generally prepared (*1–3, 5*). Segments close together in space were linked irrespective of their locations along the chain trajectories; the result is a highly random network structure in which the number and locations of the cross-links were essentially unknown (Figure 2). New synthetic techniques are now available, however, for the preparation of "model" polymer networks of known structure (*9, 42–49*). An example is the reaction shown in Figure 17, in which hydroxyl-terminated chains of poly(dimethylsiloxane) (PDMS) are end linked with tetraethyl orthosilicate. Characterizing the uncross-linked chains with respect to molecular weight (M_n) and molecular weight distribution and then running the specified reaction to completion give elastomers in which the network chains have these characteristics, in particular a molecular weight between cross-links (M_c) equal to M_n, and the cross-links have the functionality of the end-linking agent.

Trifunctional and tetrafunctional PDMS networks prepared in this way have been used to test the molecular theories of rubber elasticity with regard to the increase in non-affineness of the network deformation with increasing elongation. Some of these results are shown in Figure 18 (*44*). The ratio of the Mooney–Rivlin constants, $2C_2/2C_1$, decreases with an increase in cross-link functionality from three to four, because cross-links connecting four chains are more constrained than those connecting only three chains. Therefore, less of a decrease in modulus is brought about by the fluctuations, which are enhanced at high deformation and give the deformation its non-affine character. The decrease in $2C_2/2C_1$ with a decrease in network chain molecular weight is due to a lower degree of configurational interpenetration for short network chains. A lower degree of configurational interpenetration decreases the firmness with which the

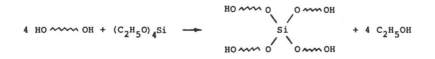

where HO⌇⌇⌇OH represents a hydroxyl-terminated PDMS chain.

Known $\overline{M}_n \rightarrow$ known $\overline{M}_c$.   Known M_n distribution $\rightarrow$ known M_c distribution.

FIGURE 17.   A typical synthetic route for preparing elastomeric networks of known structure (*42*).

cross-links are embedded, and thus, the deformation is already highly non-affine even at relatively small deformations.

A more-thorough investigation of the effects of cross-link functionality requires use of the more-versatile chemical reaction illustrated in Figure 19. Specifically, vinyl-terminated PDMS chains are end linked with a multifunctional silane. In the study summarized in Figure 20 (45), this reaction was used to prepare PDMS model networks having functionalities of 3–11, with a relatively unsuccessful attempt to achieve a functionality of 37. As shown in the figure, the modulus ($2C_1$) increases with an increase in functionality (ϕ), as expected from the increased constraints on the cross-links and as predicted by equations 17 and 18. Similarly, $2C_2$ and its value relative to $2C_1$ both decrease, for the reasons described earlier in relation to Figure 18.

Such model networks also may be used to test directly the molecular predictions of the modulus of a network of known degree of cross-linking. Some experiments on model networks (42, 44, 45) have yielded values of the elastic modulus in good agreement with theory. Others (46–48) have yielded values significantly higher than predicted, and the increases in modulus have been attributed to contributions from "permanent" chain entanglements of the type shown in the lower right portion of Figure 2. Disagreements exist, and the issue has not been resolved. Because the

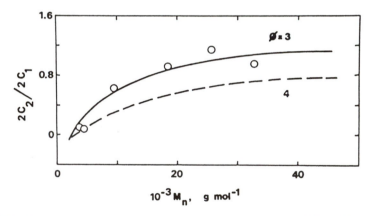

FIGURE 18. Experimental data showing values of the ratio $2C_2 / 2C_1$, which is a measure of the increase in non-affineness of the deformation as the elongation increases. The ratio decreases with an increase in junction functionality and with a decrease in network chain molecular weight, as predicted by theory (14–16).

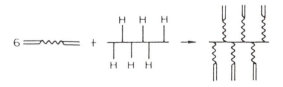

FIGURE 19. A typical reaction in which vinyl-terminated PDMS chains are end linked with a multifunctional silane.

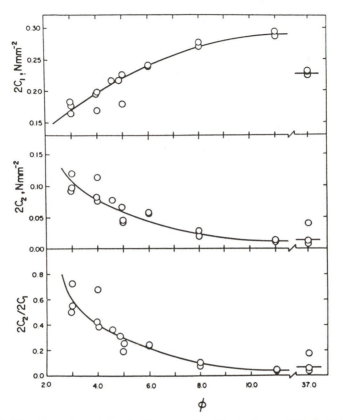

FIGURE 20. Experimental data showing the effect of cross-link functionality on $2C_1$ (a measure of the high-deformation modulus), $2C_2$, and $2C_2/2C_1$ (*45*).

relationship of modulus to structure is of such fundamental importance, a great deal of research activity is currently being done in this area (*3*).

These same very specific chemical reactions also can be used to prepare networks containing known numbers and lengths of dangling-chain irregularities, as illustrated in Figure 21 (*50*). If more chain ends are present than reactive groups on the end-linking molecules, then dangling ends will be produced, and their number is directly determined by the extent of the stoichiometric imbalance. Their lengths, however, are of necessity the same as those of the elastically effective chains, as shown in Figure 21a. This constraint can be removed by separately preparing monofunctionally terminated chains of the desired lengths and attaching them, as shown in Figure 21b. Results from some studies of this type are presented in the section *Networks at Very High Deformations*.

Some Unusual Networks

Networks Prepared in Solution or in a State of Strain

Two techniques that may be used to prepare networks having simpler topologies are illustrated in Figure 22 (*51, 52*). Basically, they involve

Excess difunctional chains

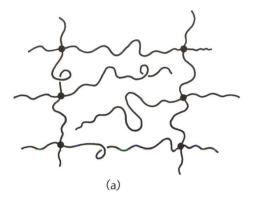

(a)

Monofunctional chains

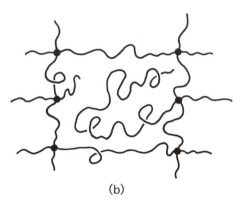

(b)

FIGURE 21. Two end-linking techniques for preparing networks with known numbers and lengths of dangling chains. (Reproduced from reference 50. Copyright 1985 American Chemical Society.)

separating the chains prior to their cross-linking by either stretching or dissolution. After the cross-linking, the stretching force or solvent is removed, and the unswollen network is studied with regard to its stress–strain properties in elongation. Some results obtained on PDMS networks cross-linked in solution by means of γ radiation (52, 53) showed a continual decrease in the time required to reach elastic equilibrium and in the extent of stress relaxation upon a decrease in the volume fraction of polymer present during the cross-linking. Also, at higher dilutions the Mooney–Rivlin constant $2C_2$ decreased as well.

These observations are qualitatively explained in Figure 23. If a network is cross-linked in solution and the solvent is subsequently removed, the chains collapse in such a way that the overlap in their configurational domains is reduced. It is primarily in this regard (namely, decreased chain–junction entangling) that samples cross-linked in solution have simpler topologies, with correspondingly simpler elastomeric behavior.

The opposite sort of experiment involves cross-linking a network in the undiluted state and then studying its stress–strain isotherms in the swollen

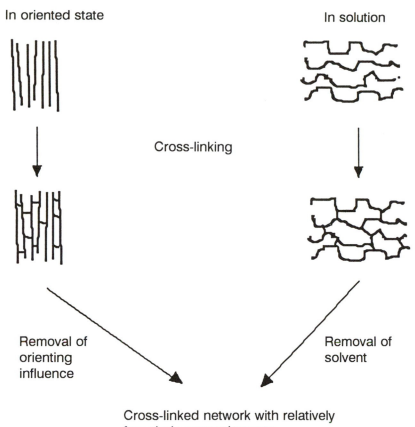

In oriented state

In solution

Cross-linking

Removal of
orienting
influence

Removal of
solvent

Cross-linked network with relatively
few chain entanglements

FIGURE 22. Two techniques that may be used to prepare networks of
simpler topology (*51, 52*).

state. A diluent might be introduced to suppress crystallization or to facilitate the approach to elastic equilibrium. A complication, however, can occur with networks of polar polymers at relatively high degrees of swelling (*53, 54*). At the same degree of polymer swelling, different solvents can have significantly different effects on the elastic force, apparently because of a "specific solvent effect" on the unperturbed dimensions, which appear in the basic relationship given in equation 9. Although frequently observed in studies of the solution properties of uncross-linked polymers, the effect is not yet well understood. It is apparently partly due to the effect of the solvent dielectric constant on the Coulombic interactions between parts of a chain and probably also to solvent–polymer segment interactions that change the conformational preferences of the chain backbone (*54*).

Unusual Diluents

End linking functionally terminated chains in the presence of chains whose ends are inert yields networks through which the unattached chains *reptate* (*55*). Networks of this type have been used to determine the efficiency with which unattached chains can be extracted from an elas-

FIGURE 23. Typical configurations of four chains emanating from a tetra-functional cross-link in a (dried) polymer network that had been prepared in solution.

tomer as a function of their lengths and the degree of cross-linking of the network (5, 56). The efficiency decreases with an increase in molecular weight of the diluent and with an increase in degree of cross-linking, as expected. Diluents present during the cross-linking also are more difficult to extract compared with the same diluents absorbed into the network after cross-linking. Such comparisons can provide valuable information on the arrangements and transport of chains within complex network structures.

If relatively large cyclic PDMS structures are present when linear PDMS chains are end linked, then some can be permanently trapped by one or more network chains threading through them, as shown by cyclic structures B, C, and D in Figure 24 (57). The amount trapped ranges from 0% for cyclic structures with fewer than ca. 30 skeletal bonds to essentially 100% for those with more than ca. 300 skeletal bonds (58). These results may be interpreted in terms of the effective "hole" sizes of the cyclic structures, as estimated from Monte Carlo simulations of their spatial configurations. The agreement between theory and experiment is very good (57).

This technique also may be used to form a network having no cross-links whatsoever. Mixing linear chains with large amounts of cyclic structures and then *di*functionally end linking them could give sufficient cyclic interlooping to yield a "chain-mail" or "Olympic" network, as depicted in Figure 25 (59). Such materials could have very unusual stress–strain isotherms.

Bimodal Networks

The end-linking reactions described earlier also can be used to make networks having unusual chain-length distributions (60–63). Networks

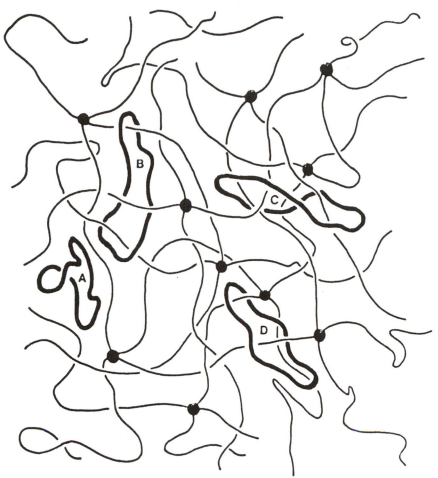

FIGURE 24. Trapping of cyclic molecules during end-linking preparation of a network. (Reproduced from reference 57. Copyright 1987 American Chemical Society.)

having a bimodal distribution are of particular interest with regard to their ultimate properties; they are discussed in the following section.

Networks at Very High Deformations

Non-Gaussian Effects

As already described in Figure 10 (*1–3*), some unfilled networks show a large and rather abrupt increase in modulus at high elongations. This increase, which is illustrated for natural rubber in Figure 26 (*64, 65*), is very important, because it corresponds to a significant toughening of the elastomer. Its molecular origin, however, has been the source of considerable controversy (*2, 5, 64, 66–72*). It had been widely attributed to the "limited extensibility" of the network chains (*70*), that is, to an inadequacy in the Gaussian distribution function. This potential inadequacy is readily

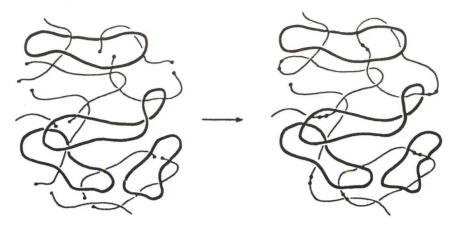

FIGURE 25. Preparation of a "chain-mail" or "Olympic" network consisting entirely of interlooped cyclic molecules. (Reproduced with permission from reference 59. Copyright 1985 Butterworth.)

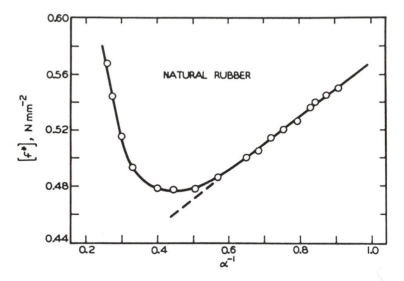

FIGURE 26. Stress–strain isotherm for an unfilled rubber network at 25 °C showing the anomalous increase in modulus at high elongation (*64*). (Reproduced with permission from reference 65. Copyright 1976 John Wiley and Sons, Inc.)

evident in the exponential in equation 1, specifically from the fact that this function does not assign a zero probability to a configuration unless its end-to-end separation (r) is infinite. This explanation in terms of limited extensibility was viewed with skepticism by some workers, because the increase in modulus was generally observed only in networks that could undergo strain-induced crystallization. Such crystallization in itself could account for the increase in modulus, primarily because the crystallites thus formed would act as additional cross-links in the network structure.

Attempts to clarify the problem by using noncrystallizable networks (*65*) were not convincing, because such networks were incapable of the large deformations required to distinguish between the two possible interpretations. The issue has now been resolved (*70, 73–76*), however, by the use of end-linked, noncrystallizable model PDMS networks. These networks have high extensibilities, presumably because of the very low incidence of dangling-chain network irregularities. They have particularly high extensibilities when they are prepared from a mixture of very short chains (around a few hundred grams per mole) with relatively long chains (around 18 000 g · mol^{-1}), as discussed in the next part of this section. Apparently, the very short chains are important because of their limited extensibilities, and the relatively long chains are important because of their ability to retard the rupture process.

Stress–strain measurements on such bimodal PDMS networks exhibit upturns in modulus that are much less pronounced than those in crystallizable polymer networks (such as natural rubber or *cis*-1,4-polybutadiene), and they are independent of temperature, as would be expected for limited chain extensibility (*53, 70*). For a crystallizable network, the upturns diminish and eventually disappear upon an increase in temperature (*73, 75*). Similarly, swelling has relatively little effect on the upturns in the case of PDMS (*53, 70*) and can even make the upturns more pronounced. In contrast, the upturns in crystallizable polymer networks disappear upon sufficient swelling, as illustrated in Figure 27 (*74, 75*).

Two additional features in Figure 27, however, merit additional comments. First, the initiation of the strain-induced crystallization (as evi-

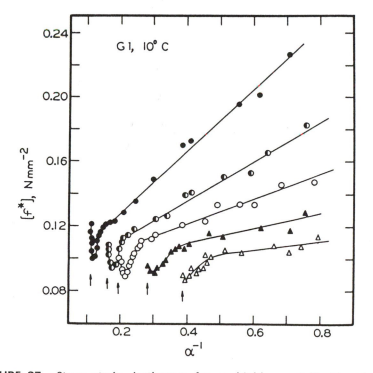

FIGURE 27. Stress–strain isotherms for a highly crystallizable *cis*-1,4-polybutadiene network swollen with 1,2-dichlorobenzene to values of the volume fraction (v_2) of polymer of 1.00 (●), 0.80 (◑), 0.60 (○), 0.40 (▲), and 0.20 (△) (*74, 75*).

denced by the departure of the isotherm from linearity) is facilitated by the presence of the low-molecular-weight diluent. In a sense, this kinetic effect opposes the thermodynamic effect, which is primarily the depression of the polymer melting point by the diluent. The second interesting point has to do with the decrease in the modulus prior to its increase. As shown schematically in Figure 28 (35, 76), this is probably due to the fact that the crystallites are oriented along the direction of stretching and that the chain sequences within a crystallite are in regular, highly extended conformations. The straightening and aligning of portions of the network chains thus decrease the deformation in the remaining amorphous regions, which decreases the stress.

In summary, the anomalous upturn in modulus observed for crystallizable polymers such as natural rubber and cis-1,4-polybutadiene is largely, if not entirely, due to strain-induced crystallization. For noncrystallizable PDMS model networks, the anomaly is clearly due to limited chain extensibility, and thus, the results on this system will be extremely useful for reliable evaluation of the various non-Gaussian theories of rubberlike elasticity.

Ultimate Properties

This section continues the discussion of unfilled elastomers at high elongations but with emphasis on ultimate properties: ultimate strength and maximum extensibility.

Some illustrative results on the effects of strain-induced crystallization on ultimate properties are given for cis-1,4-polybutadiene networks in Table III (73). The higher the temperature, the lower is the extent of crystallization, and correspondingly, the lower are the ultimate properties. The effects of increase in swelling parallel those of increase in temperature, because the diluent also suppresses network crystallization. For noncrystallizable networks, however, neither change is very important, as illustrated by the results shown for PDMS networks in Table IV (77).

For such noncrystallizable, unfilled elastomers, the mechanism for network rupture has been elucidated to a great extent by studies of model networks similar to those described in the preceding section. For example,

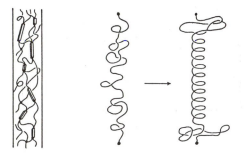

FIGURE 28. Strain-induced crystallization in a polymer network that has been elongated by a force along the vertical direction (35, 76).

**Table III. *cis*-1,4-Polybutadiene Networks
at High Elongation**

| | | Ultimate Properties | |
| | | Maximum Upturn | |
T, °C	α at Upturn	in [$f*$], %	α at Rupture
5	3.27	54.2	6.64
10	3.48	30.1	6.22
25	4.03	4.3	5.85
40	—[a]	0.0	5.68

[a] None observed.

Table IV. Ultimate Properties of PDMS Networks

υ_2	$\lambda_r{}^a$	$[f*]_r{}^b$
1.00	4.90	0.0362
0.80	4.42	0.0342
0.60	4.12	0.0338
0.40	4.16	0.0336

[a] λ_r is the value of the total elongation at the rupture point (with L_i being the initial length, unswollen).
[b] $[f*]_r$ is the modulus at rupture.

values of the modulus of bimodal networks formed by end linking mixtures of very short chains and relatively long chains, as illustrated in Figure 29 (*42*), were used to test the "weakest link" theory, in which rupture was thought to be initiated by the shortest chains (because of their very limited extensibility). Increasing the number of very short chains did *not* significantly decrease the ultimate properties. The reason, given schematically in Figure 30 (*70*), is the very non-affine nature of the deformation at such high elongations. The network simply reapportions the increasing strain among the polymer chains until no further reapportioning is possible. Generally, chain scission leading the rupture of the elastomer begins only at this point. The weakest link theory implicitly assumes an affine deformation, which predicts that the elongation at which the modulus increases should be independent of the number of short chains in the network. This assumption is contradicted by relevant experimental results, which show very different behavior (*70*): the smaller the number of short chains, the easier is the reapportioning, and the higher is the elongation required to bring about the upturn in modulus.

An exciting bonus is gained if a very large number of short chains is introduced into the bimodal network. The ultimate properties are then actually improved, as shown in Figure 31 (*78*), in which data on PDMS networks are plotted so that the area under a stress–strain isotherm corresponds to the energy required to rupture the network. If the network is made up exclusively of short chains, it is brittle and its maximum extensibility is very small. If the network is made up exclusively of long chains, the ultimate strength is very low. In neither case is the material a tough elastomer. As can readily be seen from Figure 31, the bimodal

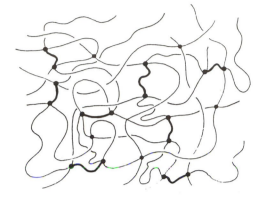

FIGURE 29. A portion of a network that is compositionally heterogeneous with respect to chain length. The very short and relatively long chains are arbitrarily shown by the thick and thin lines, respectively. (Reproduced with permission from reference 42. Copyright 1979 Hüthig & Wepf Verlag, Berlin.)

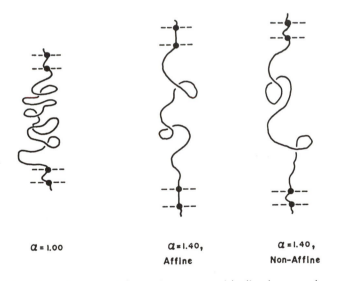

FIGURE 30. The effect of deformation on an idealized network segment consisting of a relatively long chain bracketed by two very short chains. (Reproduced with permission from reference 70. Copyright 1980 American Institute of Physics.)

networks are much improved elastomers in that they can have a high ultimate strength without the usual decrease in maximum extensibility.

A series of experiments were carried out to determine if this reinforcing effect in bimodal PDMS networks could be due to some intermolecular effect, such as strain-induced crystallization. The first such experiment showed that temperature had little effect on the isotherms, as illustrated in Figure 32 (62). This result strongly argues against the presence of any crystallization or other type of intermolecular ordering. So also do the results of stress–temperature and birefringence–temperature measure-

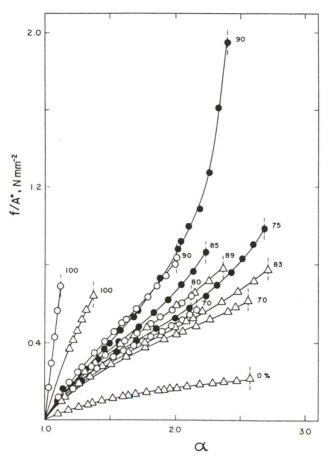

FIGURE 31.   Typical plots of nominal stress against elongation for (unswollen) bimodal PDMS networks consisting of relatively long chains [$M_c = 18\,500$ g · mol^{-1}] and very short chains [$M_c = 1100$ (△), 660 (○), and 220 (●) g · mol^{-1}]. Each curve is labeled with the mole percent of short chains it contains, and the area under each curve represents the rupture energy (a measure of the "toughness" of the elastomer). (Reproduced with permission from reference *78*. Copyright 1981 John Wiley and Sons, Inc.)

ments (*62*). In a final experiment, the short chains were prereacted in a two-step preparative technique to segregate them in the network structure (*53*, *60*), as might occur in a network cross-linked by an incompletely soluble peroxide. This segregation had very little effect on elastomeric properties, again a result arguing against any type of intermolecular organization as the origin for the reinforcing effects. Apparently, the observed increases in modulus are due to the limited chain extensibility of the short chains, with the long chains retarding the rupture process.

With the molecular origin of the unusual properties of bimodal PDMS networks having been elucidated at least to some extent, it is now possible to use these materials in various applications. The first involves the interpretation of limited chain extensibility in terms of the configurational characteristics of the PDMS chains making up the network structure (*4*).

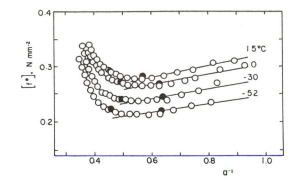

FIGURE 32. Typical results obtained to determine the effect of temperature on the bimodal PDMS stress–strain isotherms; they pertain to networks containing 75 mol % very short chains. (Reproduced with permission from reference 62. Copyright 1982 John Wiley and Sons, Inc.)

The first important characteristic of limited chain extensibility is the elongation at which the increase in modulus first becomes discernible, α_u. Although the deformation is non-affine in the vicinity of the upturn, at least a semiquantitative interpretation of such results may be made in terms of the network chain dimensions (*4*, *70*). At the beginning of the upturn, the average extension (r) of a network chain having its end-to-end vector along the direction of stretching is simply the product of the unperturbed dimension $\langle r^2 \rangle_0^{1/2}$ and α_u (*70*). Similarly, the maximum extensibility (r_m) is the product of the number of skeletal bonds (n) and the factor 1.34 Å, which gives the axial component of a skeletal bond in the most extended helical form of PDMS, as obtained from the geometric analysis of the PDMS chain (*53*, *70*). The ratio r/r_m at α_u thus represents the fraction of the maximum extensibility occurring at this point in the deformation. The values obtained indicate that the upturn in modulus generally begins at approximately 60–70% of the maximum chain extensibility (*70*). This value is approximately twice the value that had been estimated previously (*2*) as a result of misinterpretation of stress–strain isotherms of elastomers undergoing strain-induced crystallization.

The values of r/r_m at the beginning of the upturn may be compared with some theoretical results by Flory and Chang (*63*, *79*) on distribution functions for PDMS chains of finite length. Of relevance here are the calculated values of r/r_m at which the Gaussian distribution function starts to overestimate the probability of extended configurations, as judged by comparisons with the results of Monte Carlo simulations. The theoretical results (*53*, *79*) suggest, for example, that the network of PDMS chains having 53 skeletal bonds ($n = 53$), which was experimentally studied, should show an upturn at a value of r/r_m of a little less than 0.80. The observed value was 0.77 (*70*), which is in excellent agreement with theory.

The second important characteristic of limited chain extensibility is the elongation at which rupture occurs, α_r. The corresponding values of r/r_m show that rupture generally occurred at approximately 80–90% of the maximum chain extensibility (*70*). These quantitative results on chain dimensions are very important but may not apply directly to other networks, in which the chains could have very different configurational characteristics and in which the chain length distribution would presumably be quite different from the very unusual bimodal distribution intentionally produced in the PDMS networks.

The Monte Carlo simulations based on the rotational isomeric state (RIS) model for the network chains have been very useful in interpreting these upturns in modulus. Some typical results calculated for polyethylene and PDMS network chains having 20 skeletal bonds ($n = 20$) are shown in Figure 33 (*25*). The Gaussian distribution function is a relatively poor approximation of the RIS distribution at this value of n, particularly in the very important region of large r, and becomes even worse as n decreases. Calculated Mooney–Rivlin isotherms for networks made up of PDMS chains of various lengths are presented in Figure 34 (*25*). As expected, the network consisting of relatively long chains ($n = 250$) gives the Gaussian result $[f^*]/vkT = 1$ (k is the Boltzmann constant). The upturns in $[f^*]$ obtained at smaller values of n are very similar to those found experimentally and illustrated in Figure 32. Also as expected, the results show that the shorter the network chains, the smaller is the elongation at which the upturn occurs.

The upturns in modulus in these isotherms also may be interpreted by using analytical expressions, for example, the Fixman–Alben modification (*80*) of the Gaussian distribution function, combined with the constrained-junction theory and reasonable values of the constraint parameter, κ (*81*).

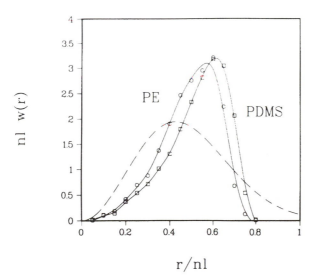

FIGURE 33. Comparisons among the rotational isomeric (RIS) radial distribution functions at 413 K for polyethylene (PE; ○) and poly(dimethylsiloxane) (□) chains having 20 skeletal bonds ($n = 20$), and the Gaussian approximation (- - -) to the PDMS distribution. The RIS curves represent cubic-spline fits to the discrete Monte Carlo data, for 80 000 chains, and each curve is normalized to an area of unity (with *l* being the skeletal bond length). (Reproduced with permission from reference 25. Copyright 1983 American Institute of Physics.)

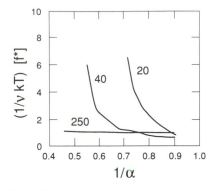

FIGURE 34.   Moduli of PDMS networks having chain lengths (n) of 20, 40, and 250 skeletal bonds. The values of [f^*] are normalized by the Gaussian prediction for the modulus, νkT, where ν is the number of network chains and kT has the usual significance. (Reproduced with permission from reference 25. Copyright 1983 American Institute of Physics.)

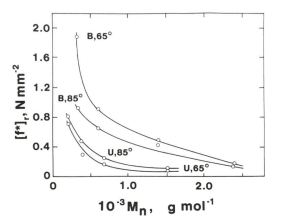

FIGURE 35.   The ultimate strength shown as a function of the molecular weight, $M_n = M_c$, between cross-links for unimodal (U) and bimodal (B) networks of crystallizable poly(ethylene oxide). (Reproduced with permission from reference 82. Copyright 1987 John Wiley and Sons, Inc.)

There is another advantage to such bimodality when the network can undergo strain-induced crystallization, the occurrence of which can provide an additional toughening effect. This is illustrated by the results for some poly(ethylene oxide) networks shown in Figure 35 (*82*). A decrease in temperature increases the extent to which the values of the ultimate strength of the bimodel networks exceed those of the corresponding unimodel networks. This trend suggests that bimodality facilitates strain-induced crystallization.

In practical terms, these results demonstrate that short chains of limited extensibility may be bonded into a long-chain network to improve its toughness. The converse effect also may be achieved; thus, bonding a small number of relatively long elastomeric chains into a short-chain PDMS thermoset greatly improves its impact resistance, as illustrated in

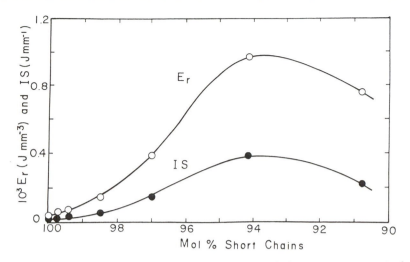

FIGURE 36. The energy required for rupture and the impact strength (as measured by the falling-dart test) shown as a function of composition for bimodal PDMS networks in the vicinity of room temperature. (Reproduced with permission from reference 83. Copyright 1984 Steinkopff Verlag.)

Figure 36 (*83*). The effects of bimodality in other types of deformation are illustrated in the following section, *Other Types of Deformation.*

Because dangling chains represent imperfections in a network structure, their presence would be expected to have a detrimental effect on the ultimate properties $(f/A^*)_r$ (ultimate strength) and α_r (maximum extensibility or elongation at which rupture occurs) of an elastomer. This expectation is confirmed by an extensive series of results obtained on PDMS networks that had been tetrafunctionally cross-linked by a variety of techniques. Some pertinent results are shown, as a function of the molecular weight between cross-links (M_c), in Figure 37 (*84*). The largest values of the ultimate strength, $(f/A^*)_r$, are obtained for the networks prepared by selectively joining functional groups occurring either as chain ends or as side groups along the chains. This result is expected because of the relatively low incidence of dangling ends in such networks. (As already described, the effects are particularly pronounced when such model networks are prepared from a mixture of relatively long chains and very short chains). Also as expected, the lowest values of the ultimate properties generally occur for networks cured by radiation (UV light, high-energy electrons, and γ radiation; *84*). The values for peroxide-cured networks are generally in between these two extremes, with the ultimate properties presumably depending on whether the free radicals generated by the peroxide are sufficiently reactive to cause some chain scission. Similar results were obtained for the maximum extensibility, α_r (*84*). These observations are at least semiquantitative and certainly interesting but are somewhat deficient in that information on the number of dangling ends in these networks is generally not available.

More definitive results have been obtained by investigation of a series of model networks prepared by end linking vinyl-terminated PDMS chains (*84*). The tetrafunctional end-linking agent was used in various amounts

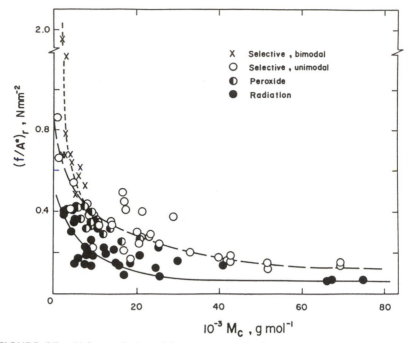

FIGURE 37. Values of the ultimate strength shown as a function of the molecular weight between cross-links (M_c) for (unfilled) tetrafunctional PDMS networks at 25 °C. (Reproduced with permission from reference 84. Copyright 1981 John Wiley and Sons, Inc.)

smaller than that corresponding to a stoichiometric balance between its active hydrogen atoms and the terminal vinyl groups of the chains. The ultimate properties of these networks, with known numbers of dangling ends, were then compared with those obtained on networks previously prepared so as to have negligible numbers of these irregularities (*84*).

Values of the ultimate strength of the networks are shown as a function of the high-deformation modulus $2C_1$ in Figure 38 (*84*). The networks containing the dangling ends have lower values of $(f/A^*)_r$, with the largest differences occurring at high proportions of dangling ends (low $2C_1$), as expected. These results confirm the less definitive results shown in Figure 37. The values of the maximum extensibility show a similar dependence (*84*).

Other Types of Deformation

Biaxial Extension

Other deformations of interest include compression, biaxial extension, shear, and torsion. The equation of state for compression ($\alpha < 1$, α is deformation ratio) is the same as that for elongation ($\alpha > 1$), and the equations for the other deformations may all be derived from equation 10

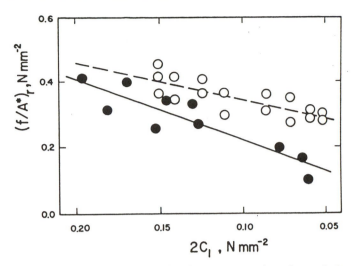

FIGURE 38. The ultimate strength shown as a function of the high-deformation modulus for tetrafunctional PDMS networks containing a negligible number of dangling ends (○) and dangling ends introduced by using less than the stoichiometrically required amount of end-linking agent (●). In the latter case, a decrease in $2C_1$ corresponds to an increase in the number of dangling ends. (Reproduced with permission from reference 84. Copyright 1981 John Wiley and Sons, Inc.)

by proper specification of the deformation ratios (*1, 2*). Some of these deformations are considerably more difficult to study than simple elongation and, unfortunately, have therefore not been investigated as extensively.

Some measurements in biaxial extension have involved the direct stretching of a sample sheet in two perpendicular directions within its plane by two independently variable amounts. In the equibiaxial case, the deformation is equivalent to compression. A good explanation of such experimental results (*85*) has been given by use of the simple molecular theory, with improvements at lower extensions by use of the constrained-junction theory (*5*).

Biaxial extension studies also can be carried out by inflating sheets of the elastomer (*2*), as illustrated in Figure 39 for some unimodal and bimodal networks of PDMS (*86*). Upturns in the modulus occur at high biaxial extensions, as expected. Also of interest, however, are the pronounced maxima preceding the upturns. This feature represents a challenge to be explained by molecular theories addressed to bimodal elastomeric networks in general.

Shear

Experimental results on natural rubber networks in shear (*87*) are not well explained by the simple molecular theory of rubberlike elasticity. The constrained-junction theory, however, gives excellent agreement with experiment (*5*).

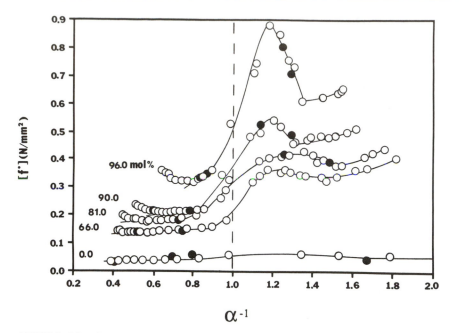

FIGURE 39. Representative stress–strain isotherms for unimodal and bimodal PDMS networks in uniaxial extension (left) and biaxial extension (right). Each curve is labeled with the mole percent short chains present in the network. The open circles represent data points measured with increasing deformations, and filled circles represent data obtained out of sequence to test for reversibility. (Reproduced with permission from reference 86. Copyright 1991 John Wiley and Sons, Inc.)

Results of shear measurements on some unimodal and bimodal networks of PDMS are shown in Figure 40 (*88*). The upturns in modulus are very similar to those obtained in elongation.

Torsion

Very little work has been done on elastomers in torsion. However, some results are available on stress–strain behavior and network thermoelasticity (*2*). More results are presumably forthcoming, particularly on the unusual bimodal networks and on networks containing some of the unusual fillers described in the section *Filled Networks*.

Swelling

Most studies of networks in swelling equilibrium give values for the cross-link density or related quantities that are in satisfactory agreement with those obtained from measurements of mechanical properties (*1, 2*).

A more interesting aspect of some swollen networks or "gels" is their abrupt collapse (decrease in volume) upon relatively minor changes in temperature, pH, solvent composition, etc. (*5, 89, 90*). Although the collapse is quite slow in large, monolithic pieces of gel, it is rapid enough

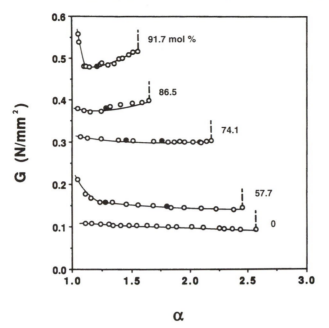

FIGURE 40. Stress–strain isotherms for unimodal and bimodal PDMS networks in shear. Each curve is labeled with the mole percent short chains present in the network, and the vertical dashed lines locate the rupture points. The open circles represent data points measured with increasing deformations, and filled circles represent data obtained out of sequence to test for reversibility. (Reproduced with permission from reference 88. Copyright 1992 John Wiley and Sons, Inc.)

in fibers and films to make the phenomenon interesting with regard to the construction of switches and related devices.

Bioelastomers

A number of cross-linked proteins are elastomeric, and investigation of their properties may be used to obtain insights into elastic behavior in general. For example, elastin (*5, 91–94*), which occurs in mammals, illustrates the relevance of several molecular characteristics to the achievement of rubberlike properties. First, a high degree of chain flexibility is achieved in elastin by its chemically irregular structure and by choices of side groups that are almost invariably very small. Because strong intermolecular interactions are generally not conducive to good elastomeric properties, the choices of side chains also are almost always restricted to nonpolar groups. Finally, elastin has a glass transition temperature of approximately 200 °C in the dry state, which means it would be elastomeric only above this temperature. Nature, however, apparently also knows about "plasticizers." Elastin, as used in the body, is invariably swollen with sufficient aqueous solutions to bring its glass transition temperature below the operating temperature of the body.

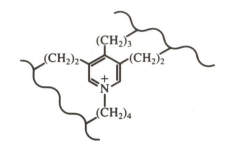

FIGURE 41. Sketch of a type of cross-link appearing in the protein elastin. (Reproduced with permission from reference 5. Copyright 1988 John Wiley and Sons, Inc.)

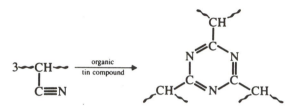

FIGURE 42. Sketch of a type of cross-link appearing in some perfluoroelastomers. (Reproduced with permission from reference 5. Copyright 1988 John Wiley and Sons, Inc.)

Elastin chains are cross-linked in vivo in a highly specific manner, by using techniques very unlike those usually used to cure commercial elastomers (*5, 95*). The cross-linking occurs through lysine repeat units, the number and placement of which along the chains are carefully controlled in the synthesis of elastin in the ribosomes. One type of resulting cross-link is shown in Figure 41 (*5*). An analogous reaction has been carried out commercially on perfluoroelastomers, which are usually very difficult to cross-link because of their inertness. Nitrile side groups placed along the chains are trimerized to triazine to give similarly stable, aromatic cross-links, as illustrated in Figure 42 (*5*).

Another bioelastomer, found in some insects, is resilin (*96*). It is an unusual material, because it is thought to be relatively highly efficient in storing elastic energy (i.e., very small losses due to viscous effects; such viscoelastic properties are discussed by W. W. Graessley in Chapter 3). A molecular understanding of this very attractive property obviously has considerable practical and fundamental importance.

Filled Networks

Elastomers, particularly those that cannot undergo strain-induced crystallization, are generally compounded with a reinforcing filler (*5*). The two most important examples are the addition of carbon black to natural rubber and to some synthetic elastomers (*97, 98*) and of silica to polysiloxane rubbers (*99, 100*). The advantages obtained include improved abrasion resistance, tear strength, and tensile strength. Disadvantages include increases in hysteresis (and thus heat buildup) and compression set (permanent deformation).

The mechanism of the reinforcement is only poorly understood. Some elucidation might be obtained by precipitating reinforcing fillers into network structures rather than blending badly agglomerated fillers into polymers prior to their cross-linking. This has, in fact, been done for a variety of fillers, for example, silica by hydrolysis of organosilicates, titania from titanates, alumina from aluminates, etc. (*5, 101, 102*). A typical and important reaction is the acid- or base-catalyzed hydrolysis of tetraethyl orthosilicate:

$$Si(OC_2H_5)_4 + 2H_2O \rightarrow SiO_2 + 4C_2H_5OH \qquad (26)$$

Reactions of this type are much used by ceramists in the new sol–gel chemical route to high-performance ceramics (*103–108*). In the ceramics area, the advantages are the possibility of using low temperatures, the purity of the products, the control of ultrastructure (at the nanometer level), and the relative ease of forming ceramic alloys. In the reinforcement of elastomers, the advantages include the avoidance of the difficult, time-consuming, and energy-intensive process of blending agglomerated filler into high-molecular-weight polymer and the ease of obtaining extremely good dispersions.

In the simplest approach to reinforcing elastomer, some of the organometallic material is absorbed into the cross-linked network, and the swollen sample is placed into water containing the catalyst, typically a volatile base such as ammonia or ethylamine. Hydrolysis to form the desired silicalike particles proceeds rapidly at room temperature to yield the order of 50 wt % filler in less than an hour (*5, 101, 102*).

A typical transmission electron micrograph of PDMS elastomer filled with approximately 30 wt % silica is shown in Figure 43 (*109*). The particles formed are well dispersed and essentially unagglomerated; such characteristics suggest that the reaction may involve simple homogeneous nucleation. This hypothesis is consistent with the fact that particles growing independently of one another and separated by cross-linked polymer would not agglomerate unless very high concentrations were reached. The particles have a relatively narrow size distribution, with almost all of them having diameters in the range 200–300 Å.

Figure 44 illustrates the reinforcing ability of such in situ generated particles (*110*). The modulus, [f^*], increases by more than an order of magnitude, and the isotherms show the upturns at high elongation that are the signature of good reinforcement. As generally occurs in filled elastomers, there is considerable irreversibility in the isotherms, which may be due to irrecoverable sliding of the chains over the surfaces of the filler particles.

If the hydrolyses in organosilicate–polymer systems are carried out with increased amounts of the silicate, bicontinuous phases can be obtained (with the silica and polymer phases interpenetrating one another) (*32*). At still higher concentrations of the silicate, the silica generated becomes the continuous phase, with the polymer dispersed in it (*111–125*).

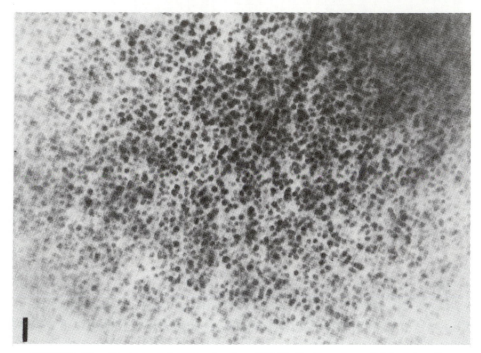

FIGURE 43. Electron micrograph of a PDMS elastomer containing in situ precipitated silica particles. (Reproduced with permission from reference 109. Copyright 1984 John Wiley and Sons, Inc.)

The result is a polymer-modified ceramic, variously called an "ormocer" (*111–113*), "ceramer" (*114–116*), or "polyceram" (*120–122*). It is obviously of considerable importance to determine how the polymeric phase, often elastomeric, modifies the ceramic in which it is dispersed.

Some typical results on such hybrid organic–inorganic composites are shown in Figure 45, which pertains to PDMS–SiO$_2$ systems (*123*). The hardness of the material can be varied greatly by changing the ratio of organic character to inorganic character, as measured by the molar ratio of organic R groups (here CH$_3$ side groups) to Si atoms. Low values of the R/Si ratio yield a brittle ceramic, and high values yield a reinforced elastomer. The most interesting range of values, R/Si ~ 1, can give a hybrid material that can be viewed as a ceramic of reduced brittleness or an elastomer of increased hardness, depending on one's point of view.

Current Problems and New Directions

Some aspects of rubberlike elasticity that are clearly in need of additional research are the following (*5*):

- Improved understanding of dependence of T_g and T_m on polymer structure
- Preparation and characterization of high-performance elastomers
- New cross-linking techniques

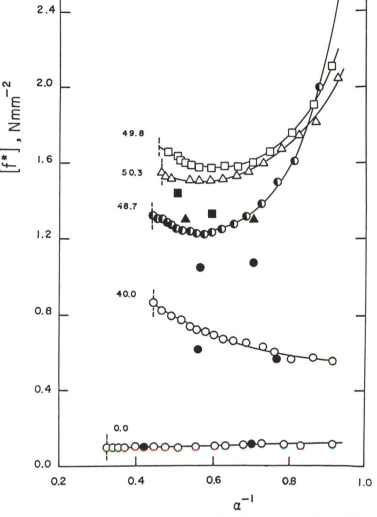

FIGURE 44. Mooney–Rivlin isotherms for PDMS elastomers filled with in situ generated silica, with each curve labeled with the amount of filler precipitated into it. Filled symbols are for results obtained out of sequence to establish the amount of elastic irreversibility, a common occurrence with reinforcing fillers. The vertical lines locate the rupture points. (Reproduced with permission from reference 110. Copyright 1984 Springer Verlag.)

- Improved understanding of network topology
- Generalization of phenomenological theory
- More experimental results for deformations other than elongation and swelling
- Better characterization of segmental orientation
- More detailed understanding of critical phenomena and gel collapse
- Additional molecular characterization by NMR spectroscopy and various scattering techniques

- Study of possibly unique properties of bioelastomers
- Improved understanding of reinforcing effects of filler particles in a network
- Quantitative interpretation of the toughening effects of elastomers in blends and in composites, particularly the polymer-modified ceramics

More high-performance elastomers are needed. These materials remain elastomeric at very low temperatures and are relatively stable at very high temperatures. Some phosphazene polymers, shown schematically in Figure 46 (*126–128*), are in this category. These polymers have rather low glass transition temperatures despite the fact that the skeletal bonds of the chains may have some double-bond character. There are thus a number of interesting problems related to the elastomeric behavior of these unusual semi-inorganic polymers. There is also increasing interest in the study of elastomers that also exhibit the type of mesomorphic behavior described by E. T. Samulski in Chapter 5.

An example of a cross-linking technique currently under development is the preparation of triblock copolymers such as those of styrene–butadiene–styrene. The phases of this system separate in such a way that

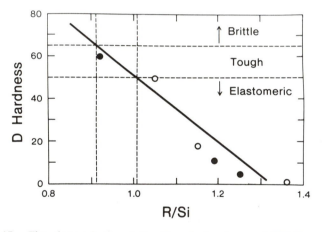

FIGURE 45. The dependence of the D-scale hardness of PDMS composites on the ratio of alkyl groups to silicon atoms. The open circles correspond to bimodal PDMS, and the filled circles correspond to unimodal PDMS. (Reproduced with permission from reference 123. Copyright 1987 Springer Verlag.)

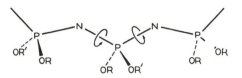

FIGURE 46. Sketch of a semi-inorganic phosphazene polymer (*127*). (Reproduced with permission from reference 126. Copyright 1977 John Wiley and Sons, Inc.)

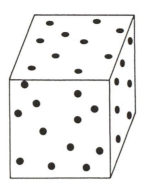

FIGURE 47. Sketch of a multiphase, thermoplastic elastomer (*129*).

relatively hard polystyrene domains act as temporary, physical cross-links, as shown in Figure 47 (*129*). The resulting elastomer is thermoplastic, and it may be reprocessed simply by heating it to above the glass transition temperature of polystyrene. It is thus a *reprocessible* elastomer. As already mentioned, more novel approaches probably could be learned by studies of the cross-linking techniques used by Nature in preparing bioelastomers.

A particularly challenging problem is the development of a more quantitative molecular understanding of the effects of filler particles, in particular carbon black in natural rubber and silica in siloxane polymers (*97, 98, 130, 131*). Such fillers provide tremendous reinforcement in elastomers in general, and how they do this is still only poorly comprehended. A related but even more complex problem involves much the same components, namely, one that is organic and one that is inorganic. When one or both components are generated in situ, however, an almost unlimited variety of structures and morphologies can be generated. How physical properties such as elastomeric behavior depend on these variables is obviously a challenging but important problem.

Numerical Problems

Some Typical Elongation or Compression Data

Suppose a network has tetrafunctional cross-links ($\phi = 4$, $A_\phi = 1/2$), a density of 0.900 g·cm^{-3}, and [f^*] ($\alpha = \infty$) = 0.100 N·mm^{-2} [10^5 N·m^{-2} (Pa) = 10^{-1} MN·m^{-2} (MPa) = 1.02 kg·cm^{-2}] at 298.2 K. Calculate the network chain density, the cross-link density, and the average molecular weight between cross-links (*5*).

Some Typical Swelling Data

A typical network studied in this regard might tetrafunctionally cross-linked in the undiluted state ($v_{2S} = 1.00$) and exhibit an equilibrium degree of swelling characterized by $v_{2m} = 0.100$ in a solvent having a molar volume (V_1) of 80 cm^3·mol^{-1} (8.00×10^4 mm^3·mol^{-1}) and an interaction parameter with the polymer (χ_1) corresponding to 0.30. Calculate the network chain density (5).

Solutions to Numerical Problems

Elongation or Compression

Use of the given data in equation 17 with k (Boltzmann constant) in units (1.381×10^{-20} $N \cdot mm \cdot K^{-1} \cdot chain^{-1}$) compatible with $[f^*]$ ($\alpha = \infty$) in units of $N \cdot mm^{-2}$ gives the following:

$$v/V = 4.86 \times 10^{16} \text{ chains} \cdot mm^{-3}$$

Use of Avogadro's number ($N_{avo} = 6.02 \times 10^{23}$ mol^{-1}) then gives the following:

$$v/V = 8.06 \times 10^{-8} \text{ mol of chains} \cdot mm^{-3}$$

As specified by the relationship $\mu = (2/\phi)v$, the density of cross-links would be one-half ($2/\phi$) of this value:

$$v/V = 4.03 \times 10^{-8} \text{ mol of cross-links} \cdot mm^{-3}$$

Because the polymer has a density (ρ) of 0.900 $g \cdot cm^{-3}$ (9.00×10^{-4} $g \cdot mm^{-3}$), the relationship $M_c = \rho/(v/V)$ indicates the following (M_c is the molecular weight between cross-links):

$$M_c = 1.12 \times 10^4 \text{ g} \cdot mol^{-1}$$

Swelling

The standard swelling relationship, equation 19, with $A'_\phi = 1$ would give the following:

$$v/V = 7.13 \times 10^{-8} \text{ mol of chains} \cdot mm^{-3}$$

Use of the improved relationship, equation 20, with the reasonable estimates (30) $\kappa = 20$ and $p = 2$ gives $K = 0.42$ (30), and thus:

$$v/V = 8.95 \times 10^{-8} \text{ mol of chains} \cdot mm^{-3}$$

This result is not very different from the value calculated with the simpler relationship given in equation 19.

Literature Cited

1. Flory, P. J. *Principles of Polymer Chemistry*; Cornell University Press: Ithaca NY, 1953.

2. Treloar, L. R. G. *The Physics of Rubber Elasticity*, 3rd ed.; Clarendon: Oxford, 1975.

3. Mark, J. E. *J. Chem. Ed.* **1981**, *58*, 898.

4. Flory, P. J. *Statistical Mechanics of Chain Molecules*; Interscience: New York, 1969.

5. Mark, J. E.; Erman, B. *Rubberlike Elasticity: A Molecular Primer*; Wiley-Interscience: New York, 1988.

6. Mason, P. *Cauchu: The Weeping Wood*; Australian Broadcasting Commission: Sydney, Australia, 1979.

7. Erman, B.; Mark, J. E. *Annu. Rev. Phys. Chem.* **1989**, *40*, 351.

8. *Characterization of Highly Cross-Linked Polymers*; Labana, S. S.; Dickie, R. A., Eds.; ACS Symposium Series 243; American Chemical Society: Washington, DC, 1984.

9. Mark, J. E. *Rubber Chem. Technol.* **1982**, *55*, 762.

10. Ogden, R. W. *Rubber Chem. Technol.* **1986**, *59*, 361.

11. James, H. M.; Guth, E. *J. Chem. Phys.* **1947**, *15*, 669.

12. James, H. M.; Guth, E. *J. Chem. Phys.* **1953**, *21*, 1039.

13. Ronca, G.; Allegra, G. *J. Chem. Phys.* **1975**, *63*, 4990.

14. Flory, P. J. *Proc. R. Soc. London A* **1976**, *351*, 351.

15. Flory, P. J. *Polymer* **1979**, *20*, 1317–1320.

16. Flory, P. J.; Erman, B. *Macromolecules* **1982**, *15*, 800.

17. Oeser, R.; Ewen, B.; Richter, D.; Farago, B. *Phys. Rev. Lett.* **1988**, *60*, 1041.

18. Ewen, B.; Richter, D. In *Elastomeric Polymer Networks*; Mark, J. E.; Erman, B., Eds.; Prentice Hall: Englewood Cliffs, NJ, 1992.

19. Erman, B.; Monnerie, L. *Macromolecules* **1989**, *22*, 3342.

20. Vilgis, T. A. *Prog. Coll. Polym. Sci.* **1987**, *75*, 4.

21. Vilgis, T. A. *Prog. Coll. Polym. Sci.* **1987**, *75*, 243.

22. Ball, R. C.; Doi, M.; Edwards, S. F. *Polymer* **1981**, *22*, 1010.

23. Gaylord, R. J. *Polym. Bull.* **1982**, *8*, 325.

24. Kilian, H.-G.; Enderle, H. F.; Unseld, K. *Colloid Polym. Sci.* **1986**, *264*, 866.

25. Mark, J. E.; Curro, J. G. *J. Chem. Phys.* **1983**, *79*, 5705.

26. Curro, J. G.; Mark, J. E. *J. Chem. Phys.* **1984**, *80*, 4521.

27. Mark, J. E.; Curro, J. G. *J. Chem. Phys.* **1984**, *80*, 5262.

28. Mark, J. E.; Curro, J. G. *J. Polym. Sci., Polym. Phys. Ed.* **1985**, *23*, 2629.

29. Finkelmann, H. In *Liquid Crystallinity in Polymers*; Ciferri, A., Ed.; VCH Publishers: New York, 1991.

30. Flory, P. J. *Macromolecules* **1979**, *12*, 119.

31. Noda, I.; Dowrey, A. E.; Marcott, C. In *Fourier Transform Infrared Characterization of Polymers*; Ishida, H., Ed.; Plenum: New York, 1987.

32. Schaefer, D. W.; Mark, J. E.; McCarthy, D.; Jian, L.; Sun, C.-C.; Farago, B. In *Polymer-Based Molecular Composites*; Schaefer, D. W.; Mark, J. E., Eds.; Materials Research Society: Pittsburgh, PA, 1990; Vol. 171.

33. Mark, J. E.; Odian, G. "Polymer Chemistry Course Manual;" American Chemical Society: Washington, DC, 1984.

34. Mark, J. E. *Rubber Chem. Technol.* **1975**, *48*, 495.

35. Mark, J. E. *Acc. Chem. Res.* **1979**, *12*, 49.

36. Flory, P. J.; Ciferri, A.; Hoeve, C. A. J. *J. Polym. Sci.* **1960**, *45*, 235.

37. Ciferri, A.; Hoeve, C. A. J.; Flory, P. J. *J. Am. Chem. Soc.* **1961**, *83*, 1015.

38. Goritz, D.; Muller, F. H. *Kolloid Z. Z. Polym.* **1973**, *251*, 679.

39. Mark, J. E. *Rubber Chem. Technol.* **1973**, *46*, 593.

40. Mark, J. E. *Macromol. Rev.* **1976**, *11*, 135.

41. Godovsky, Y. K. *Adv. Polym. Sci.* **1986**, *76*, 31.

42. Mark, J. E. *Makromol. Chem. Suppl.*, **1979**, *2*, 87.

43. Rempp, P.; Herz, J. E. *Angew. Makromol. Chemie* **1979**, *76/77*, 373.

44. Mark, J. E.; Rahalkar, R. R.; Sullivan, J. L. *J. Chem. Phys.* **1979**, *70*, 1794.
45. Llorente, M. A.; Mark, J. E. *Macromolecules* **1980**, *13*, 681.
46. Gottlieb, M.; Macosko, C.; Benjamin, G. S.; Meyers, K. O.; Merrill, E. W. *Macromolecules* **1981**, *14*, 1039.
47. Stadler, R.; Jacobi, M. M.; Gronski, W. *Makromol. Chem., Rapid Commun.* **1983**, *4*, 129.
48. Miller, D. R.; Macosko, C. W. *J. Polym. Sci. Polym. Phys. Ed*. **1987**, *25*, 2441.
49. Kornfield, J. A.; Spiess, H. W.; Nefzger, H.; Hayen, H.; Eisenbach, C. D. *Macromolecules* **1991**, *24*, 4787.
50. Mark, J. E. In *Silicon-Based Polymer Science: A Comprehensive Resource*; Zeigler, J. M.; Fearon, F. W. G., Eds.; Advances in Chemistry Series 224; American Chemical Society: Washington, DC, 1990.
51. Langley, N. R.; Dickie, R. A.; Wong, C.; Ferry, J. D.; Chasset, R.; Thirion, P. *J. Polym. Sci., A-2* **1968**, *6*, 1371.
52. Johnson, R. M.; Mark, J. E. *Macromolecules* **1972**, *5*, 41.
53. Mark, J. E.; Eisenberg, A.; Graessley, W. W.; Mandelkern, L.; Koenig, J. L. *Physical Properties of Polymers*, 1st ed.; American Chemical Society: Washington, DC, 1984.
54. Yu, C. U.; Mark, J. E. *Macromolecules* **1974**, *7*, 229.
55. De Gennes, P. G. *Scaling Concepts in Polymer Physics*; Cornell University Press: Ithaca, NY, 1979.
56. Mark, J. E.; Zhang, Z.-M. *J. Polym. Sci. Polym. Phys. Ed*. **1983**, *21*, 1971.
57. DeBolt, L. C.; Mark, J. E. *Macromolecules* **1987**, *20*, 2369.
58. Clarson, S. J.; Mark, J. E.; Semlyen, J. A. *Polym. Commun*. **1986**, *27*, 244.
59. Garrido, L.; Mark, J. E.; Clarson, S. J.; Semlyen, J. A. *Polym. Commun*. **1985**, *26*, 53.
60. Mark, J. E.; Andrady, A. L. *Rubber Chem. Technol*. **1981**, *54*, 366.
61. Llorente, M. A.; Andrady, A. L.; Mark, J. E. *Colloid Polym. Sci*. **1981**, *259*, 1056.
62. Zhang, Z.-M.; Mark, J. E. *J. Polym. Sci. Polym. Phys. Ed*. **1982**, *20*, 473.
63. Mark, J. E. In *Elastomers and Rubber Elasticity*; Mark, J. E.; Lal, J., Eds.; ACS Symposium Series 193; American Chemical Society: Washington, DC, 1982.
64. Mullins, L. *J. Appl. Polym. Sci*. **1959**, *2*, 257.
65. Mark, J. E.; Kato, M.; Ko, J. H. *J. Polym. Sci. C* **1976**, *54*, 217.
66. Smith, K. J., Jr.; Greene, A.; Ciferri, A. *Kolloid Z. Z. Polym*. **1964**, *194*, 49.
67. Morris, M. C. *J. Appl. Polym. Sci*. **1964**, *8*, 545.
68. Treloar, L. R. G. *Rep. Prog. Phys*. **1973**, *36*, 755.
69. Chan, B. L.; Elliott, D. J.; Holley, M.; Smith, J. F. *J. Polym. Sci. C* **1974**, *48*, 61.
70. Andrady, A. L.; Llorente, M. A.; Mark, J. E. *J. Chem. Phys*. **1980**, *72*, 2282.
71. Doherty, W. O. S.; Lee, K. L.; Treloar, L. R. G. *Br. Polym. J*. **1980**, *15*, 19.
72. Furukawa, J.; Onouchi, Y.; Inagaki, S.; Okamoto, H. *Polym. Bull*. **1981**, *6*, 381.
73. Su, T.-K.; Mark, J. E. *Macromolecules* **1977**, *10*, 120.
74. Chiu, D. S.; Su, T.-K.; Mark, J. E. *Macromolecules* **1977**, *10*, 1110.
75. Mark, J. E. *Polym. Eng. Sci*. **1979**, *19*, 254.
76. Mark, J. E. *Polym. Eng. Sci*. **1979**, *19*, 409.
77. Chiu, D. S.; Mark, J. E. *Colloid Polym. Sci*. **1977**, *225*, 644.
78. Llorente, M. A.; Andrady, A. L.; Mark, J. E. *J. Polym. Sci. Polym. Phys. Ed*. **1981**, *19*, 621.
79. Flory, P. J.; Chang, W. C. *Macromolecules* **1976**, *9*, 33.
80. Fixman, M.; Alben, R. *J. Chem. Phys*. **1973**, *58*, 1553.
81. Erman, B.; Mark, J. E. *J. Chem. Phys*. **1988**, *89*, 3314.
82. Sun, C.-C.; Mark, J. E. *J. Polym. Sci. Polym. Phys. Ed*. **1987**, *25*, 2073.

83. Tang, M.-Y.; Letton, A.; Mark, J. E. *Colloid Polym. Sci.* **1984**, *262*, 990.

84. Andrady, A. L.; Llorente, M. A.; Sharaf, M. A.; Rahalkar, R. R.; Mark, J. E.; Sullivan, J. L.; Yu, C. U.; Falender, J. R. *J. Appl. Polym. Sci.* **1981**, *26*, 1829.

85. Obata, Y.; Kawabata, S.; Kawai, H. *J. Polym. Sci. A-2* **1970**, *8*, 903.

86. Xu, P.; Mark, J. E. *J. Polym. Sci., Polym. Phys. Ed.* **1991**, *29*, 355.

87. Rivlin, R. S.; Saunders, D. W. *Philos. Trans. R. Soc. London A* **1951**, *243*, 251.

88. Wang, S; Mark, J. E. *J. Polym. Sci. Polym. Phys. Ed.* **1992**, *30*, 801.

89. Tanaka, T. *Phys. Rev. Lett.* **1978**, *40*, 820.

90. Tanaka, T. *Sci. Am.* **1981**, *244*, 124.

91. Ross, R.; Bornstein, P. *Sci. Am.* **1971**, *224*, 44.

92. Hoeve, C. A.; Flory, P. J. *Biopolymers* **1974**, *13*, 677.

93. *Elastin and Elastic Tissue*; Sandberg, L. B.; Gray, W. R.; Franzblau, C., Eds.; Plenum: New York, 1977.

94. Andrady, A. L.; Mark, J. E. *Biopolymers* **1980**, *19*, 849.

95. Gosline, J. M. In *The Mechanical Properties of Biological Materials*; Vincent, J. F. V.; Currey, J. D., Eds.; Cambridge University Press: Cambridge, 1980.

96. Jensen, M.; Weis-Fogh, T. *Philos. Trans. R. Soc. London Ser. B.* **1962**, *245*, 137.

97. Boonstra, B. B. *Polymer* **1979**, *20*, 691.

98. Rigbi, Z. *Adv. Polym. Sci.* **1980**, *36*, 21.

99. Warrick, E. L.; Pierce, O. R.; Polmanteer, K. E.; Saam, J. C. *Rubber Chem. Technol.* **1979**, *52*, 437.

100. Wolff, S.; Donnet, J.-B. *Rubber Chem. Technol.* **1990**, *63*, 32.

101. Mark, J. E. *Chemtech* **1989**, *19*, 230.

102. Mark, J. E.; Schaefer, D. W. In *Polymer-Based Molecular Composites*; Schaefer, D. W.; Mark, J. E., Eds.; Materials Research Society: Pittsburgh, PA, 1990; Vol. 171.

103. *Ultrastructure Processing of Ceramics, Glasses, and Composites*; Hench, L. L.; Ulrich, D. R., Eds.; Wiley: New York, 1984.

104. *Ultrastructure Processing of Advanced Ceramics*; Mackenzie, J. D.; Ulrich, D. R., Eds.; Wiley: New York, 1988.

105. Ulrich, D. R. *J. Non-Cryst. Solids* **1988**, *100*, 174.

106. Ulrich, D. R. *Chemtech* **1988**, *18*, 242.

107. Mackenzie, J. D. *J. Non-Cryst. Solids* **1988**, *100*, 162.

108. Ulrich, D. R. *J. Non-Cryst. Solids* **1990**, *121*, 465.

109. Ning, Y.-P.; Tang, M.-Y.; Jiang, C.-Y.; Mark, J. E.; Roth, W. C.; *J. Appl. Polym. Sci.* **1984**, *29*, 3209.

110. Mark, J. E.; Ning, Y.-P. *Polym. Bull.* **1984**, *12*, 413.

111. Schmidt, H. In *Inorganic and Organometallic Polymers*; Zeldin, M.; Wynne, K. J.; Allcock, H. R., Eds.; ACS Symposium Series 360; American Chemical Society: Washington, DC, 1988; p 333.

112. Schmidt, H.; Wolter, H. *J. Non-Cryst. Solids* **1990**, *121*, 428.

113. Nass, R.; Arpac, E.; Glaubitt, W.; Schmidt, H. *J. Non-Cryst. Solids* **1990**, *121*, 370.

114. Wang, B.; Wilkes, G. L. *J. Polym. Sci. Polym. Chem. Ed.* **1991**, *29*, 905.

115. Wilkes, G. L.; Huang, H.-H.; Glaser, R. H. In *Silicon-Based Polymer Science*; Zeigler, J. M.; Fearon, F. W. G., Eds.; Advances in Chemistry Series 224; American Chemical Society: Washington, DC, 1990; Vol. 224; pp 207.

116. Brennan, A. B.; Wang, B.; Rodrigues, D. E.; Wilkes, G. L. *J. Inorg. Organomet. Polym.* **1991**, *1*, 167.

117. Sobon, C. A.; Bowen, H. K.; Broad, A.; Calvert, P. D. *J. Mater. Sci. Lett.* **1987**, *6*, 901.

118. Calvert, P.; Mann, S. *J. Mater. Sci.* **1988**, *23*, 3801.

119. Azoz, A.; Calvert, P. D.; Kadim, M.; McCaffery, A. J.; Seddon, K. R. *Nature* (*London*) **1990**, *344*, 49.
120. Doyle, W. F.; Uhlmann, D. R. In *Ultrastructure Processing of Advanced Ceramics*; Mackenzie, J. D., Uhrich, D. R., Eds.; Wiley-Interscience: New York, 1988; p 795.
121. Doyle, W. F.; Fabes, B. D.; Root, J. C.; Simmons, K. D.; Chiang, Y. M.; Uhlmann, D. R. In *Ultrastructure Processing of Advanced Ceramics*; Mackenzie, J. D.; Ulrich, D. R., Eds.; Wiley-Interscience: New York, 1988; p 953.
122. Boulton, J. M.; Fox, H. H.; Neilson, G. F.; Uhlmann, D. R. In *Better Ceramics Through Chemistry IV*; Zelinski, B. J. J.; Brinker, C. J.; Clark, D. E.; Ulrich, D. R., Eds.; Materials Research Society: Pittsburgh, PA, 1990; Vol. 180, p 773.
123. Mark, J. E.; Sun, C.-C. *Polym. Bull.* **1987**, *18*, 259.
124. Ning, Y. P.; Zhao, M. X.; Mark, J. E. In *Chemical Processing of Advanced Materials*; Hench, L. L.; West, J. K., Eds.; Wiley: New York, 1992.
125. Zhao, M. X.; Ning, Y. P.; Mark, J. E. In *Proceedings of the Symposium on Composites*: *Processing, Microstructure, and Properties*; Sacks, M. D., Ed.; American Ceramics Society: Westerville, OH, 1993.
126. Mark, J. E.; Yu, C. U. *J. Polym. Sci. Polym. Phys. Ed.* **1977**, *15*, 371.
127. Andrady, A. L.; Mark, J. E. *Eur. Polym. J.* **1981**, *17*, 323.
128. Mark, J. E.; Allcock, H. R.; West, R. *Inorganic Polymers*, Prentice Hall: Englewood Cliffs, NJ, 1992.
129. Aggarwal, S. L. *Polymer* **1976**, *17*, 938.
130. Polmanteer, K. E.; Lentz, C. W. *Rubber Chem. Technol.* **1975**, *48*, 795.
131. Kraus, G. *Rubber Chem. Technol.* **1978**, *51*, 297.

Selected General Bibliography

Flory, P. J. *Principles of Polymer Chemistry*; Cornell University Press: Ithaca NY, 1953.

Meares, P. *Polymers*: *Structure and Bulk Properties*; Van Nostrand: New York, 1965.

Dusek, K.; Prins, W. *Adv. Polym. Sci.* **1969**, *6*, 1.

Polymer Networks: *Structure and Mechanical Properties*; Chompff, A. J.; Newman, S., Eds.; Plenum: New York, 1971.

Smith, K. J. Jr., In *Polymer Science*; Jenkins, A. D., Ed.; North-Holland: Amsterdam, Netherlands, 1972.

Rubber Technology; Morton, M., Ed.; Van Nostrand Reinhold: New York, 1973.

Wall, F. T. *Chemical Thermodynamics*, 3rd ed.; Freeman: San Francisco, 1974.

Rubber and Rubber Elasticity; Dunn, A. S., Ed.; Wiley-Interscience: New York, 1974; Polymer Symposium 48.

Treloar, L. R. G. *The Physics of Rubber Elasticity*; Clarendon: Oxford 1975.

Chemistry and Properties of Crosslinked Polymers; Labana, S. S., Ed.; Academic: New York, 1977.

Science and Technology of Rubber; Eirich, F. R., Ed.; Academic: New York, 1978.

Nash, L. K.; *J. Chem. Ed.* **1979**, *56*, 363.

Mark, J. E. *J. Chem. Ed.* **1981**, *58*, 898.

Elastomers and Rubber Elasticity; Mark, J. E.; Lal, J., Eds.; ACS Symposium Series 193; American Chemical Society: Washington, DC, 1982.

Eichinger, B. E. *Annu. Rev. Phys. Chem.* **1983**, *34*, 359.

Characterization of Highly Cross-Linked Polymers; Labana, S. S.; Dickie, R. A., Eds.; ACS Symposium Series 243; American Chemical Society: Washington, DC, 1984.

Furukawa, J. *Makromol. Chem. Suppl.* **1985**, *14*, 3.

Kilian, H. G.; Enderle, H. F.; Unseld, K. *Colloid Polym. Sci.* **1986**, *264*, 866.

Advances in Elastomers and Rubber Elasticity; Lal, J.; Mark, J. E., Eds.; Plenum: New York, 1986.

Queslel, J. P.; Mark, J. E. In *Encyclopedia of Polymer Science and Engineering*, 2nd ed.; Meyers, R. A., Ed.; Wiley-Interscience: New York, 1986.

Elastomers and Rubber Technology; Singler, R. E.; Byrne, C. A., Eds.;   U.S. Government Printing Office: Washington, DC, 1987.

Queslel, J. P.; Mark, J. E. In *Encyclopedia of Physical Science and Technology*; Meyers, R. A., Ed.; Academic: New York, 1987.

Queslel, J. P.; Mark, J. E. *J. Chem. Ed.* **1987**, *64*, 491.

Edwards, S. F.; Vilgis, T. A. *Rep. Prog. Phys.* **1988**, *51*, 243.

Biological and Synthetic Polymer Networks; Kramer, O., Ed.; Elsevier: London, 1988.

Mark, J. E.; Erman, B. *Rubberlike Elasticity: A Molecular Primer*; Wiley-Interscience: New York, 1988.

Heinrich, G.; Straube, E.; Helmis, G. *Adv. Polym. Sci.* **1988**, *85*, 33.

Erman, B.; Mark, J. E. *Annu. Rev. Phys. Chem.* **1989**, *40*, 351.

Queslel, J. P.; Mark, J. E. In *Comprehensive Polymer Science*; Allen, G., Ed.; Pergamon: Oxford, 1989.

Mark, J. E. *Kautsch. Gummi Kunstst.* **1989**, *42*, 191.

Queslel, J. P.; Mark, J. E. In *Determination of Molecular Weight*; Cooper, A. R., Ed.; Wiley-Interscience: New York, 1989.

Mark, J. E. In *Frontiers of Macromolecular Science*; Saegusa, T.; Higashimura, T.; Abe, A., Eds.; Blackwell Scientific Publishers: Oxford, 1989.

International Seminar on Elastomers, Proceedings; White, J. L.; Murakami, K., Eds.; Wiley: New York, 1989.

Physical Networks: Polymers and Gels; Burchard, W.; Ross-Murphy, S. B., Eds.; Elsevier: London, 1990.

Queslel, J. P.; Mark, J. E. In *Encyclopedia of Physical Science and Technology*, 2nd ed.; Meyers, R. A., Ed.; Academic: New York, 1992.

Elastomeric Polymer Networks; Mark, J. E.; Erman, B., Eds.; Prentice-Hall: Englewood Cliffs, NJ, 1992.

Chapter 2

The Glassy State and the Glass Transition

Adi Eisenberg

**Department of Chemistry, McGill University,
801 Sherbrooke St. W., Montreal, Quebec,
Canada H3A 2KG**

The glass transition is perhaps the most important single parameter that determines the application of many noncrystalline polymers now available. For example, if the glass transition temperature of a polymeric material is approximately -70 °C, that particular polymer would not be suitable for a window application but would be appropriate for a rubber tire or tubing application. Conversely, if the glass transition temperature is $+100$ °C, use as a rubber will be completely excluded, but any of the hard-plastic applications may be considered.

In structural terms, glasses are characterized by the absence of long-range order. Order along a polymer chain in polymeric glasses is, naturally, to be expected, as well as some short-range liquidlike order, which is encountered not only in polymeric glasses but also in regular liquids. Order in noncrystalline materials, in general, has been the subject of continuing interest and the topic of specialized publications (*1*). In addition, many polymers are partly crystalline, and the crystalline regions are subject to the same behavior as crystals in general.

The noncrystalline regions exhibit a common range of phenomena that are the subject of this chapter. These phenomena are exhibited not only by polymers but also by many other materials, and the study of glass transition phenomena is relatively new. The phenomena accompanying vitrification were not really explored until the 1920s, and a complete theoretical understanding of the glass transition temperature is still not available. The area is receiving considerable attention at this time, despite the fact that many original publications are available on the glass transition specifically

2505–2/93/0061 $9.75/1

and on glass transition phenomena in general. In addition, several reviews or compendia have also been published (2–13).

Although most of the topics (e.g., viscoelasticity, polymer spectroscopy, and rubber elasticity) discussed in this volume have been reviewed extensively in monographs or even single-topic textbooks, the glass transition has been reviewed only in much shorter review articles. Furthermore, reviews devoted to the fundamentals of the field date back a number of years. Therefore, this chapter will be concerned to a large extent with an introductory review of the glass transition. The inclusion of both the fundamental aspects and a detailed treatment of the more advanced topics would lengthen the presentation beyond the limits justified by the relative importance of the topic. However, for the interested reader who wishes to delve more deeply into the subject, recent relevant references are provided.

The first part of the discussion will be devoted to a presentation of glass transition phenomena, stressing particularly volumetric and mechanical properties. The second part will be a brief introduction to glass transition theories, stressing a very straightforward free volume theory but also mentioning the thermodynamic and kinetic theories. The most modern treatments will be mentioned very briefly, along with the relevant references; an extensive discussion of recent theories is beyond the scope of this chapter. The third section will be devoted to a discussion of the molecular parameters (e.g., chain stiffness, internal plasticization, and intramolecular forces) and their effect on the glass transition. The discussion will answer such questions as why polystyrene has a glass transition temperature very different from those of poly(n-butyl acrylate) and poly(dimethylsiloxane). The fourth section will explore controllable parameters and their effect on the glass transition. The factors that can change the glass transition temperature of a particular polymer will be discussed. Factors such as the pressure, the presence of diluent, the molecular weight, and the presence of comonomers will be explored.

The fifth section will deal very briefly with selected methods of determining the glass transition temperature. Many methods are available, and only some of the most important ones will be described briefly to introduce the most common techniques. The chapter will conclude with a discussion of molecular motions in the glassy state and methods of measuring them, specifically such methods that allow one to determine which piece of a molecule in a glassy material moves at what frequency as a function of temperature. The experimental techniques for measuring molecular motion are quite well established; however, the results of the experiments (i.e., the molecular aspects) are still being interpreted and, therefore, this field is very challenging and interesting. This area also has many practical implications, in that the presence of these low-temperature molecular motions occasionally renders the material ductile, as opposed to brittle, and improves a wide range of other physical properties.

Many types of materials exhibit glass transition phenomena. The focus of this chapter is on polymeric materials, the so-called linear, synthetic, high-molecular-weight polymers. Some of the other materials that exhibit the glass transition phenomenon can be prepared either as crystalline or noncrystalline solids. Therefore, a discussion of the contrasting behavior of crystalline and glassy materials is given in the section on glass transition phenomena. The vitrification process is similar for this entire range of materials.

Glass Transition Phenomena

Materials Exhibiting Glass Transition Phenomena

The list of materials exhibiting glass transition phenomena (Table I) starts with the network glasses (i.e., materials like SiO_2, B_2O_3, P_2O_5, etc.). Network glasses are three-dimensional polymers in which the repeat units are tri- or tetrafunctional moieties like those found in P_2O_5 or SiO_2. If network modifiers such Na_2O or K_2O are introduced into these networks, some network points are ruptured, and an Si–O–Si bridge, for example, is converted to two $-SiO^-Na^+$ groups. Some linear segments are to be expected in these modified networks; for example, for an equimolar mixture of Na_2O and P_2O_5, a predominantly linear polymer, $(NaPO_3)_x$, is found. In contrast to the pure network glasses (e.g., SiO_2), for which the glass transition temperature is about 1200 °C, the modified networks based on silica (e.g., soda–lime–silica glasses) have a considerably lower glass transition temperature of about 500 °C.

The next category, linear or branched polymers, is the main subject of this chapter. Included in this category are all the commercially available organic polymers, as well as some of the inorganic polymers (e.g., those based on sulfur, selenium, or even the phosphates). Under other classification schemes, the inorganic polymers might be included in the category of modified networks. Numerically, the category of linear or branched polymers is perhaps the largest category of all the materials listed in Table I.

The fourth group comprises the hydrogen-bonded systems (e.g., glycerine). The glass transition temperatures for these materials are still lower (for example, glycerine has a glass transition temperature of approximately -80 °C). The fifth category includes salts or salt mixtures of relatively low molecular weight (e.g., zinc chloride, beryllium chloride, potassium carbonate–magnesium carbonate mixtures, and potassium nitrate–calcium nitrate mixtures). These materials have glass transition temperatures that are considerably lower than those of the modified

Table I. Materials Exhibiting Glass Transition Phenomena

Category	Example
Network glasses	SiO_2, B_2O_3, P_2O_5, and As_2S_3
Modified networks	$SiO_2 + Na_2O$ and $P_2O_5 + K_2O$
Linear or branched polymers	Polystyrene and polyethylene
H-bonded glasses	Glycerine
Salts or salt mixtures	$ZnCl_2$, BeF_2, and $K_2CO_3 \cdot MgCO_3$
Electrolyte solutions	H_2O–HCl eutectic
Metals	$Pd_{80}Si_{20}$ and $Fe_{40}Ni_{40}P_{14}B_6$
Low-molecular-weight materials	2-Methylpentane

Source: Reference 6.

networks but usually somewhat higher than those of the normal organic polymers. Electrolyte solutions are still another category (for example, the water–HCl eutectic, which has a glass transition temperature well below $-100\ °C$).

Perhaps the most surprising category of all is that of the glassy metals. Included in this group are materials that are prepared from liquid metals, usually mixtures of several elements, by rapid cooling at rates of $100\,000–1\,000\,000\ °C/s$. Glassy metals have glass transition temperatures that span a fairly wide region, $0–200\ °C$ being typical.

The last category includes low-molecular-weight materials (e.g., 2-methylpentane, mixtures of methyltetrahydrofuran and toluene, and various other nonlinear hydrocarbons). These materials, when cooled at reasonably rapid rates, vitrify quite easily to give materials with glass transition temperatures ranging from -150 to $-200\ °C$.

A useful classification scheme has been suggested by Angell et al. (14); in their scheme, glass formers are grouped into families ranging from "strong" to "fragile" depending on their relaxation behavior. Networks such as SiO_2 are classified as strong, whereas the low-molecular-weight liquids are classified as fragile. Polymer would be typical of materials exhibiting intermediate relaxation behavior.

Changes in Thermodynamic Properties Accompanying the Glass Transition

A number of reviews describe the phenomenology of the glass transition (3–13). The discussion of glass transition phenomena can begin most conveniently with a study of the volumetric properties of polymeric materials, especially those that can either crystallize or vitrify, depending on the handling method.

Volume and Enthalpy Changes

Figure 1A is a plot of the volume as a function of temperature for a crystalline material. As the material is heated, the volume expands in a perfectly normal fashion until the temperature approaches the melting point. If the material is completely crystalline and has a sharp melting point, the volume changes discontinuously and increases appreciably over very small temperature range. For normal polymeric materials, however,

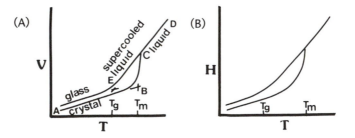

FIGURE 1. (A) Volume–temperature and (B) enthalpy–temperature relations for glass-forming materials.

the melting process takes place over a fairly wide temperature range. For this reason, the volume increases in a manner indicated by the solid line in Figure 1A rather than discontinuously, as is observed with low-molecular-weight materials. At the melting point, the material has lost all traces of crystallinity; it no longer shows the sharp X-ray bands expected from partly or completely crystalline materials; only an amorphous halo is observed. As the material is heated beyond the melting point (in the region between points C and D in Figure 1A), a much more rapid expansion than that observed below the melting point takes place. The rapid expansion after the melting point suggests that the packing is much less orderly than in the crystal, as expected, and that much more free volume is introduced as the material is heated. (This concept of free volume, although not very quantitative, is extremely useful and will be referred to repeatedly in the subsequent discussion). Far above the melting point (for example, at point D in Figure 1A), the material is a liquid of a relatively high viscosity, depending on the molecular weight.

During cooling from point D, most polymeric materials may be cooled beyond point C without crystallization. Some polymers (e.g., polyethylene) cannot be prevented from crystallizing, but many others (e.g., isotactic polystyrene) can be cooled below the crystalline melting point without the slightest danger of crystallization. As a matter of fact, in some cases, it is extremely difficult or even impossible to crystallize polymers from the melt. Beyond point C, cooling cannot continue indefinitely while still maintaining a more or less linear relationship between the volume and the temperature. Within a relatively narrow temperature range, indicated by point E in Figure 1A, the slope of the volume–temperature plot changes dramatically, and below that temperature range, the expansion coefficient, which is the slope of this plot, becomes very much smaller than it was for the liquid.

The glass transition temperature has been defined as the intersection of the straight line segments of the volume–temperature plot in the vicinity of point E. Experimentally, however, the glass transition is not a precise point but a relatively narrow temperature range. In volume–temperature plots, the glass transition temperature is defined by a discontinuity in the rate of change of volume with respect to temperature, that is, by a change in the expansion coefficient. A very similar behavior is observed if the enthalpy is plotted as a function of temperature (Figure 1B).

The kinetics of the vitrification process can be explored by examining the phenomena in the vicinity of point E accompanying the cooling of a polymer at different cooling rates (5). Far above the glass transition the material is at its equilibrium volume; the material then is cooled rapidly until a change in slope is observed in the volume–temperature plots, as shown in Figure 2. The inflection point, or the glass transition temperature, is observed at a reasonably high temperature (point A). When cooling is done at a very slow rate, the inflection point is observed at a considerably lower temperature (point B). Intermediate cooling rates lead to inflection points at intermediate temperatures. This type of experiment shows clearly that the glass transition temperature, as determined by straightforward cooling experiments, is a function of the cooling rate. The magnitude of the effect of cooling rate on the glass transition temperature is approximately a 3 °C change per order of magnitude change in the cooling rate. Thus, a material that has been cooled at 100 °C/min will have a glass transition temperature that is approximately 6 °C higher than the same material cooled at 1 °C/min. Therefore, the glass transition temperature of a rapidly quenched sample is quite different from that of a slowly cooled sample.

Another interesting aspect of the glass transition can be seen when a material that is cooled very rapidly to a point near A in Figure 2 is kept at a constant temperature and the volume is monitored as a function of time at this constant temperature. Depending on the actual cooling rate and the temperature to which the sample is cooled, behavior of the type shown in Figure 3 will be encountered (5). In Figure 3, the volume at any time t [$V(t)$] minus the equilibrium volume [i.e., the volume attained on storage of the sample for a very long time; $V(\infty)$] is plotted as a function of the logarithm of time. Different curves are obtained, depending on the temperature to which a sample is quenched. The glass transition temperature (T_g) in Figure 3 is taken to be the normal T_g, that is, the value obtained at a cooling rate of approximately 1 °C/min by the usual technique. Thus, if a material is cooled very rapidly to a temperature of $T_g - 10$ °C and stored at that temperature for some time, the volume shrinks along the line shown by the curve labeled $T_g - 10$ °C. At $T_g - 5$ °C, the shrinkage is somewhat more rapid; at T_g or $T_g + 2$ °C, the shrinkage is still more rapid. In any case, the volume of a material that is quenched very rapidly to a particular temperature in the vicinity of the glass transition and maintained at that temperature will continue shrinking. This behavior is of great commercial importance if the sample is expected to retain its structural integrity and dimensional stability over a long period. The detailed aspects of the

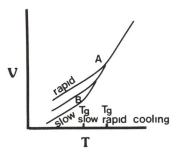

FIGURE 2.   Volume–temperature behavior as a function of cooling rate.

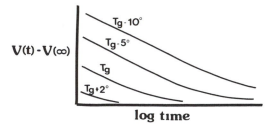

FIGURE 3.   Isothermal volume contraction near the glass transition temperature. (Adapted from reference 5.)

volume shrinkage process have been investigated by a number of authors. Kovacs (*5*) provided the first detailed treatment, which was followed by more recent approaches (*15, 16*). The physical aging process of amorphous materials is well explored and is the subject of a monograph (*17*).

Because enthalpy responds to temperature changes in the same manner as volume, it is not surprising that enthalpy relaxation also takes place and that the thermal properties of a polymer change just as its volume changes with storage in the vicinity of the glass transition temperature. Enthalpy relaxation, specifically, has received considerable attention in view of its importance in thermal measurements of the glass transition (*18*).

Hysteresis Effects

Sometimes unusual hysteresis effects are evident in volume–temperature plots, as shown by two examples in Figure 4 (*19*). Consider what happens when the sample is cooled very rapidly along line 1 in Figure 4 and subsequently heated, very much more slowly than the rate at which was cooled, along the dotted line 2. The sample, as it warms up, has much more time to adjust to its environment than it had while it was cooling. This longer time for adjustment results in a volume shrinkage so that the volume of the sample near T_g on heating is considerably smaller than the volume at the corresponding temperature on cooling. Storing the sample for long periods, as was shown in Figure 3, results in a volume shrinkage; a slow heating after a rapid cooling thus has the same effect. When thermal tests are performed, the same phenomena are, obviously, also operative. Thus, in a DSC (differential scanning calorimetry; *see Methods Based on Specific Heat Changes*) experiment performed at some heating rate on a sample that had been quenched at a much more rapid rate, artifacts in the thermogram can appear that are strictly the result of hysteresis or relaxation phenomena of the type discussed here.

The opposite effect is observed when the sample is cooled very slowly and heated rapidly along lines 3 and 4 in Figure 4. As expected, slow cooling results in a very low apparent glass transition temperature. However, a very rapid heating of a slowly cooled sample results in an overshooting of the original glass transition temperature and leads to a very rapid expansion considerably above the inflection point of the cooling

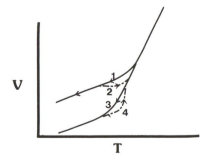

FIGURE 4. Hysteresis effects in volume–temperature plots near T_g. (Adapted from reference 19.)

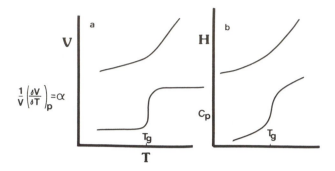

FIGURE 5.   Temperature dependence of (a) volume and expansion coefficient (α) and (b) enthalpy and heat capacity (C_p) near T_g.

curve. This rapid expansion again leads to a serious hysteresis effect, which makes the determination of the glass transition temperature from volumetric properties alone somewhat uncertain if the heating and cooling rates are unequal. Again, thermal measurements also reflect these phenomena. These effects emphasize the need to maintain equal heating and cooling rates when determining glass transition temperatures by these techniques.

Expansion Coefficient and Heat Capacity

Finally, it is worth looking at the first derivative of either the volume–temperature or the enthalpy–temperature curve. The first derivative of the volume–temperature curve divided by the volume, $(1/V)(\delta V/\delta T)_p$ (where V is volume, T is temperature, and P is pressure), is simply the expansion coefficient, α. Plots of V versus T and α versus T, shown in Figure 5a, indicate clearly a discontinuity in the plot of first derivative of volume versus time, not just a change in slope. As a result of this discontinuity, the glass transition temperature has been referred to as a second-order transition in some of the early literature. However, the glass transition temperature, as it is observed, is not a second-order thermodynamic transition but a kinetic phenomenon that may have underlying thermodynamic reasons, as will be shown later. Similar curves of the first derivative of enthalpy versus temperature are shown in Figure 5b. The heat capacity (C_p), which is the first derivative of the enthalpy with respect to temperature at constant pressure $(\delta H/\delta T)_p$, shows a discontinuity similar to that observed in the expansion coefficient, again explaining the historical reference to the glass transition as a second-order transition.

Changes in Mechanical Properties Accompanying the Glass Transition

Viscosity

Figure 6 shows a plot of the logarithm of viscosity as a function of temperature for polymeric materials (*19*). Some disagreement exists with respect to the shape of the curve at very high viscosities, with some authors suggesting a continuous increase rather than a sigmoidal character. Whatever the shape of the curves at viscosities (η) of $\geq 10^{18}$ poises, in the vicinity of the glass transition temperature, no dramatic discontinuities of the type observed in volume–temperature plots are seen; however, for an extremely wide range of noncrystalline materials (including linear organic polymers, the splat-cooled metals, and organic or inorganic low-molecular-weight-materials), the viscosity at the glass transition temperature is of the order of 10^{13} poises.

Modulus

A similar absence of discontinuities is observed in a plot of the modulus (G or E) as a function of temperature. For viscoelastic materials, the modulus is a function not only of the temperature of the measurement but also of the time. Therefore, a time scale of experiment must be fixed. Any convenient point could be adopted, and for the present discussion, the 10-s modulus will be adopted. The plot of the 10-s modulus as a function of temperature, as shown in Figure 7, shows four distinct regions of behavior (*20a*). At very low temperatures, the modulus has a value of the order of 10^{10} dynes/cm^2, typical of organic glasses. Over a fairly narrow temperature range between $T_g - 20$ °C and $T_g + 20$ °C, the modulus drops by approximately 3 orders of magnitude, and sometimes even more, to a value of the order of $10^{6.5}$ dynes/cm^2, a value typical of lightly cross-linked rubbers or of physically entangled, long chains. The modulus can

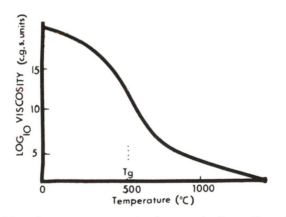

FIGURE 6. Viscosity–temperature plot for a soda–lime–silica glass. (Reproduced with permission from reference 19. Copyright 1956 Methuen.)

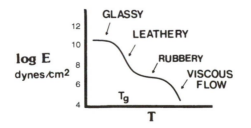

FIGURE 7. Modulus–temperature plot for a typical linear polymer. (Adapted from reference 20.)

drop very rapidly beyond that point as the temperature increases, if the material is not cross-linked. However, in the presence of cross-links, the material can exhibit a plateau characteristic of rubbery materials at a value of the Young's modulus of $\sim 10^{6.5}$ dynes/cm^2 (or higher), which can extend over considerable temperature ranges. If the material is not cross-linked, the modulus drops further at a temperature depending on the molecular weight. In the glass transition region, for a very wide range of materials, the modulus has a value of the order of 10^9 dynes/cm^2; however, if the material contains very strong intermolecular forces (for example, electrostatic forces due to ionic groups), the modulus would have a much higher value at the glass transition temperature. However, for linear, branched, or slightly cross-linked organic polymers such as polystyrene, poly(methyl methacrylate), or polybutadiene, a modulus of 10^9 dynes/cm^2 is typical in the vicinity of the glass transition temperature.

Loss Tangent or Energy Dissipation

The loss tangent is a measure of the energy dissipation, and many experimental techniques are available for its determination. For example, if a rubber ball is dropped from a particular height and it bounces back to almost the same height, the loss tangent of that material would be fairly low; on the other hand, if the ball rebounds only slightly, the loss tangent would be quite high. The higher the loss tangent, or the greater the amount of energy that is dissipated on impact, the lower the bounce of the ball. Many quantitative methods of determining the loss tangent are available (20b), but these are not of importance at this point.

When the energy dissipation of a polymer is investigated as a function of temperature at constant frequency, a plot of the type found in Figure 8 is observed (21). A number of low-temperature energy dissipation peaks are seen (see Molecular Motions below the Glass Transition) as well as a very dramatic peak at the glass transition temperature. In many cases, this loss tangent maximum is taken as the glass transition temperature. A cautionary comment, however, is called for. The position of the peak in dynamic experiments is a function of the frequency at which the experiment is done; therefore, when results obtained from volume–temperature studies are compared with those from dynamic mechanical studies, comparable frequencies or comparable cooling rates must be involved. The volumetric or thermodynamic results for a material studied at a cooling rate of 1 °C/min should be compared to the loss tangent peak obtained at a frequency of about 1 cycle/min; comparisons over drastically different frequencies or cooling rate ranges may lead to erroneous results. Also, peak positions in the loss modulus (E'' or G'') are not usually observed at the same temperature as peaks in the loss tangent (22). The loss tangent will be encountered again when molecular motions are discussed later in this chapter.

Theories of the Glass Transition

The theories of the glass transition are the subject of several reviews (6, 10, 11). A complete theoretical understanding of the glass transition phe-

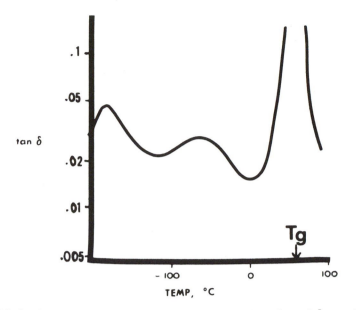

FIGURE 8.   Loss tangent versus temperature at ca. 1 Hz for poly[2-methyl-6-(2-octyl)phenylene oxide].   (Adapted from reference 21.)

nomenon is not yet available. The theories that have been proposed can be divided into three categories: free volume theories, thermodynamic theories, and kinetic theories. Most recently, statistical mechanical treatments have received considerable attention. In the present discussion, only one free volume theory will be described briefly, to be followed by a brief mention of a thermodynamic and a kinetic theory. References will be provided to useful reviews of modern approaches.

Free Volume Theory

The free volume theory to be discussed here starts with the Doolittle equation (*23*), which is given as equation 1:

$$\eta = A \exp(BV_o/V_f) \tag{1}$$

In equation 1, η is the viscosity; V_o and V_f are the occupied and the free volumes, respectively; and A and B are constants. This equation provides the theoretical background for the so-called Williams–Landell–Ferry equation (commonly known as a WLF equation; *24*), which is of enormous importance in the study of polymer viscoelasticity. Taking the logarithm of equation 1 gives equation 2:

$$\ln \eta = \ln A + BV_o/V_f \tag{2}$$

If f (fractional free volume) is defined as $V_f/(V_o + V_f) \simeq V_f/V_o$ (because $V_o \gg V_f$), then equation 2 can be rewritten as follows:

$$\ln \eta = \ln A + B \cdot 1/f \tag{3}$$

If the free volume at the glass transition temperature is f_g and if f_g increases above the glass transition temperature with the expansion coefficient α_f, which is approximately equal to the liquid-state expansion coefficient minus the glassy-state expansion coefficient (or to the difference in expansion coefficients above and below the glass transition temperature), then the fractional free volume at any temperature (T) above the glass transition, f_T, is given by equation 4:

$$f_T = f_g + \alpha_f(T - T_g) \tag{4}$$

By substituting equation 4 into equation 3, the ratio of the viscosities (a_T), that is η_T, the viscosity at any temperature T, divided by η_{T_g}, the viscosity at T_g, can be solved from equation 5:

$$\ln(\eta_T/\eta_{T_g}) = \ln a_T = B(1/f_T - 1/f_g) \tag{5}$$

After substituting equation 4 into equation 5 and appropriate manipulations, equation 6 is obtained:

$$\ln a_T = B\left[\frac{1}{f_g + \alpha_f(T - T_g)} - \frac{1}{f_g}\right]$$

$$= \frac{B}{f_g}\left[\frac{f_g - f_g - \alpha_f(T - T_g)}{f_g + \alpha_f(T - T_g)}\right]$$

$$= -\frac{B}{f_g}\left[\frac{T - T_g}{(f_g/\alpha_f) + T - T_g}\right]$$

$$\log a_T = -\frac{B}{2.3f_g}\left[\frac{T - T_g}{(f_g/\alpha_f) + T - T_g}\right] \tag{6}$$

The constant B is very close to unity. Equation 6 is a form of the WLF equation. The WLF equation in terms of the so-called universal parameters is given in equation 7 (24):

$$\log a_T = -17.4(T - T_g)/(51.6 + T - T_g) \tag{7}$$

The constants 17.4 and 51.6 are nearly universal, being valid, within a reasonable degree of approximation, for a very wide range of materials. The value 17.4 for the first constant implies that $1/2.3f_g = 17.4$ or that $f_g \simeq 0.25$. The fractional free volume at the glass transition temperature is, therefore, around 2.5% for most known materials. The value 51.6, the second so-called universal constant, suggests, in turn, that $f_g/\alpha_f \simeq 51.6$. With $f_g \simeq 0.25$, the free volume expansion coefficient (α_f) would be about 4.8×10^{-4}/K. Both of these parameters, although not universal, are valid for a wide range of materials, although by no means for all glass formers.

This development forms the core of the so-called free volume theory, which suggests that the free volume at the glass transition temperature

(f_g) is around 2.5% for most materials. This relationship is very useful in predicting various effects, for example, those of molecular weight and plasticizer content. Several other approaches to free volume theories exist (*25*), and a review has been provided by Grest and Cohen (*26*). The presentation in this chapter is meant to provide a conceptual introduction to the field.

Thermodynamic Theory

The most sophisticated example of a thermodynamic theory is that due to Gibbs and Di Marzio (*see* reference 27), which is concerned with the configurational entropy of a polymer as a function of temperature. The theory suggests that if the configurational entropy approaches 0 at a temperature above 0 K then a thermodynamic glass transition should exist; this transition is called T_2. The theory suggests that the behavior of the experimental glass transition, as observed under normal kinetic conditions, is very similar to that of T_2. Thus, although it is impossible to reach T_2, conclusions about the glass transition temperature can be drawn from the theory. Within this framework, a wide range of predictions can and have been made (*27*) about the behavior of the glass transition as a function of cross-link density, plasticizer content, molecular weight, etc. The agreement between theory and experiment is excellent. (The experimental variation of the glass transition with the various parameters mentioned is the subject of the next section.) The Gibbs–Di Marzio theory is most insightful, but because of its mathematical complexity, it will not be discussed further. Recent work by Sillinger (*28*) has suggested, however, that such an "ideal" thermodynamic glass transition cannot occur for substances of limited molecular weight and with conventional intermolecular interactions. That paper (*28*) also contains extensive further references to earlier theoretical treatments by Stillinger and co-workers.

Kinetic Theory

Several kinetic theories of the glass transition have been proposed. Rather than describe any of them in detail in this short presentation, only a conceptual approach will be given, which is based on the rate of volume contraction of the type shown in Figure 3 and which has been treated in some detail theoretically (*5, 15*). A very simple assumption concerning the rate of volume shrinkage that can be made for illustrative purposes is that it is a first-order process. As expressed in equation 8, the rate of volume contraction is proportional to the volume at time t (V_t) minus the volume at infinity (V_∞):

$$dV/dt = -(1/\tau_v)(V_t - V_\infty) \tag{8}$$

The more excess free volume there is in a polymer, the faster it will shrink, the shrinkage being indicated by the minus sign and the rate constant by $1/\tau_v$. This description is not very accurate. First-order kinetics are not exactly obeyed, and τ_v, which is the reciprocal of the first-order rate constant, is dependent on the time itself. A somewhat better approximation is given in equation 9:

$$dV/dt = -[1/(b + at)](V_t - V_\infty) \tag{9}$$

The equation suggests that the rate of volume shrinkage is proportional to $1/(b + at)$, a time-dependent rate constant, times $V_t - V_\infty$. In this equation, a and b are constants that can be determined experimentally from measurements of the rate of volume shrinkage in the vicinity of the glass transition. Once the constants a and b have been determined, the time-dependent τ_v, the volume relaxation time (i.e., the time needed for $1/2.72$ of the excess free volume to be excluded), can be determined.

The results of this approach as applied to polystyrene are shown in Table II. At approximately 100 °C, relaxation time is of the order of 10^{-2} s; at 95 °C, it is approximately 1 s. The relaxation time increases progressively until it reaches 60 h at 79 °C and approximately 1 year at 77 °C. The glass transition for this particular sample has been determined, presumably at a cooling rate of 1 °C/min, as approximately 90 °C. The glass transition temperature is, therefore, clearly observed physically at a point at which the volume relaxation time is of the order of 1–5 min.

The significance of these observations can best be understood in the following way: if the sample is cooled at a constant rate of approximately 1 °C/min, as is done in most experiments, the volume equilibrium will be reached if the volume relaxation time is very much shorter than 1 min. At very high temperatures ($\gg$100 °C), the volume relaxation time is short, of the order of nanoseconds or microseconds at high enough temperatures. Suppose that a sample is given 1 min to cool from 100 to 99 °C, whereas the volume relaxation time is of the order of 10 ms. In the first 10 ms of this cooling process, $1/2.7$ of the excess free volume will have been squeezed out, and in the next 10 ms again $1/2.7$ of the remaining excess free volume will have followed. After three or four of these relaxation time periods, that is, after 30 of 40 ms, almost all the excess free volume that was present in the sample, by virtue of its temperature, will have been squeezed out. In effect, the sample is given 1 min to cool from 100 to 99 °C, whereas the sample requires only about 30 or 40 ms to squeeze out all the free volume consistent with this temperature decrease. Therefore, we can be absolutely sure that when the sample reaches 99 °C, it will be at the equilibrium free volume.

Table II. Volume Relaxation Times at Various Temperatures

Temperature (°C)	τ_v
100	0.01 s
95	1 s
91	40 s
90	2 min
89	5 min
88	18 min
85	5 h
79	60 h
77	1 year

The same is true in going from 95 to 94 °C; the sample has 1 min for a 1 °C change, whereas the volume relaxation time is of the order of 1 s. In 4 or 5 s, thermodynamic equilibrium will be reached, and again, we can be sure that the sample is at the equilibrium free volume. In going from 90 to 89 °C, the sample, again, has 1 min; the volume relaxation time at that point, however, is 2 min. Therefore, the sample just simply does not have enough time to reach volume equilibrium. A certain amount of free volume is squeezed out, but not enough for the sample to reach equilibrium. If the sample is kept at 89 °C for a long period, it would continue squeezing out free volume, but because the temperature of the sample is continually lowered at the rate of 1 °C/min, the sample would be at a temperature < 89 °C (88, 87, 86, or even 85 °C) before it gets rid of all the excess free volume consistent with the temperature change from 90 to 89 °C. The sample will never get rid of that free volume completely, because by the time it reaches, say, 77 °C, the volume relaxation time is of the order of 1 year, whereas it has only 1 min to reach 76 °C. Thus, a certain amount of excess free volume has been frozen in, approximately 2.5% as suggested by the WLF equation, and if the sample is stored at low enough temperatures, it can never get rid of this excess free volume. However, if the sample is stored close to the glass transition temperature, say $T_g - 5$ °C or $T_g - 10$ °C, that is, temperatures at which the volume relaxation times are of the order of minutes or hours, then at least some of the excess free volume can be removed, and this removal of excess free volume is observed as a volume shrinkage. This type of consideration explains the curves shown in Figure 3.

What has been presented here is obviously not a true kinetic theory of the glass transition but a kinetic approach to the glass transition. Several formal kinetic theories exist, and readers are referred to the reviews or the original literature (*12*). An excellent review of recent developments in dynamical theories of the liquid–glass transition has also been provided by Frederickson (*29*). Physical aging also has been reviewed (*17*).

Molecular Parameters Affecting the Glass Transition

This section considers why a particular polymer has a particular glass transition temperature, or viewed another way, why the glass transition temperature of polystyrene differs, for example, from that of polypropylene. Several factors affecting the glass transition temperature will be considered, starting with chain stiffness.

Chain Stiffness

Two concepts are frequently combined in the term chain stiffness: linearity plus rigidity (sticklike behavior) and high rotational barriers. An inflexible polymer chain can be completely linear [e.g., poly(phenylene) or the linear polyimides], or it can have a few kinks with long linear segments between kinks. If such materials are noncrystalline, they usually exhibit very high glass transition temperatures (*8*). The other concept related to chain stiffness is the presence of high rotational barriers, with the chain maintaining a random coil shape. As might be expected, the mobility of such polymer chains is primarily affected by the barrier to rotation around backbone carbon–carbon bonds. This barrier, in turn, is determined primarily by the size of the substituent groups on the carbon atoms. For example, as shown in Table III (7), if the substituent group is a methyl group, the barrier to rotation around the carbon–carbon bond is relatively low. Thus, the glass transition temperature of the material containing methyl groups on every second carbon atom (i.e., polypropylene) is −10 °C. If instead of a methyl group on every second carbon atom a phenyl ring is placed in the same position (i.e., polystyrene), the glass transition temperature rises to +100 °C, an increase of about 110 °C in changing from a methyl to a phenyl group on alternate carbon atoms of the main chain. If the substituent ring is further enlarged to *ortho*-methylbenzene, the glass transition temperature of this new material (*ortho*-methylstyrene) goes up to 115 °C. A naphthyl group attached in the alpha position as a substituent on every second carbon atom raises the glass transition temperature to 135 °C (for α-vinylnaphthalene). Finally, for vinylbiphenyl, the glass transition temperature is 145 °C. This series shows clearly that larger substituents (provided the substituent is rigid) result in higher glass transitions. If two substituents are on the same carbon atom as for styrene versus α-methylstyrene, the glass transition temperature goes up from 100 °C for polystyrene to 175 °C for poly(α-methylstyrene). If a naphthalene ring is attached at two positions of the backbone rather than at one, then the glass transition temperature increases from 135 °C for the alpha-substituted naphthalene to 264 °C for poly(acenaphthylene). Clearly, the larger the substituents, or the more hindered the rotations because of multiple anchoring, the higher the glass transitions.

Although these relationships are very clear cut, they cannot be extrapolated to those of flexible side chains or to flexible pendant groups in general. What happens if flexible pendant groups (e.g., alkyl side chains) are attached to a polymer to yield series such as the acrylates, methacrylates, α-olefins, or *para*-alkylstyrenes? Judging from the preceding discussion, we would expect the glass transition temperature to increase initially with the size of the side chain. However, as shown in Figure 9, the opposite takes place: longer substituents (i.e., longer side chains) result in lower glass transition temperatures.

Table III. Effect of Rigid Substituents on T_g in $(CH_2–CHX)_n$ or $(CH_2–CXY)_n$ Polymers

X	T_g (°C)
Methyl	−10
Phenyl	100
ortho-Methylphenyl	115
α-Naphthyl	135
Biphenyl	145
Methyl and phenyl on same carbon	175
α-Substituted naphthyl attached to two backbone carbons	264

The reason for this behavior becomes clear on short deliberation. As pointed out before, an increase in the rotational barrier increases the glass transition. However, for flexible alkyl side chains, only the first unit of the side chain, which is attached rigidly to the main chain, has the effect of increasing the glass transition temperature. In the poly(alkyl methacrylate), only the first CH_2 group has this effect. Any additional alkyl units on the side chain would have a different effect, the reason for the different behavior being that the methylene units beyond the first alkyl group can get out of the way of the rotating backbone units. Looked at another way, the addition of low-molecular-weight materials such as hexane, octane, or decane to a polymer plasticizes the polymer because of the low glass transition temperatures of the alkanes. Low-molecular-weight materials have glass transition temperatures ranging from −150 to −200 °C. The addition of these materials as side chains to polystyrene or other polymers has the effect of depressing the glass transition temperature. Thus, the longer the side chain, the more plasticizers are added to a polymer, and the lower the glass transition temperature should be, as is, indeed, observed. As is shown quantitatively for four polymer series in Figure 9, longer alkyl side chains result in lower glass transition temperatures. Not unexpectedly, the attachment of side chains may complicate the mechanical behavior by introducing new relaxation mechanisms connected with the side chain motions (*see* reference 21 for an example).

At this point then, we have a much better appreciation for the structure of the side chain as it affects the glass transition temperature. Let us

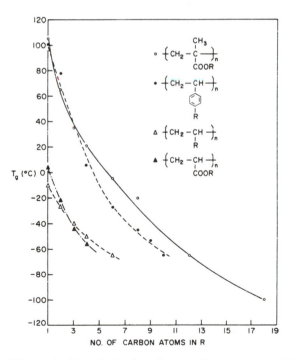

FIGURE 9. Effect of side chain length on the glass transition of (○) methacrylates, (●) *p*-alkylstyrenes, (△) α-olefins, and (▲) acrylates. (Adapted from reference 6.)

Table IV. Glass Transition Temperatures of Butyl Methacrylates

Butyl Group	T_g of Methacrylate (°C)
Tertiary butyl	43
Secondary butyl	−22
Normal butyl	−56

Source: Reference 30.

look in some detail at butyl methacrylate and inquire what happens as we change the structure of the butyl substituent. The expected order of glass transition temperatures is tertiary butyl > secondary butyl > normal butyl, corresponding to highest hindrance to rotation with the tertiary butyl group and lowest hindrance to rotation with the normal butyl group. This order is confirmed by the data in Table IV (30) for butyl methacrylate.

The same trends also are observed with α-olefins, acrylates, and *para*-alkylstyrenes. The behavior is exactly what we expect on the basis of both the bulkiness and the internal plasticizing effect in this series of materials.

Intermolecular Forces

Intermolecular forces affect the glass transition temperature profoundly. The cohesive energy density (31), which, for low-molecular-weight materials, is equal to the vaporization energy divided by the molar volume, is a measure of these intermolecular forces. In polymers, because vaporization is impossible without destruction of the chain, the cohesive energy density can be determined by swelling measurements, the cohesive energy density of the polymer being equal to that of the low-molecular-weight liquid that swells the polymer to the greatest extent. Polymers frequently have to be cross-linked to prevent complete dissolution in some solvents. A relation that has been proposed to correlate the cohesive energy density (CED) with the glass transition temperature is given in equation 10:

$$CED = 0.5mRT_g - 25m \tag{10}$$

In this equation, m is a parameter analogous to the number of degrees of freedom of a molecule (or the ability of atoms or groups of atoms in the molecule to rotate), and R is the universal gas constant. This equation is useful, because it shows that intermolecular forces, indeed, raise the glass transition temperature of a polymer.

Ion-containing polymers provide perhaps an ideal system for the study of intermolecular forces, because the interactions between ions can be quantified most precisely. A very extensive series of studies on the poly-phosphate system (32) (*see* structure on next page), has revealed that the glass transition temperature, indeed, increases dramatically as the strength of the ionic interactions increases. Table V lists the glass transition temperatures of a series of phosphate polymers. HPO_3 is a nonionic polymer with a glass transition temperature of −10 °C. All the other systems are ionic. Clearly, the smaller the size or the higher the charge of the cation, the higher the glass transition temperature. The 530 °C difference in the

$$\begin{array}{ccc} O & & O \\ \parallel & & \parallel \\ \sim P - O - & P - O \sim \\ \mid & & \mid \\ O & & O \end{array}$$

Repeat unit of polyphosphate chain.

Table V. Effect of Ionic Forces on the Glass Transition

Material	T_g (°C)	Pauling Radius (Å)	q^a	q/a^a
HPO_3	−10	—[b]	—	0.00
$LiPO_3$	+335	0.60	1	0.50
$NaPO_3$	+280	0.95	1	0.42
$Ca(PO_3)_2$	+520	0.99	2	0.84
$Sr(PO_3)_2$	+485	1.13	2	0.79
$Ba(PO_3)_2$	+470	1.35	2	0.73
$Zn(PO_3)_2$	+520	0.74	2	0.93
$Cd(PO_3)_2$	+450	0.97	2	0.84

Source: Reproduced with permission from reference 32. Copyright 1971 John Wiley and Sons.
[a] q is charge; a is cation radius plus oxygen anion radius.
[b] —, Not determined.

glass transition temperatures of HPO_3 and calcium phosphate is greater than that spanning the entire range of the common synthetic organic polymers of commercial importance. In addition to these homopolymers, many mixed systems can be envisaged, that is, materials with identical backbones but different counterions (lithium in some and sodium in others, or calcium in some and lithium in others).

How is the glass transition temperature specifically related to the various parameters of the ions? The glass transition temperature may be considered as an isokinetic point, that is, a point at which the rate of the squeezing out of the excess free volume is constant. In terms of molecular mobility, this means that the rate of molecular motions that determines the squeezing out of the free volume must be constant, and from the specific materials under consideration here, we can conclude that this must be determined by the strength of the interactions between anions and cations. Thus, the rate-determining step for these materials is the removal of the anion from the coordination sphere of the cation or the reverse. As shown in equation 11a the glass transition should be proportional to the electrostatic work (W_{el}) necessary to remove an anion from the coordination sphere of the cation:

$$T_g \propto W_{el} \tag{11a}$$

In turn, W_{el} is proportional to the electrostatic force (F_{el}) integrated over the distance from the distance of closest approach, a, to infinity (obviously, the removal is not to infinity, but the mathematical treatment is very much simpler if the upper limit is taken as infinity, and the approximation does not introduce serious errors for distances greater than 5–10 Å):

$$T_g \propto \int_a^\infty F_{el}\, dx \qquad (11b)$$

The electrostatic force, in turn, is equal to the anion charge (q_a) times the cation charge (q_c) divided by the square of the distance (x) between them:

$$T_g \propto \int_a^\infty \frac{q_a q_c}{x^2}\, dx \qquad (11c)$$

On integration, equation 11c yields the following:

$$T_g \propto q_a q_c / a \qquad (11d)$$

If q_a, the anion charge, is lumped into the proportionality constant, the result is equation 12:

$$T_g \propto q_c / a \qquad (12)$$

Figure 10 shows the results of studies on silicates, phosphates, and acrylates (33), the plot being of T_g versus q_c/a (a is the distance between centers of charge). Linear relationships, indeed, are observed for all three polymer systems. With transition metal ions, the situation is more complex.

For organic polymers the situation is still more complicated (34). Figure 11 shows the results for the ethyl acrylate–acrylic acid copolymer system both in the acid form and in the form of various salts. Several families of curves are observed, all of which are sigmoidal except for that corresponding to the acid. For the salts, the glass transition temperature rises rapidly at first with anion content, followed by a sigmoid with an accelerated rise of the glass transition and then by another relatively lower rate of increase at still higher ion concentrations. The onset of the sigmoid is related to the changes that occur in ionomers as a function of ion concentration, in this specific case the onset of cluster-dominated behavior (35). At very low ion contents, the ions are present primarily in the form of small aggregates, called multiplets, with the higher ion contents being characterized by spatial coordination of the multiplets (which reduces chain mobility over appreciable regions), that is, clustering. The onset of the sigmoid coincides with a point at which the clusters begin to dominate a wide range of physical properties. As might be expected, if the glass transition temperature is plotted not against the ion concentration itself but against the normalized ion concentration (i.e., the ion concentration multiplied by the q/a factor, which is so useful for the inorganic systems), all of these lines coalesce, as shown in Figure 12, into a single line that is valid for the entire family of ethyl acrylate–acrylic acid copolymer salts.

In general, in low-ion-content regions, the rate of change of the glass

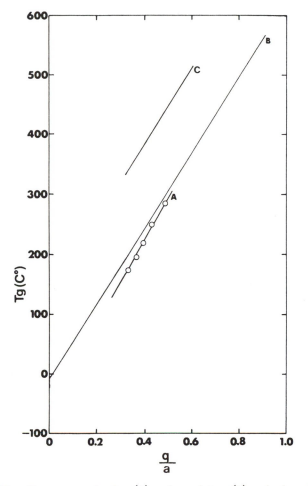

FIGURE 10. T_g versus q/a for (A) polyacrylates, (B) polyphosphates, and (C) polysilicates. (Adapted from reference 33.)

transition with ion concentration (C) lies between 2 and 10 °C/mol % (equation 13):

$$2\,°C/mol\,\% < \partial T_g/\partial C < 10\,°C/mol\,\% \tag{13}$$

Lower matrix T_g values, higher charges, and smaller cation sizes result in a greater effect. Thus, for magnesium methacrylate in polyethylene, the effect is expected to be larger than for cesium acrylate in polystyrene. A list is shown in Table VI. As can be seen, ionic interactions in polymers lead to major effects on the glass transition.

Controllable Parameters Affecting the Glass Transition

This section will consider how the glass transition temperature of a particular polymer is affected by various parameters such as pressure,

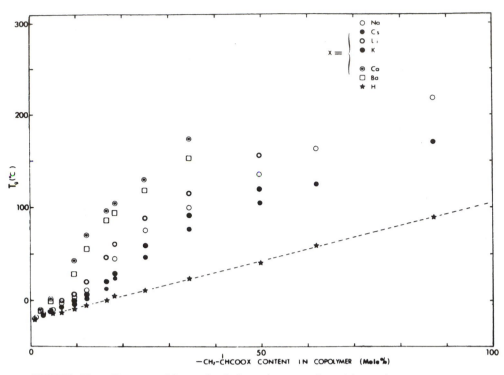

FIGURE 11. Glass transitions of ethyl acrylate–acrylic acid copolymers neutralized with various cations versus ion content. (Adapted from reference 34.)

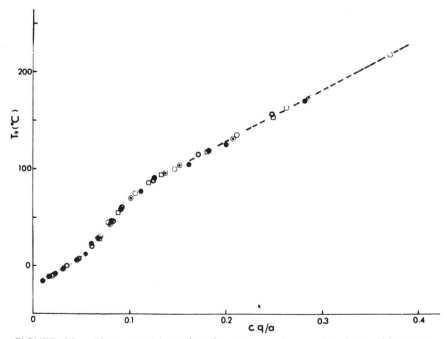

FIGURE 12. Glass transitions for the same polymers in Figure 12 versus Cq/a. (Adapted from reference 34.)

Table VI. $\partial T_g/\partial C$ for Various Ionomers

Nonionic Component[a]	Ionic Component[b]	Conc. Range (mol %)	T_g (°C)	dT_g/dC (°C/mol %)	Reference[c]
Phenoxy	Ca(SCN)$_2$	0–20	~ 100	1.8	45
PPO	ZnCl$_2$	12–32	−65	4.0	47
PPO	LiClO$_4$	0–10	−70	5.5	44
Polysulfone	Sulfone–				
	SO$_3^-$Na$^+$	0–100	≈175	1.3	57
EA	NaA	0–12	−20	2.7	41
S	NaMA	0–10	100	3.2	28
B	LiMA	0–8	−90	5.4	15
E	NaMA	0–3	−20	5.7	35
B	MVPI	0–8	−90	8.9	15
E	Mg$_{1/2}$MA	0–2	−20	9.7	35

Source: Reproduced with permission from reference 36. Copyright 1977 Academic.
[a] PPO, poly(propylene oxide); EA, ethyl acrylate; S, styrene; B, butadiene; E, ethylene.
[b] NaA, sodium acrylate; NaMA, sodium methacrylate; MVPI, methylvinylpyridinium iodide.
[c] Numbers refer to references in chapter 2 of reference 36.

molecular weight, and diluent concentration. As mentioned before, the application of the simple free volume theory provides a useful semiquantitative guide to the phenomena.

Glass Transition Pressure

The free volume approach considers what happens to a polymer above its glass transition temperature when it is subjected to hydrostatic pressure and provides a qualitative prediction of the effect (*37*). As expected, the free volume is squeezed out when pressure is applied at temperatures above T_g; therefore, because the free volume is lower, the polymer is closer to its glass transition temperature.

One also can speak of a glass transition pressure at constant temperature. For example, as shown in Figure 13, one can look at the glass transition as a function of pressure; thus, volume–pressure experiments at constant temperature can be performed to obtain the glass transition pressure at various temperatures. Alternatively, glass transition temperatures can be determined at various constant pressures. Typically, the glass transition temperature increases with pressure at the rate of approximately 20 °C/1000 atm (1 atm = 101.325 kPa), as is shown in Figure 14 for a range of materials. Thus, the pressure effect is a realistic problem when one searches for polymers for high-pressure applications (for example, in applications at the bottom of the ocean or other environments where pressures of the order of few hundred or few thousand atmospheres are involved). For small pressure changes of the order of 1 atm or so, the effects are clearly negligible.

Plasticizers

As might be expected, the presence of diluents or plasticizers decreases the glass transition temperature, an example being shown in Figure 15 for

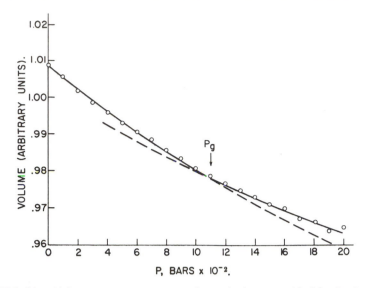

FIGURE 13. Volume versus pressure for selenium at 40 °C. P_g is glass transition pressure. (Adapted from reference 37.)

polystyrene (*38*). Various plasticizers have different effects on the glass transition; mostly they depress the T_g, although some additives raise the glass transition temperature.

As before, the free volume approach provides an excellent, semiquantitative treatment of the plasticizer effect. Equation 14 describes the fractional free volume as a function of temperature, suggesting, as before, that the fractional free volume (f_T) is equal to 0.025 (the value at the glass transition temperature) plus the free volume expansion coefficient, α_f, times ($T - T_g$):

$$f_T = 0.025 + \alpha_f(T - T_g) \tag{14}$$

This equation applies both to the polymer and the diluent. If the subscripts p and d indicate polymer and diluent, respectively, and V_p and V_d indicate volume fraction of polymer and diluent, respectively, then equation 15 describes the fractional free volume at any temperature T for a polymer–diluent system:

$$f_T = 0.025 + \alpha_{fp}(T - T_{gp})V_p + \alpha_{fd}(T - T_{gd})V_d \tag{15}$$

At the glass transition temperature, $f_T = 0.025$ and $T = T_g$. Rearranging the equation yields equation 16, which shows the behavior of T_g as a function of plasticizer content (*39*):

$$T_g = \frac{\alpha_{fp}V_pT_{gp} + \alpha_{fd}(1 - V_p)T_{gd}}{\alpha_{fp}V_p + \alpha_{fd}(1 - V_p)} \tag{16}$$

T_{gd} and α_{fd}, the glass transition temperature and the free volume expansion coefficient of the diluent, respectively, are not usually available and

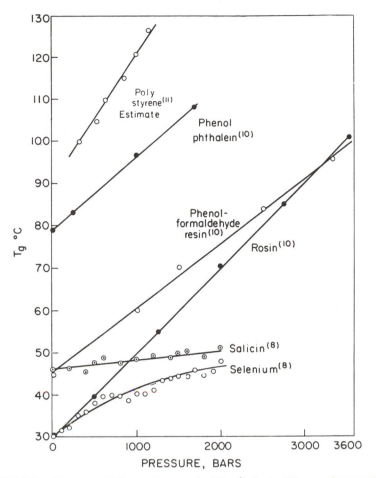

FIGURE 14. T_g versus P for various materials. (Adapted from reference 37. References are for those given in reference 37.)

sometimes not even measurable experimentally because of crystallization. If they are taken as adjustable parameters, the fit of the equation can be excellent. Many other quantitative treatments of the glass transition temperature as a function of plasticizer content are available, and the reader is referred to the original literature or the reviews (*40–44*), reference 44 being perhaps the most general.

Molecular Weight

The molecular weight is another parameter affecting the glass transition temperature of polymers. Many systems have been studied, with similar results. Again, the free volume approach gives the most useful quantitative introduction to the phenomena. It is reasonable to assume that each chain end, at any temperature, moves more rapidly than a chain middle, because a chain end has only one chain attached to it, whereas a chain middle has two. By virtue of its greater mobility, a chain end therefore has associated with it a greater excess free volume. If θ is the excess free volume per chain end, then 2θ is the excess free volume per chain, and $2\theta N_{av}$ is the

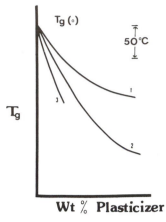

FIGURE 15. Glass transition temperature of polystyrene versus plasticizer content for (1) β-naphthyl salicylate, (2) nitrobenzene, and (3) carbon disulfide. (Adapted from reference 38.)

excess free volume per mole of chains (N_{av} is Avogadro's number). If $2\theta N_{av}$ is divided by M, the molecular weight of the polymer, the quotient is the excess free volume per gram of chains:

$$\text{Excess free volume per gram of chains} = 2\theta N_{av}/M \qquad (17)$$

Finally, multiplication of the excess free volume per gram of chains by the density (ρ) gives the excess free volume per unit volume of chains, that is, the excess free volume (per unit volume) that is introduced if a chain of initially infinite molecular weight is cut down to a molecular weight of M:

$$\text{Excess free volume per cm}^3 \text{ of chains} = (2\theta N_{av}/M)\rho \qquad (18)$$

Thus, at the glass transition temperature for the infinite polymer, the fractional free volume, f_g, is 0.025. Cutting of the chains introduces more free volume. To get back to the glass transition temperature, this free volume must be eliminated by cooling the entire polymer sample to the glass transition temperature of the polymer of molecular weight M. For a unit volume of material, the total free volume that must be squeezed out is simply the free volume expansion coefficient (α_f) times the difference between the glass transition temperature of the infinite polymer [$T_g(\infty)$] and the glass transition temperature of a polymer of molecular weight M [$T_g(M)$]:

$$2\rho\theta N_{av}/M = \alpha_f\left[T_g(\infty) - T_g(M)\right] \qquad (19)$$

Rearranging equation 19 gives equation 20:

$$T_g(M) = T_g(\infty) - 2\rho\theta N_{av}/\alpha_f M = T_g(\infty) - K/M \qquad (20)$$

Thus, the glass transition temperature of a polymer of molecular weight M is equal to the glass transition temperature of polymer of infinite molecu-

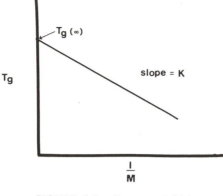

FIGURE 16. T_g versus $1/M$.

lar weight minus K/M, where K is simply $2\rho\theta N_{av}/\alpha_f$. Thus, a plot of the glass transition temperature versus the reciprocal molecular weight should give a linear relationship with a slope K and intercept of $T_g(\infty)$, as shown in Figure 16. Because in this constant K the only unknown is θ, the excess free volume per chain end, θ can be calculated if K is known. These calculations reveal that θ is of the order of the volume of a repeat unit, that is, 20 to 50 Å^3/chain end. This value can be considerably smaller if, for example, the chain end contains ionic groups, which show a higher degree of attraction for each other than do chain middles in the absence of ionic interactions. A recent study of the molecular weight effect specifically is given in reference 45, and reference 44 brings this work into a broader context.

Copolymers

The introduction of a comonomer into a polymer is another method of changing the glass transition temperature. Usually, a random copolymer of two monomers with different glass transition temperatures has an intermediate glass transition temperature. These random copolymers behave like regular polymers in that they exhibit only one glass transition temperature. If the glass transition is studied as a function of comonomer concentration, very simple relations between the glass transition temperatures of the homopolymers and those of the copolymers are found in ideal situations. One example is given in equation 21 (46), in which the subscripts 1 and 2 refer to pure polymers 1 and 2, and W is the mass of pure polymer.

$$1/T_g = W_1/T_{g1} + W_2/T_{g2} \tag{21}$$

More complicated equations exist:

$$T_g = \left[T_{g1} + k(T_{g2} - T_{g1})W_2\right]\big/\left[1 - (1-k)W_2\right] \tag{22}$$

$$1/T_g = \left[1/(W_1 + RW_2)\right](W_1/T_{g1} + RW_2/T_{g2}) \tag{23}$$

In these equations, k and R are constants. A complete listing of copolymer T_g equations is considerably beyond the scope of the present discussion. Relevant studies are given in references 47–49; reference 13 provides a good summary.

Both positive and negative deviations from ideality can be found in copolymer systems. For a copolymer the homopolymers of which have identical glass transition temperatures, negative or positive deviations are sometimes found, that is, systems in which the glass transition temperatures of copolymers are lower or higher than those of either of the homopolymers, the crucial factor here being the rotational barriers between AB monomer pairs as compared with AA or BB monomer pairs.

So far, the discussion has centered on random copolymers. If the copolymers are not random, many complications can arise. The most extreme example occurs in block copolymers, that is, copolymers in which one sequence of a homopolymer A is chemically attached to one or two sequences of pure homopolymer B. If the sequences are incompatible, the materials will form phase-separated polymer systems in which two glass transition temperatures are observed. This behavior is also true of blends of two homopolymers. In this sense, the glass transition temperature can be regarded as one of the tests of compatibility for blends or blocks. Naturally, a material will manifest its own glass transition temperature only if its dimensions exceed a certain value (i.e., ca. 100 Å), and the use of a single glass transition temperature to probe homogeneity is valid only down to that limit. If only one glass transition temperature is found, then the systems must be compatible, provided that the glass transition temperatures of the homopolymers are far enough apart. If two glass transition temperatures are found, then presumably the systems are not compatible. Intermediate situations suggest intermediate degrees of compatibility. In this connection, the study of dynamic mechanical properties, specially the loss tangents, are quite useful. ten Brinke and co-workers (*18*), recently described a very useful thermal method for determining the glass transition temperatures of heterogeneous materials with two very close glass transitions. Usually, mechanical or thermal methods (*see Methods of Determining the Glass Transition*) can be applied to mixtures only if their glass transitions differ by more than ca. 30 °C.

Cross-linking

Cross-linking is still another method of influencing the glass transition. A cross-link has the opposite effect of a decrease in molecular weight: as the cross-link density increases, the free volume of a sample decreases, and the glass transition temperature increases correspondingly. This behavior is ideal; normally, the introduction of cross-links into a polymer system occurs not just by removal of two hydrogen atoms from the chain backbone and attachment of the resulting free radicals. Cross-linking is usually accomplished by the addition of a specific cross-linking agent, which can be considered a comonomer in addition to being a cross-link. Therefore, two different effects must be considered: a copolymer effect, resulting from the incorporation of a second unit, and a cross-linking effect. Both of these effects are taken into account in equation 24 (*50*):

$$T_g = T_g(\infty) - K/M + K_x \rho \qquad (24)$$

This equation suggests that the glass transition temperature of the cross-linked polymer is equal to the glass transition temperature of the polymer at infinite molecular weight minus K/M (which is used here in exactly the same sense as it was in connection with molecular weight effects) to which the $K_x \rho$ product is added; K_x is a constant and ρ is the number of cross-links per gram. Other equations have been proposed, but equation 24 illustrates the complexities involved in the study of cross-linking effects. Other examples are given in reference 51. Nielsen treated the problem more recently (*52*).

Crystallinity

Crystallinity also affects the glass transition temperature, but the results are not clear cut. In some cases, for example, with poly(ethylene terephthalate), as the crystallinity increases from 2 to 65%, the glass transition temperature increases from 81 to 125 °C. In other cases, for example, with poly(4-methylpentene), as the crystallinity increases from 0 to 76%, the glass transition temperature decreases from 29 to 18 °C. Many factors are responsible for this type of behavior, including changes in tacticity of polymers that can be crystallized to different extents and molecular-weight effects. The effect of these secondary parameters on the degree of crystallinity, in turn, influences the effect of the degree of crystallinity on the glass transition temperature. Also, in highly crystalline matcrials, the amorphous regions are primarily the surfaces of the crystal lamellae as well as the interlammelar regions, and the structure of those amorphous regions will be very different from that of the bulk of the same material in the melt.

Polyethylene is one example for which this situation is particularly complex. Thus, for polyethylene, T_g values of -128 °C (*53*), -80 °C (*54*), and -30 °C (*55*) have been reported on the basis of different experiments. Boyer (*56*) discusses the phenomenon extensively.

Tacticity

The tacticity of the polymers is still another property through which the glass transition may be modified. Most polymers that possess only one substituent on every second carbon atom do not show any tacticity effects; however, polymers such as poly(methyl methacrylate), which contain both a methyl group and a pendant ester on the same carbon atom, show profound effects of tacticity. Thus, syndiotactic poly(methyl methacrylate) has a glass transition temperature of 115 °C, whereas the isotactic material has a glass transition temperature of 45 °C. This tacticity effect is quite general for unsymmetrically disubstituted polymers and has been discussed by Karasz and MacKnight (*57*). It is not observed in monosubstituted species (for example, the acrylates or the styrenes).

Elongation

Still another phenomenon that affects the glass transition is the degree of elongation or the percentage of strain. Two conflicting effects are in operation here. On extension, the slight increase in the free volume tends to depress the glass transition temperature; on the other hand, the decrease in the entropy of the chains on extension tends to increase the glass transition temperature. Both of these effects have been observed in different cases. The effects are not large, but they are present.

Methods of Determining the Glass Transition

Methods Based on Volume Changes

A wide range of methods have become available during the past four decades for determining the glass transition temperature. Only a few can be mentioned here. Dilatometry (6) is perhaps the most widely accepted classical method for the determination of the glass transition temperature, and all the early determinations were performed by that method. A dilatometer consists of a glass bulb with an attached small capillary. A polymer sample is placed in the large bulb, as shown in Figure 17 (left) and mercury or some other confining liquid is introduced up to a convenient level of the capillary. The dilatometer is then placed in a temperature bath, and the temperature is changed at a uniform rate so that a volume–temperature plot is obtained. The inflection point on the plot indicates the position of the glass transition. If the experiment is done carefully and quantitatively, the specific volume of the polymer as a function of temperature can be obtained.

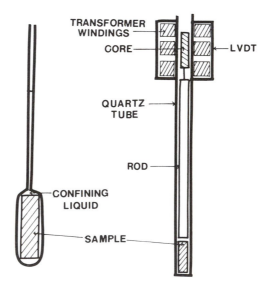

FIGURE 17.   LVDT, Linear variable differential transformer. (Left) Volume dilatomer and (Right) linear dilatometer.

The procedure as described is quite laborious; a far simpler method involves the determination of only one dimension of the sample in a so-called one-dimensional dilatometer as shown in Figure 17 (right). In this method, the sample is placed at the bottom of a quartz tube that is attached at the other end to a linear variable differential transformer (LVDT). A quartz rod is placed on top of the sample, and to the other end of the quartz rod (in the middle of the LVDT) is attached the LVDT core. As the sample expands, the LVDT core is moved relative to the windings of the variable differential transformer, and the length of the sample is thus translated into a voltage. Extremely high resolution is possible, and very precise determinations of the expansion coefficient can be performed by this method with calibrated instrumentation. Both of these methods depend on the volume of the sample, and many variations are possible, including measurements of the refractive index and the transmission of ionizing radiation (for example, beta rays) (9).

Methods Based on Specific Heat Changes

The specific heat of the polymer changes as the polymer passes through the glass transition, and this effect is the basis of one of the most convenient methods of determining the glass transition: differential thermal analysis or differential scanning calorimetry (9). In differential thermal analysis, two containers are placed in identical environments (for example, in a metal block). One container is filled with the polymer to be measured, and the other container is filled with some inert material, such as sand, that does not undergo any transitions in the temperature range of interest. Thermocouples are attached to each container, and the temperature of the aluminum block, which contains both containers, is changed. If the temperature is changed at a constant rate, a discontinuity in any thermal property of one of the samples will be manifested as a change in the temperature difference between the two samples. Thus, because the heat capacity changes as the polymer goes through the glass transition temperature and because no comparable change occurs in the blank (or the sand) sample, a change in ΔT, the difference in the temperatures of the sand and the polymer, will be observed. This method is extremely convenient and suitable for automation, and several commercial instruments are available.

A slight modification of this method is differential scanning calorimetry, which measures the amount of heat that must be supplied to one or the other of the two samples to keep them at a constant temperature. Thus, one can measure exothermic or endothermic phenomena in the sample. Again, commercial instruments are available, and the method is rapid and completely automated.

Estimation of Unmeasured Glass Transitions

Methods have been proposed to estimate glass transition temperatures if measurements cannot be made. A very convenient procedure has been described by Van Krevelen (58), which is based on the concept of independent group contributions. The group contributions are tabulated, and the glass transition of a new polymer can be estimated by combining the appropriate group contribution values.

Molecular Motions below the Glass Transition

Only one example of a study will be presented here, because a more thorough discussion is beyond the scope of this presentation. A number of review articles have been written on this topic, and the reader is referred to these reviews or the original literature (*22, 59, 60*).

As was mentioned in connection with Figure 8, low-temperature peaks are observed for all polymers when the loss tangent is measured as a function of temperature. These low-temperature peaks generally reflect a molecular mechanism, for example, a movement of a benzene ring, a cyclohexyl group, a methyl group, etc., depending on the groups that are present in the polymer. The field of study, known as mechanical spectroscopy, is concerned with the elucidation of which segment of the molecule moves with which frequency as a function of temperature. In a typical study, the loss tangent is measured at a constant frequency as a function of temperature. Various frequencies require a very wide range of instruments, and the instrument that is used at 1 Hz (for example, a torsion pendulum) differs dramatically from that used at around 1 kHz (for example, a vibrating reed), which in turn differs drastically from the instrument used in the ultrasonic region at 1 MHz.

An example of a mechanical spectroscopy study is that by Heijboer (*59*). In this study, whenever a cyclohexyl group is attached to a polymer chain, peaks of the type shown in Figure 18 are observed. The investigation was done over a frequency range of 10^{-4} cycles/s to 8×10^5 cycles/s. The results showed a remarkable constancy of the peak height as a func-

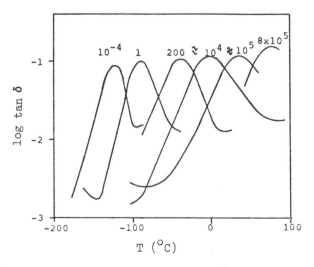

FIGURE 18.   Log tan δ versus temperature for poly(cyclohexyl methacrylate). (Adapted from reference 59.)

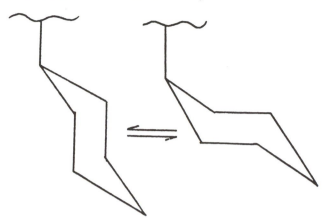

SCHEME I. Chair–chair flip in pendant cyclohexane groups.

tion of temperature and frequency. Mechanistically, the peak was assigned to a chair–chair flip within the cyclohexyl group of the type shown in Scheme I. The activation energy for the motion was obtained from the plot of the logarithm of the frequency at maximum loss versus the reciprocal temperature (i.e., from an Arrhenius plot), which gives a straight line consistent with a slope of 11.2 kcal. In a subsequent NMR investigation (*61*), the barrier for the chair–chair flip was, indeed, found to be approximately 11.2 kcal, in excellent agreement with the earlier mechanical study. It seems unimportant whether the cyclohexyl group was attached to a methacrylate polymer or to a different type of polymer or whether the group was at all attached to a polymer chain. Under most conditions, the barrier to the chair–chair flip is constant and uniquely assignable to this specific motion of the cyclohexyl group.

Since the study of Heijboer (*59*), many other materials have been investigated by similar techniques. In some cases, as mentioned before, correlations have been found between brittleness of glasses and the absence of these low-temperature relaxations, but these studies are by no means conclusive, and much work remains to be done. Suffice it to say that materials below the glass transition temperature are not, by any stretch of the imagination, rigid, unmoving materials; the molecules can move in various ways, and if the material is close enough to the glass transition temperature, considerable volume shrinkage can occur, as described earlier.

Conclusion

The glass transition phenomenon represents a relatively new field, having been explored only since the 1920s. The phenomenon is well explored, but no thorough and unique understanding of the theoretical basis of the glass transition has emerged, although many attempts have been made, and some have been quite successful, at least in part. The field, however, represents one of enormous industrial importance, because the ability to control the glass transition temperature, and to understand it, aids in the industrial utilization of polymeric materials. Finally, materials in the glassy state are very much "alive," in that they are subject to specific molecular motions on a small scale; in the vicinity of the glass transition, the material can shrink considerably with time.

References

1. See, for example, *Order in the Amorphous "State" of Polymers;* Keinath, S. K.; Miller, R. L.; Reike, J. K., Eds.; Plenum: New York, 1987.
2. Tammann, G. *Der Glazustand;* Leopold Voss: Leipzig, 1933.
3. Kauzmann, W. *Chem. Rev.* **1948,** *43,* 219.
4. Boyer, R. F. *Rubber Chem. Technol.* **1963,** *36,* 1303.
5. Kovacs, A. J. *Fortschr. Hochpolym. Forsch.* **1963,** *3,* 394.
6. Shen, M. C.; Eisenberg, A. In *Progress in Solid State Chemistry;* Reiss, H., Ed.; Pergamon: Oxford, 1967; Vol. 2, Chapter 9. Reprinted in *Rubber Chem. Technol.* **1970,** *43,* 95.
7. Eisenberg, A.; Shen, M. C. *Rubber Chem. Technol.* **1970,** *43,* 165.
8. Wrasidlo, W. In *Thermal Analysis of Polymers; Adv. Polym. Sci.* **1974,** *13,* 3 pp. 29–53.
9. *Polymer Handbook*; 3rd ed.; Brandrup J.: Immergut, E. H., Eds.; John Wiley and Sons: New York; pp. VI-209–VI-277. [The second edition of the Polymer Handbook (1975) gives a particularly good summary of the methods of determining T_g].
10. "The Glassy Transition and the Nature of the Glassy State;" Goldstein, M.; Simha, R., Eds.; *Ann. N.Y. Acad. Sci.* **1976,** *270.*
11. "Structure and Mobility in Molecular and Atomic Glasses;" O'Reilly, J. M.; Goldstein, M., Eds.; *Ann. N.Y. Acad. Sci.* **1981,** *371.*
12. "Dynamic Aspects of Structural Change in Liquids and Glasses;" Angell, C. A.; Goldstein, M. Eds.; *Ann. N.Y. Acad. Sci.* **1986,** *484.*
13. "Glass-Rubber Transition," *Physical Polymer Science;* Sperling, L. H.; John Wiley and Sons: New York, 1986; Chapter 6.
14. Angell, C. A.; Dworkin, A.; Figuere, P.; Fuchs, A.; Szwarc, H. J. *Chim. Phys.* **1985,** *82,* 773. Angell, C. A. *J. Non-Cryst. Solids* **1985,** *73,* 1. Martin, S. W.; Angell, C. A. *J. Phys. Chem.* **1986** *90,* 6736. Angell, C. A. *J. Phys. Chem. Solids* **1988,** *49,* 863.
15. Kovacs, A. J.; Aklonis, J. J.; Hutchinson, J. M.; Ramos, A. R. *J. Polym. Sci. Polym. Phys. Ed.* **1979,** *17,* 1097.
16. Moynihan, C. T.; Macedo, P. B.; Montrose, C. J.; Gupta, P. K.; De Bolt, M. A.; Dill, J. F.; Dom, B. E.; Drake, P. W.; Easteal, A. J.; Etterman, P. B.; Moeller, R. P.; Sasabe, H.; Wilder, J. A. In "The Glassy Transition and the Nature of the Glassy State;" Goldstein, M.; Simha, R., Eds.; *Ann. N.Y. Acad. Sci.* **1976,** *270,* 15.
17. Struick, E. C. E. *Physical Aging in Amorphous Polymers and Other Materials;* Elsevier: Amsterdam, 1978.
18. For a recent reference, see, for example Bosma, M.; ten Brinke, G.; Ellis, T. S. *Macromolecules* **1988,** *21,* 1465.
19. Jones, G. O. *Glass;* Methuen: England, 1956.
20. a. Tobolsky, A. V. *Properties and Structure of Polymers;* John Wiley: New York, 1960; b. Ferry, J. D. *Viscoelastic Properties of Polymers,* 2nd ed.; John Wiley: New York, 1970.
21. Cayrol, B.; Eisenberg, A.; Harrod, J. F.; Rocaniere, P. *Macromolecules* **1972,** *5,* 676.

22. McCrum, N. G.; Read, B. E.; Williams, G. *Anelastic and Dielectric Effects in Polymeric Solids;* John Wiley and Sons: New York, 1967.
23. Doolittle, A. K. *J. Appl. Phys.* **1951,** *22,* 1471.
24. Williams, M. L.; Landel, R. F.; Ferry, J. D. *J. Am. Chem. Soc.* **1955,** *77,* 3701.
25. *Order in the Amorphous "State" of Polymers;* Keinath, S. K.; Miller, R. L.; Reike, J. K., Eds.; Plenum: New York, 1987; p 199.
26. Grest, G. S.; Cohen, M. H. *Adv. Chem. Phys. 48,* 455.
27. "Structure and Mobility in Molecular and Atomic Glasses;" O'Reilly, J. M.; Goldstein, M., Eds.; *Ann. N.Y. Acad. Sci.* **1981,** 371; p 1.
28. Stillinger, F. H. *J. Chem. Phys.* **1988,** *88,* 7818.
29. Frederickson, G. H. *Ann. Rev. Phys. Chem.* **1988,** *39,* 149.
30. Shetter, J. A. *Polym. Lett.* **1963,** *1,* 209.
31. Hayes, R. A. *J. Appl. Polym. Sci.* **1961,** *5,* 318.
32. Eisenberg, A.; Farb, H.; Cool, L. G. *J. Polym. Sci. A-2* **1966,** *4,* 855.
33. Eisenberg, A.; Matsuura, H.; Yokoyama, T. *J. Polym. Sci. A-2* **1971,** *9,* 2131.
34. Matsuura, H.; Eisenberg, A. *J. Polym. Sci. Polym. Phys. Ed.* **1976,** *14,* 1201.
35. Eisenberg, A.; Hird, B.; Moore, A. B. *Macromolecules* **1990,** *23,* 4098.
36. Eisenberg, A.; King, M. *Ion-Containing Polymers;* Academic: New York, 1977.
37. Eisenberg, A. *J. Phys. Chem.* **1967,** *67,* 1333.
38. Jenckel, E.; Heusch, R. *Kolloid Z* **1953,** *130,* 89.
39. Kelley, F. N.; Bueche, F. *J. Polym. Sci.* **1961,** *50,* 549.
40. Couchman, P. R.; Karasz, F. E. *Macromolecules* **1978,** *11,* 117.
41. Chow, T. S. *Macromolecules* **1980,** *13,* 362.
42. Czekaj, T.; Kapko, J. *Eur. Polym. J.* **1981,** *17,* 1227.
43. ten Brinke, G.; Karasz, F. E.; Ellis, T. S. *Macromolecules* **1983,** *16,* 244.
44. Couchman, P. R. *Polym. Eng. Sci.* **1984,** *24,* 135.
45. Couchman, P. R. *Polym. Eng. Sci.* **1981,** *21,* 377.
46. Fox, T. G. *Bull. Am. Phys. Soc.* **1956,** *1,* 123.
47. Gordon, M.; Taylor, J. S. *J. Appl. Chem.* **1952,** *2,* 493.
48. Mandelkern, L.; Martin, G. M.; Quinn, F. A. *J. Res. Nat. Bur. Stand.* **1957,** *58,* 137.
49. Couchman, P. R. *Nature* **1982,** *268,* 720. Also *Macromolecules* **1982,** *15,* 770.
50. Fox, T. G.; Loshaek, S. *J. Polym. Sci.* **1955,** *15,* 371.
51. Loshaek, S. *J. Polym. Sci.* **1955,** *15,* 391.
52. Nielsen, L. E. *J. Macromol. Sci.* **1969,** *C3,* 69.
53. Stehling, F. C.; Mandelkern, L. *Macromolecules* **1970,** *3,* 242.
54. Illers, K. M. *Koll. Z. Z. Polym.* **1972,** *250,* 426.
55. Davis, G. T.; Eby, R. K. *J. Appl. Phys.* **1973,** *44,* 4274.
56. Boyer, R. F. In *Encyclopedia of Polymer Science and Technology, Suppl.;* Bikales, N. M., Ed.; Interscience: New York, 1977; Vol. 2.
57. Karasz, F. E.; MacKnight, W. J. *Macromolecules* **1968,** *1,* 537.
58. Van Krevelen, D. M. *Properties of Polymers,* 3rd ed.; Elsevier: London, 1990; pp 130–151. This reference provides a review of extensive work on the system.
59. Heijboer, J. Ph.D. Thesis, University of Leiden, Netherlands, 1972.
60. Eisenberg, A.; Eu, B. C. *Annu. Rev. Mater. Sci.* **1976,** *6,* 335.
61. Anet, F. A. L.; Bourn, A. J. R. *J. Am. Chem. Soc.,* **1967,** *89,* 760.

Chapter 3

Viscoelasticity and Flow in Polymer Melts and Concentrated Solutions

William W. Graessley

Department of Chemical Engineering, Princeton University,
Princeton, NJ 08544

This chapter deals with viscoelasticity, with particular emphasis on aspects related to the flow properties of polymer melts and concentrated solutions. The time-dependent response of polymers in the glassy state and near the glass transition was discussed in Chapter 2. The concern in this chapter is with the response at long times for temperatures well above the glass transition. The elastic behavior of polymer networks well above the glass transition was discussed in Chapter 1. The conditions here are similar, and elastic effects may be very important, but steady-state flow can now occur, because the chains are not linked to form a permanent network. All the molecules have finite sizes, and for flexible-chain polymers, which are the focus of this chapter, they have random coil conformations at equilibrium (*see* Chapters 1 and 7).

The discussion in this chapter will cover linear viscoelasticity (*1*), a primary means of rheological characterization for polymer liquids, and steady shear flow (*2*), a relatively well-understood bridge into nonlinear viscoelastic behavior. The effects of large-scale chain structure (molecular weight, molecular weight distribution, and long-chain branching) will be discussed, and some current theoretical ideas about molecular aspects will be described. The general mathematical framework of the subject (*3*) and applications to the solution of practical flow problems (*4, 5*) are more advanced topics and will not be discussed here. Some other important topics have been omitted or considered only briefly, but sources of information on these are included in the *References*. The aim of this chapter is to provide some physical understanding of the viscoelastic behavior in polymer liquids from both the macroscopic and molecular viewpoints.

2505–2/93/0097 $12.75/1

General Considerations

Deformation and Stress

Deformation means a change in shape (6). A liquid deforms as it flows in a tube (Figure 1), the flow being driven by pressure or gravity. The liquid particles move in straight lines parallel to the tube axis. Particles at the center line have the maximum velocity. The liquid particles at the wall do not move at all. In long tubes, each particle moves at constant velocity if the driving force is constant; a particle moves slightly slower than adjacent particles that are nearer the center line and slightly faster than those that are nearer the wall. Tube flow exemplifies simple shear deformation. The layers of liquid along the flow direction slide over one another without

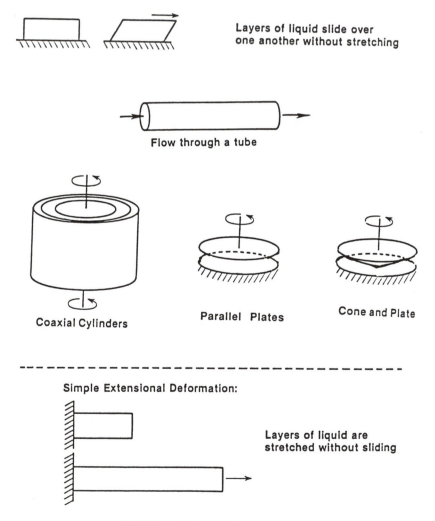

FIGURE 1. Types of deformation.

stretching. Other examples of simple shear are the flows induced by relative rotation of coaxial cylinders, coaxial parallel plates, and a coaxial cone and plate (Figure 1).

Simple extension belongs to a different class of deformations. Layers of particles along the flow direction are stretched without sliding relative to adjacent layers (Figure 1). Extensional flows, sometimes called elongational flows, are important in many polymer-processing operations, for example, in fiber spinning or film formation. These flows are difficult to generate and sustain in a controlled way, however. Most laboratory methods used to characterize polymer flow properties involve simple shear flows. The effects of these two classes of deformation on chain conformations, caused by the respective relative motions, are sketched in Figure 2. For simplicity and brevity only the response in shear flow will be discussed in this chapter.

Deformation always involves a change in the distance between particles, which induces a resisting force. Constant particle velocities in flow through a tube (Figure 1), for example, are the result of a balance of two forces. The force from the applied pressure difference that drives the flow is opposed by a force from the shear stress, generated at the molecular level and acting between the adjacent layers of deforming liquid along the flow direction. The applied torque in the various coaxial geometries is opposed similarly. Stress, in general, describes the force per unit area transmitted by contact between adjacent layers of particles. The relationship between stress and deformation is a property of the material itself.

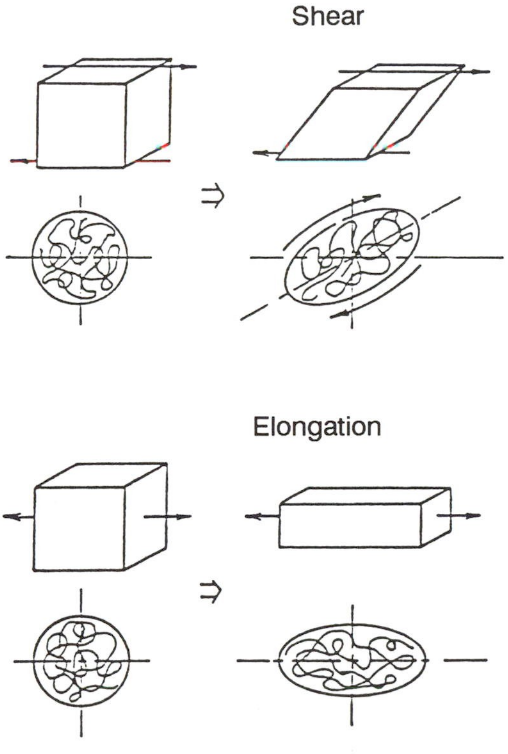

Shear

Elongation

FIGURE 2. Effects of shear and elongational flow on chain conformations.

Rheology is the study of stress–deformation relationships, although that term is usually reserved for discussions of materials that are more complicated in behavior than ordinary liquids and solids.

Viscoelasticity

Liquids have no preferred shape. The stress in a perfectly viscous liquid, except for a pressure contribution that acts equally in all directions, depends only on the rate of deformation. The history of deformation is irrelevant in a perfectly viscous liquid. The stress at each moment depends only on how rapidly the liquid is being deformed at that moment. A perfectly viscous liquid has no memory. All the mechanical work expended in producing the deformation is dissipated instantaneously.

A solid body, on the other hand, has a preferred shape, the shape it assumes spontaneously when no force is applied. In a perfectly elastic solid, the stress (again, except for the pressure) depends only on the deformation from that preferred shape. All the mechanical work expended in producing the deformation is stored as elastic energy. A substance is viscoelastic if it exhibits both energy dissipation and energy storage in its mechanical behavior. In a viscoelastic liquid, the stress depends on the history of the deformation. Some finite time must elapse for the liquid to "forget" the sequence of shapes that it had in the past.

All real substances are viscoelastic. How they respond in particular situations depends on the rate of testing compared with the rate of spontaneous structural reorganization at the molecular level (*1*). In ordinary liquids that are well above the glass transition temperature, T_g, the local arrangements of the molecules relax quickly through the action of Brownian motion (Figure 3). Structural "memory" is very short ($\sim 10^{-10}$ s, perhaps) so that any changes in the distance between molecules induced by the deformation are quickly relaxed, and the response is essentially viscous unless the testing rate is extraordinarily rapid. In ordinary solids, on the other hand, the relaxation of structure is extremely slow ($\sim 10^{10}$ s, perhaps). Structural memory is very long, and the response, at least for small deformations, is essentially elastic unless the testing rate is extraordinarily slow. One property that sets polymeric liquids apart is the enormously wide range for the period from the onset of spontaneous structural rearrangement until its completion.

The distribution of structural relaxation over many orders of magnitude in time is a natural consequence of macromolecular structure. Thus, rearrangement of flexible chains well above T_g is very rapid on the scale of a chain unit. The local-rearrangement time for macromolecules is affected by the nature of the chain unit itself and, to some extent, by the need for local cooperation with adjacent units in the same chain. It is independent of chain length for long chains, and it decreases with increasing temperature. It is probably not different in any significant way from the local-rearrangement time for ordinary liquids at the same temperature relative to the glass transition ($\sim 10^{-9}$ s, perhaps). However, the complete rearrangement of chain conformation requires much longer times ($\sim 10^{1}$ s is not uncommon, for example). The chain units must not only rearrange locally but also diffuse over progressively longer distances to rearrange the conformations of progressively longer segments of the chain. The time re-

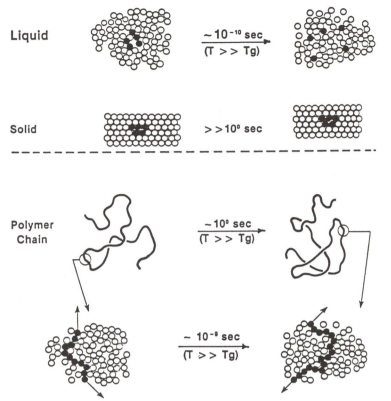

FIGURE 3. Molecular rearrangement and time scales for typical liquids, solids, and polymer melts. T_g is glass transition temperature.

quired for complete rearrangement, therefore, strongly depends on the large-scale chain architecture. These relatively sluggish processes, called the terminal-zone relaxations or the slow dynamics of the chains, strongly influence the flow properties. For this reason, molecular weight, molecular weight distribution, and long-chain branching play such important roles in determining the rheological properties of polymers.

Polymer networks can be classified as true solids, because they have a preferred shape to which they eventually return in the absence of external forces. Polymer melts and solutions are true liquids in the sense of having no such preferred shape. However, both networks and liquid polymers may require a significant time to reach their final state of stress when deformed to a new shape, for exactly the reasons discussed earlier. Beyond that, however, the chain conformations in both polymer networks and polymer liquids can be displaced significantly from equilibrium by deformation. Nonlinear elastic effects, produced by these large conformational deformations, are readily demonstrated in polymer liquids, just as in networks (7). This ease of evoking finite deformation effects is a second property that distinguishes polymeric liquids from monomeric liquids. The combination of both properties gives rise to nonlinear viscoelastic behavior, which is discussed later in the chapter. If the deformation is small, or applied sufficiently slowly, the molecular arrangements are never far from equilibrium. The mechanical response is then just a reflection of dynamic processes at the molecular level, which go on constantly, even for the system at full mechanical and thermal equilibrium. This situation is the domain of linear viscoelasticity. The magnitudes of stress and strain in this case are related linearly, as will be explained later, and the behavior for any liquid can then be described completely by a single function of time. The properties of this function can be obtained by a variety of experimental procedures, as discussed in the following section.

Linear Viscoelasticity

Consider the case of a simple shear deformation, depicted in Figure 4. The shear stress, σ, is F/A, the shear force (F) per unit area of the surface (A) that it acts upon (the shear plane). Deformation is specified by the shear strain γ which is Δ/H, the displacement (Δ) per unit distance normal to the shear planes (H). Deformation rate is specified by the shear rate $\dot{\gamma}$ which is $(d\Delta/dt)/H$, the rate of change of shear strain with time (t). For the example shown, σ, γ and $\dot{\gamma}$ are the same everywhere in the material, and the deformation is said to be homogeneous.

Stress–Strain Relationships

The stress–strain relationship in simple shear for Hookean solids and Newtonian liquids, the classical models for purely elastic response and for purely viscous response, respectively, are as follows:

$$\sigma(t) = G\gamma(t) \tag{1}$$

$$\sigma(t) = \eta\dot{\gamma}(t) \tag{2}$$

Equation 1 is Hooke's law, and equation 2 is Newton's law. In these equations, $\gamma(t)$ and $\dot{\gamma}(t)$ are the shear strain and strain rate, respectively, at any time t, and $\sigma(t)$ is the shear stress at the same time. A single

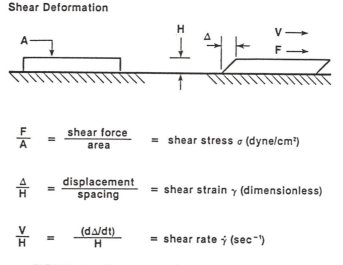

Shear Deformation

$$\frac{F}{A} = \frac{\text{shear force}}{\text{area}} = \text{shear stress } \sigma \text{ (dyne/cm}^2\text{)}$$

$$\frac{\Delta}{H} = \frac{\text{displacement}}{\text{spacing}} = \text{shear strain } \gamma \text{ (dimensionless)}$$

$$\frac{V}{H} = \frac{(d\Delta/dt)}{H} = \text{shear rate } \dot{\gamma} \text{ (sec}^{-1}\text{)}$$

FIGURE 4.   Shear stress, shear strain, and shear rate.

constant completely defines the mechanical response in each case: the shear modulus, G, for the solid and the shear viscosity, η, for the liquid. The stress depends only on the current value of strain for the solid and only on the current value of strain rate for the liquid. The history of loading is not a factor in either case. Hooke's law accurately describes the small-strain behavior of many solid materials, and Newton's law is broadly applicable to small-molecule liquids except near the glass transition.

The history of loading is crucial for a viscoelastic substance. The response to a sudden deformation moves from solidlike at short times to liquidlike at long times, and the history of loading, in this case the time elapsed since the deformation was imposed, is crucial. In a typical stress relaxation experiment (Figure 5), some small but otherwise arbitrary shear strain γ_0, is imposed instantaneously (in principle), and the stress at subsequent times is monitored while the strain is held fixed. The stress would be constant for a Hookean solid, because the strain is constant. The stress would be zero for a Newtonian liquid (except for an initial spike), because the strain rate is zero. The stress for a viscoelastic substance begins at some initial value that decreases with time and finally reaches an equilibrium value for a solid or zero for a liquid. If the strain is small enough, the ratio of stress to strain is a function of time alone. The response is said to be linear, and that ratio is a linear viscoelastic property of the material called the shear stress relaxation modulus, $G(t)$.

$$G(t) = \sigma(t)/\gamma_0 \qquad \text{(small strain)} \qquad (3)$$

In a typical shear creep experiment (Figure 6), a constant shear stress, σ_0, is imposed, and the shear strain, $\gamma(t)$, at subsequent times is monitored; in the creep recovery phase, the sample is unloaded (the shear stress

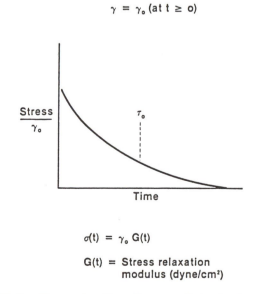

$$\gamma = \gamma_0 \text{ (at } t \geq 0)$$

Stress / γ_0

τ_0

Time

$$\sigma(t) = \gamma_0 \, G(t)$$

$$G(t) = \text{Stress relaxation modulus (dyne/cm}^2\text{)}$$

FIGURE 5. Stress relaxation after a small step in shear strain.

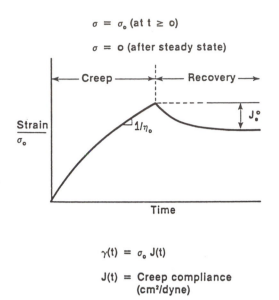

$\sigma = \sigma_0$ (at $t \geq 0$)

$\sigma = 0$ (after steady state)

$\gamma(t) = \sigma_0 \, J(t)$

$J(t) = $ Creep compliance
(cm²/dyne)

FIGURE 6.   Shear creep and creep recovery from steady state for small shear stresses.

is set to zero), and the strain at subsequent times is monitored. The creep strain would be constant (independent of time after loading) for the Hookean solid, and the creep strain rate would be constant for the Newtonian liquid (the total strain would increase at a constant rate), because the stress is constant. The strain (for the solid) and strain rate (for the liquid) would go immediately to zero with unloading (the creep recovery phase), because the stress is now zero.

For a viscoelastic liquid in the creep experiment, the strain rate decreases with time and finally reaches a steady state for which the strain rate is constant. With unloading (the creep recovery phase) the liquid recoils and eventually reaches equilibrium at some total strain smaller than that at the time of unloading. If the shear stress is small enough, the strain is always a simple shear deformation, and the response is linear. The ratio of shear strain to shear stress in the creep phase is a function of time alone and is called the shear creep compliance, $J(t)$:

$$J(t) = \gamma(t)/\sigma_0 \qquad \text{(small stress)} \qquad (4)$$

The linear viscoelastic properties $G(t)$ and $J(t)$ are not independent of one another. Both are manifestations of the same dynamic processes at the molecular level in the liquid at equilibrium. They are related through the Boltzmann superposition principle (1), which links all such linear viscoelastic functions.

Two quantities that play an important role in flow behavior are the steady-state viscosity at zero shear rate, η_0 and the steady-state recoverable shear compliance, J_e°. Both are obtained quite directly from creep results, η_0 from σ_0 and $\dot{\gamma}$ in the steady-state region of the creep phase ($\dot{\gamma}_{ss}$)

and J_e^o from the total recoil strain (γ_r) in the recovery phase:

$$\eta_0 = \sigma_0/\dot{\gamma}_{ss} \qquad \text{(zero shear viscosity)} \qquad (5)$$

$$J_e^o = \gamma_r/\sigma_0 \qquad \text{(recoverable compliance)} \qquad (6)$$

The Boltzmann superposition principle relates η_0 and J_e^o to the properties of $G(t)$:

$$\eta_0 = \int_0^\infty G(t)\,dt \qquad (7)$$

$$J_e^o = \frac{1}{\eta_0^2} \int_0^\infty tG(t)\,dt \qquad (8)$$

For a Newtonian liquid, J_e^o is zero. All liquids have a viscosity, but a nonzero value for J_e^o is one clear indication of a viscoelastic nature. The value of J_e^o also characterizes the elasticlike features of the response in steady shearing flows, as will be seen shortly.

The product $\eta_0 J_e^o$ is an average relaxation time, τ_0, which measures the time required for final equilibration of stress in the liquid:

$$\tau_0 = \eta_0 J_e^o \qquad (9)$$

The magnitude of τ_0 also plays an important role in locating the onset of nonlinear viscoelastic response in steady shear flows (*see Nonlinear Viscoelasticity*).

The characteristics of $G(t)$ for melts of nearly monodisperse linear polymers, illustrating the glassy, transition, plateau, and terminal zones of response, are sketched in Figure 7. Deformation carries the chains into distorted conformations (Figure 8). At very short times, the response is glassy. The modulus for glass response, G_g, is large ($\sim 10^{10}$ dynes/cm^2 is typical for liquids) and essentially constant. The modulus begins to decrease in the same range of times that the chains begin to relax locally, and it continues to decrease as the relaxation propagates over progressively longer chain distances. For short chains, the relaxation simply proceeds

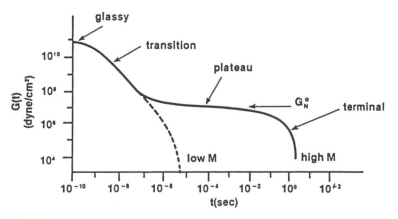

FIGURE 7. Shear stress relaxation function for a typical polymer melt.

smoothly and uneventfully to zero. For long chains, however, the relaxation rate, $d \log G(t)/d \log t$, then slows perceptibly, and the modulus remains relatively constant for some range of times before resuming a more rapid relaxation rate to full equilibrium.

This intermediate zone, or plateau, separates the short-time-relaxation region, called the transition zone, where the large-scale chain architecture has little effect, and the long-time-relaxation region, called the terminal zone, where such architectural features as molecular weight, molecular weight distribution, and long-chain branching have a profound effect. Response in the plateau zone resembles that of a rubber network. The width of the plateau zone increases rapidly with chain length, but the plateau modulus, G_N^o (typically 10^6–10^7 dynes/cm^2 and depending on the polymer species and concentration), is independent of chain length.

Entanglements

This "interruption" of relaxation at intermediate times (or chain distances) is attributed to chain entanglement or, more precisely, to chain uncrossability. At high polymer concentrations, the domains of individual chains overlap extensively (Figure 9). Long chains equilibrate in the deformed state up to a certain average distance along the chains, called the entanglement spacing (corresponding to the end of the transition zone), but further conformational equilibration is delayed until the chains extricate themselves from the constraining mesh of surrounding chains (the onset of the terminal zone). The chains must somehow diffuse around the contours of their neighbors to return to an equilibrium distribution of conformations

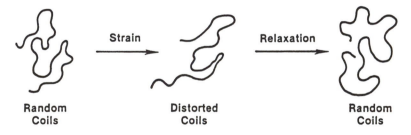

FIGURE 8. Distortion and relaxation of chain conformation after a step shear strain.

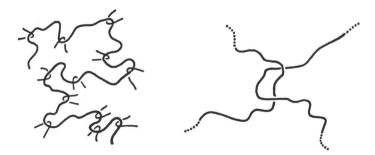

FIGURE 9. Chain entanglement interactions.

and, thereby, to obliterate all memory of the original shape. These terminal relaxation processes dominate the flow behavior. They are governed by entanglement interactions, which are essentially geometrical in nature and are universally observed in the melt dynamics of long, flexible chains. The plateau modulus is important in flow behavior, because it sets the modulus scale for the terminal response. The details of shape in the transition and glassy zones, on the other hand, have little influence on flow behavior.

Storage and Loss Moduli and Related Properties

Although stress relaxation and creep experiments are used extensively, the oscillatory shear experiment (Figure 10) is the most commonly used method for determining the linear viscoelastic properties of polymer melts and concentrated solutions. The liquid is strained sinusoidally, and the in-phase and out-of-phase components of the shear stress at steady state are measured as functions of the frequency, ω.

$$\gamma(t) = \gamma_0 \sin \omega t \qquad \text{(input)} \qquad (10)$$

$$\sigma(t)/\gamma_0 = G'(\omega)\sin \omega t + G''(\omega)\cos \omega t \qquad \text{(response)} \qquad (11)$$

$G'(\omega)$ is the dynamic storage modulus, and $G''(\omega)$ is the dynamic loss modulus. The strain amplitude γ_0, is kept small enough to evoke only a linear response. For a perfectly elastic solid (equation 1), the stress would be in phase with the strain to give $G'(\omega) = G$ and $G''(\omega) = 0$ at all frequencies. For a perfectly viscous liquid (equation 2), the stress would be 90° out of phase with the strain (but in phase with the strain rate, $\gamma_0 \omega \cos \omega t$) to give $G'(\omega) = 0$ and $G''(\omega) = \eta \omega$. The phase angle, $\delta(\omega) = \arctan[G''(\omega)/G'(\omega)]$, depends on frequency for a viscoelastic substance, as do the individual values of $G'(\omega)$ and $G''(\omega)/\omega$. Related quantities, such as the dynamic viscosity, $\eta'(\omega)[\eta'(\omega) = G''/\omega]$, and the loss tangent, $G''(\omega)/G'(\omega)$, are sometimes used as well.

The frequency dependence of $G'(\omega)$ and $G''(\omega)$ for a melt of long, nearly monodisperse linear chains is sketched in Figure 11. Compared with $G(t)$ (Figure 7), the order of appearance of the various viscoelastic zones is reversed. Low frequencies correspond to long times, and high

OSCILLATORY STRAIN RESPONSE

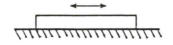

Strain: $\gamma(t) = \gamma_0 \sin \omega t; \dot{\gamma}(t) = \dfrac{d\gamma}{dt} = \gamma_0 \omega \cos \omega t$

γ_0 = strain amplitude

ω = frequency (radian/sec, sec^{-1})

FIGURE 10. Steady-state response to a small-amplitude, sinusoidal shear deformation.

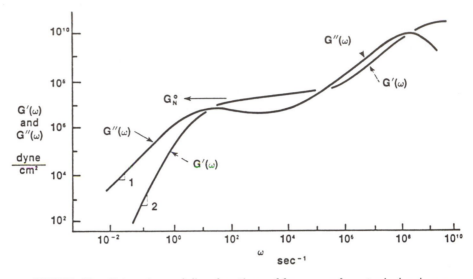

FIGURE 11.   Dynamic moduli as functions of frequency for a typical polymer melt.

frequencies correspond to short times. At the lowest frequencies, $G'(\omega)$ is much smaller than $G''(\omega)$: the viscous response dominates. The curves eventually cross, however, and at intermediate frequencies, $G'(\omega)$ is larger than $G''(\omega)$: the elastic response dominates in the plateau zone. The relative magnitudes reverse again on entering the transition zone. Eventually $G'(\omega)$ levels off at the glassy modulus to G_g, and $G''(\omega)$ falls again in the glassy zone. The loss modulus has two peaks, corresponding in location to the terminal zone (low-frequency) and the transition zone (high-frequency) relaxation processes.

The behaviors of both $G'(\omega)$ and $G''(\omega)$ are linked to $G(t)$ through the Boltzmann superposition principle, and the zero shear viscosity and recoverable shear compliance can be obtained from their properties in the low-frequency limit:

$$\eta_0 = \lim_{\omega \to 0} G''(\omega)/\omega \tag{12}$$

$$J_e^o = \frac{1}{\eta_0^2} \lim_{\omega \to 0} G'(\omega)/\omega^2 \tag{13}$$

The plateau modulus can be estimated from the relatively constant values of $G'(\omega)$ in the plateau zone for nearly monodisperse samples, but more refined methods are also available (1).

Temperature Dependence

The stress relaxation modulus and dynamic moduli, as set forth in Figures 7 and 11, span many orders of magnitude in modulus, as well as in time or frequency scale. No single experiment can cover the entire span; 5 orders of magnitude is a typical dynamic range for even the best instruments. Those sketches in fact represent master curves—composites of data for the same material measured at different temperatures. With very few exceptions, conventional low-density polyethylene being the most prominent (8), homogeneous polymer liquids obey the principle of time–temperature superposition (1, 9). A change in temperature shifts the viscoelastic functions along the modulus and time (or frequency) scales without changing their shapes.

$$G(t, T) = G(t/a_T, T_0)/b_T \qquad (14)$$

$$G'(\omega, T) = G'(\omega a_T, T_0)/b_T \qquad (15)$$

$$G''(\omega, T) = G''(\omega a_T, T_0)/b_T \qquad (16)$$

In equations 14–16, T_0 is some convenient but arbitrary reference temperature, and a_T and b_T are empirically determined ratios of the time and modulus scales for another temperature $T(a_T = b_T = 1$ at $T_0)$. Moreover, the modulus shift is usually very small so that the major effect of a temperature change is to rescale the time (or frequency). Raising the temperature shifts the response curves to smaller times (higher frequencies). The rate of molecular rearrangement is increased, but the molecular organization (i.e., the physical structure of the liquid) is practically unchanged. Measurements at different temperatures can thus be assembled to form a master curve, covering many more decades than is possible by measurements at any single temperature.

Typical behavior is shown in Figure 12, where storage modulus is plotted versus frequency in the plateau and terminal regions at several temperatures for a polystyrene sample (10). A reference temperature (T_0) is selected, in this case $T_0 = 160$ °C, and "best fit" scale factors for data obtained at other temperatures are determined empirically. The time scale can shift very rapidly, as indicated by the plot of a_T versus T in Figure 13 (10). The Williams–Landel–Ferry (WLF) equation, introduced in Chapter 2, describes rather well the temperature dependence of a_T for most polymer melts and concentrated solutions. This important subject is discussed in some detail in the treatise by Ferry (1), which also provides a wealth of data for many polymer species. For the purposes of this chapter, temperature dependence is primarily a function of the local liquid composition. Thus, except for rather short chains, the values of a_T are independent of molecular weight and molecular weight distribution. Moreover, the temperature dependence of viscosity depends directly on a_T. Thus, with the approximation of no shift in modulus scale with temperature, the following relation applies:

$$\eta_0(T) = \eta_0(T_0) a_T \qquad (17)$$

Both G_N° and J_e° are independent of temperature. These results follow directly from the combination of Boltzmann and time–temperature superposition and are extremely useful for extrapolation purposes. Thus, for example, flow-related viscoelastic properties can be measured at experimentally convenient temperatures and estimated at other temperatures rather easily.

Effects of Chain Architecture

Master curves for the dynamic moduli of linear polystyrenes with different molecular weights and narrow distributions (10) are shown in Figure 14.

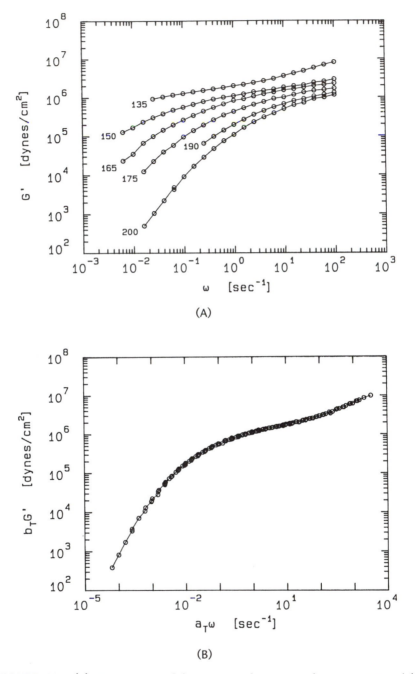

FIGURE 12. (A) Storage modulus versus frequency for a commercial polystyrene at several temperatures. (B) Master curve for data in A for a reference temperature, $T_0 = 150$ °C. (Courtesy of S. H. Wasserman, Princeton University.)

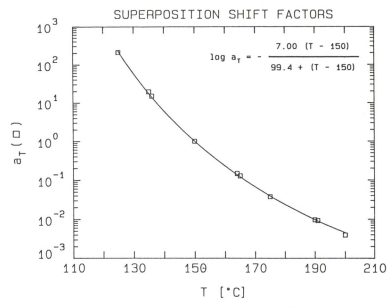

FIGURE 13. Temperature shift factor, a_T ($T_0 = 150$ °C), and the WLF equation fit for the data shown in Figure 12. (Courtesy of S. H. Wasserman, Princeton University.)

Notable in this figure are the increase in plateau width with increasing molecular weight and the similarity in shapes of the terminal regions for different molecular weights. The plateau modulus, $G_N^o = 2.0 \times 10^6$ dynes/cm^2 for undiluted polystyrene, is independent of chain length.

Master curves for polystyrene samples with different molecular weight distributions (*11*) are shown in Figure 15. The samples have similar viscosities, because their loss moduli merge at low frequencies (*see* equation 12). However, their recoverable compliances are quite different. For samples with the same viscosity, J_e^o depends only on $G'(\omega)$ at low frequencies (equation 13), and those values are much higher for the sample with broader distribution. Indeed, $G'(\omega)$ for that sample has still not reached the limiting behavior ($G' \propto \omega^2$) at the lowest accessible frequencies. This result illustrates the general point that J_e^o is extremely sensitive to distribution breadth and particularly to the presence of a high-molecular-weight tail. Molecular weight distribution strongly affects the shape of the terminal region. The response is "smeared out", so to speak, because chains of different sizes relax to equilibrium at different rates. For the example in Figure 15, the terminal zone for the polydisperse sample is so broad that the terminal loss peak in G'', rather well defined in the narrow-distribution sample, is merely a broad shoulder on the transition loss peak. Modulus values for the two samples merge at high frequencies. Response at high frequencies depends only on local chain motions; the effects of chain length and distribution are gone.

Nonlinear Viscoelasticity

This section considers the behavior of polymeric liquids in steady, simple shear flows—the shear rate dependence of viscosity and the development

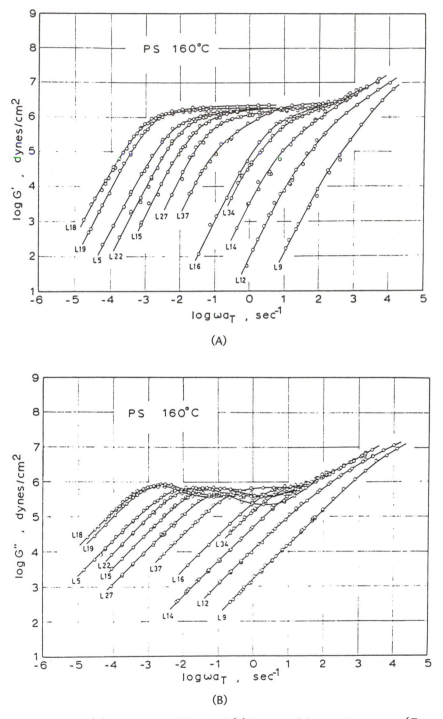

FIGURE 14. (A) Storage modulus and (B) loss modulus master curves ($T_0 =$ 160 °C) for a series of nearly monodisperse polystyrenes. Weight-average molecular weights ($\overline{M}_w$) range from 8900 (L9) to 581 000 (L18). (Reproduced from reference 10. Copyright 1970 American Chemical Society.)

of normal stress differences. Also considered in this section is an elastic recoil phenomenon, called die swell, that is important in melt processing. These properties belong to the realm of nonlinear viscoelastic behavior. In contrast with linear viscoelasticity, neither strain nor strain rate is always small, Boltzmann superposition no longer applies, and the chains are displaced significantly from their equilibrium conformations (Figure 16). The large-scale organization of the chains (i.e., the physical structure of the liquid, so to speak) is altered by the flow. The effects of finite strain appear, much as they do when a polymer network is deformed appreciably.

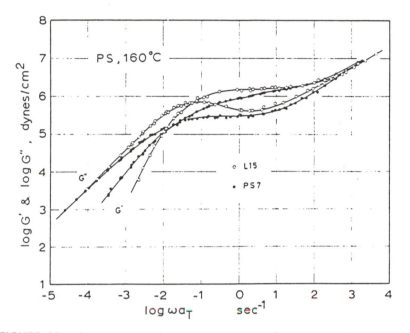

FIGURE 15. Comparison of dynamic moduli ($T_0 = 160$ °C) for nearly monodisperse (L15; $\overline{M}_w / \overline{M}_n < 1.1$) and broad-distribution (PS7; $\overline{M}_w / \overline{M}_n \sim$ 1.6) polystyrenes. (Reproduced from reference 11. Copyright 1970 American Chemical Society.)

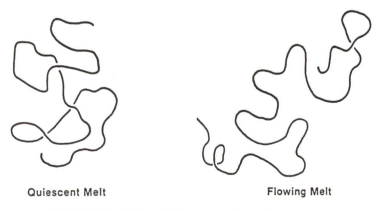

FIGURE 16. Effect of flow on chain conformation.

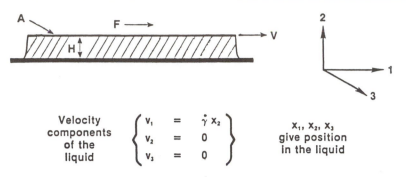

$$v_1 = \dot{\gamma} x_2, \qquad v_2 = 0, \qquad v_3 = 0 \tag{18}$$

FIGURE 17.   Schematic of steady shear flow.

If any liquid is sheared at a constant shear rate $\dot{\gamma}$, the stress that results will eventually reach a steady state. In the parallel-plate illustration in Figure 17, the upper plate moves at constant velocity, V, in direction 1, and a constant shear stress, $\sigma = F/A$, acts in direction 1 on all planes of the liquid that are normal to direction 2. The deformation illustrated in Figure 17 is homogeneous: the shear rate, $\dot{\gamma} = V/H$, is the same everywhere in the liquid, and the components of velocity are as follows:

$$v_1 = \dot{\gamma} x_2, \qquad v_2 = 0, \qquad v_3 = 0 \tag{18}$$

In equation 18, the subscripts 1, 2, and 3 refer to the flow, velocity gradient, and neutral directions, respectively, and x_2 is the vertical distance measured from the fixed plate. Aside from pressure, the forces acting on each element of the liquid are also the same everywhere.

Viscosity and Normal Stress

Suppose we could isolate a small cubical element of the liquid at some instant (with faces parallel to the three coordinate directions) and examine the forces acting upon it. For a Newtonian liquid, the components of force that act normal to the faces would all have the same magnitude, which would depend only on the pressure. A shear component also would exist in the force acting on certain of the faces (Figure 18). Again all equal in magnitude, these shear forces would scale with the viscosity and would be directly proportional to the shear rate. The situation is changed in two important ways for a viscoelastic liquid. First, the normal components of force acting in each of the three directions are no longer equal in magnitude. The major result of these differences in the normal components of force is to produce a tension in the liquid along the line of flow (Figure 18). The normal force differences depend on shear rate and are zero only in the limit of $\dot{\gamma} = 0$. Second, the shear forces, although still all equal in magnitude and as shown in Figure 18, are not directly proportional to shear rate except in the limit of sufficiently small $\dot{\gamma}$.

The components of stress for this chosen orientation are simply the various force components divided by the face areas of the elemental cube. Aside from pressure, the components of stress for any viscoelastic liquid in simple shear flow are completely specified by three shear rate functions, the shear stress function, and two normal stress difference functions. The

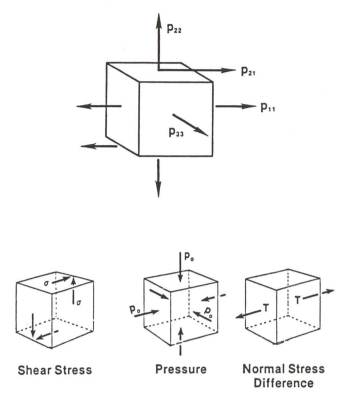

FIGURE 18.   Stress components in shear flow. (Bottom diagram reproduced with permission from reference 6. Copyright 1964 Academic Press.)

shear stress is denoted by $\sigma(\dot{\gamma})$ (p_{21} in Figure 18), and the first and second normal stress differences are denoted by $N_1(\dot{\gamma})$ and $N_2(\dot{\gamma})$, respectively ($p_{11}-p_{22}$ and $p_{22}-p_{33}$ in Figure 18). All three become zero for $\dot{\gamma} = 0$. The shear stress is proportional to shear rate at low enough shear rates (the linear viscoelastic regime). The normal stress differences, small compared with the shear stress in this range of shear rates, are proportional to the square of shear rate. Thus,

$$\sigma(\dot{\gamma}) = \eta(\dot{\gamma})\dot{\gamma} \tag{19}$$

where $\eta(\dot{\gamma})$ is the steady-state viscosity, and

$$N_1(\dot{\gamma}) = \theta_1(\dot{\gamma})\dot{\gamma}^2 \tag{20}$$

$$N_2(\dot{\gamma}) = \theta_2(\dot{\gamma})\dot{\gamma}^2 \tag{21}$$

where $\theta_1(\dot{\gamma})$ and $\theta_2(\dot{\gamma})$ are the first and second normal stress coefficients, respectively. From general arguments (*12*) it has been shown that in the low shear rate limit the following relationships exist:

$$\eta(0) = \eta_0 \tag{22}$$

$$\theta_1(0) = 2J_e^o\eta_0^2 \tag{23}$$

Equations 22 and 23 establish direct, and now well-verified, connections to the linear viscoelastic response.

Some general facts are known about these properties for simple, nonassociating polymeric liquids. Both $\eta(\dot{\gamma})$ and $\theta_1(\dot{\gamma})$ decrease with increasing shear rate, and both begin to depart from η_0 and $2J_e^{\circ}\eta_0^2$ near $\dot{\gamma} = 1/\tau_0$ (equation 9). The ratio $N_1(\dot{\gamma})/\sigma^2(\dot{\gamma})$, however, is relatively insensitive to both shear rate and temperature, an observation that is sometimes useful for extrapolation and estimation purposes. Much less is known about N_2. However, it has been shown to be a negative quantity for homogeneous polymer liquids and closely related to, but smaller in magnitude than, N_1. Thus, on the basis of limited data, $-N_2/N_1$ is in the range 0.1–0.3 and insensitive to the shear rate (13).

Viscosity–shear rate behavior is relatively easy to measure (as will be discussed later). Measurement of first normal stress difference is more difficult, especially at high shear rates ($\dot{\gamma}\tau_0 \gg 1$), and data on N_2 are very scarce indeed. Figure 19 shows the working parts of the cone and plate rheometer, a device that is commonly used to measure both $\sigma(\dot{\gamma})$ and $N_1(\dot{\gamma})$ at relatively low shear rates. The liquid is placed in the gap, one of the fixtures is held fixed, and the other is rotated at constant angular velocity, $\dot{\phi}$. The generated torque, M, and total thrusting force, F, in the axial direction are measured. For a small gap angle α, the shear rate is $\dot{\gamma} = \dot{\phi}/\alpha$ and is the same everywhere in the liquid. At steady state, the following relationships apply (14):

$$\sigma(\dot{\gamma}) = 3M(\dot{\gamma})/2\pi R^2 \qquad (24)$$
$$N_1(\dot{\gamma}) = 2F(\dot{\gamma})/\pi R^2 \qquad (25)$$

In these equations, M is the torque, and R is the radius.

The relationship between the axial force F and N_1 is not too difficult to understand. The flow-induced tension along the lines of flow (from N_1) causes the outer liquid elements to squeeze inward upon the inner elements; the result is a buildup of pressure on the cone and plate surfaces from near zero at the outer edge to a maximum at the center. The total axial force, F, is thus just the sum of contributions from this pressure, and equation 25 gives its precise connection with N_1.

Data on σ and N_1 versus $\dot{\gamma}$ for concentrated solutions of nearly monodisperse polyisoprenes (15) are shown at the bottom of Figure 20. At

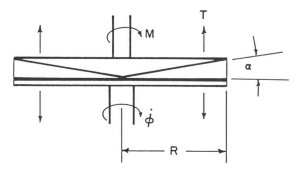

FIGURE 19. Schematic of cone and plate rheometer. (Reproduced with permission from reference 16. Copyright 1975 Syracuse University Press.)

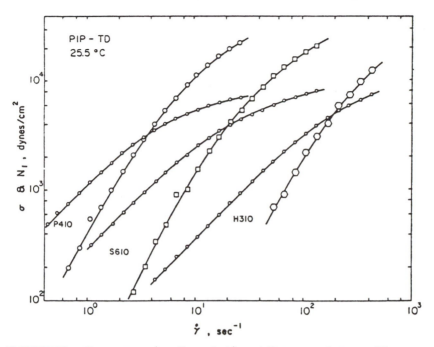

FIGURE 20. Shear stress (small symbols) and first normal stress difference (large symbols) versus shear rate for 10% solutions of three polyisoprenes (P410, S610, and H310) in tetradecane. (Reproduced from reference 15. Copyright 1976 American Chemical Society.)

low shear rates, σ is indeed much larger than N_1, but because σ grows as $\dot{\gamma}$ and N_1 as $\dot{\gamma}^2$, the curves eventually cross, and N_1 is larger than σ at high shear rates. Near the crossing point ($\sigma \sim N_1$), σ begins to depart from its direct proportionality to $\dot{\gamma}$; that is, the steady-state viscosity, $\eta(\dot{\gamma})$, begins to decrease from its low-shear-rate limit, η_0. That qualitative characteristic appears to be quite general for flexible polymer liquids. The onset of non-Newtonian viscosity behavior occurs near the shear rate at which σ and N_1 become equal, and N_1 grows increasingly large compared with σ at higher shear rates. Figure 21 shows the same data when plotted as $\eta(\dot{\gamma})$ and $J_s(\dot{\gamma}) = N_1/2\sigma^2$; the figure demonstrates the near constancy of N_1/σ^2 even well beyond the knee ($\dot{\gamma}\tau_0 \sim 1$) of the viscosity–shear rate curve.

The use of cone and plate rheometers is limited to relatively low shear rates by the onset of flow instabilities, typically occurring not far beyond the σ–N_1 crossing point in polymer melts. A capillary rheometer is sketched in Figure 22. Operation at much higher shear rates is possible, but N_1 cannot be determined directly in such instruments. The steady-state viscosity, however, can be obtained from measurements of the volumetric flow rate, Q, and the pressure drop, $\Delta P = P - P_0$ (where P_0 is the ambient pressure). For long tubes ($L/D \gg 1$), the following equation applies for Newtonian liquids:

$$\eta = \frac{\pi D^4}{128 L} \cdot \frac{\Delta P}{Q} \tag{26}$$

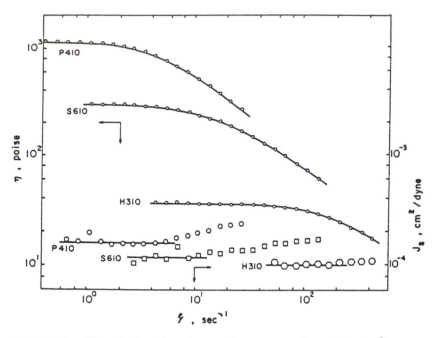

FIGURE 21. Viscosity ($\sigma/\dot{\gamma}$) and normal stress compliance ($N_1/2\sigma^2$) versus shear rate for 10% solutions of three polyisoprenes (P410, S610, and H310) in tetradecane (*see* Figure 20). (Reproduced from reference 15. Copyright 1976 American Chemical Society.)

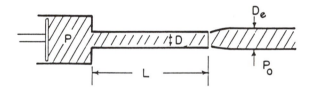

FIGURE 22. Schematic of capillary rheometer. (Reproduced with permission from reference 16. Copyright 1975 Syracuse University Press.)

Equation 26 is not generally correct for viscoelastic liquids, because it is based on the assumption of a constant viscosity. Unlike cone and plate flow, the shear rate in tubes varies with location, in this case with distance from the tube centerline, so that any shear rate dependence of $\eta(\dot{\gamma})$ rules out the use of equation 26. However, regardless of liquid properties, the shear stress at the wall (σ_w) always be calculated from the pressure drop (2):

$$\sigma_w = \frac{D}{4L} \cdot \Delta P \qquad (L/D \gg 1) \tag{27}$$

Likewise, the shear rate at the wall ($\dot{\gamma}_w$) can be obtained from an appropriate numerical differentiation of Q versus ΔP data (2):

$$\dot{\gamma}_w = \frac{8Q}{\pi D^3}\left[3 + \frac{d \log Q}{d \log \Delta P}\right] \tag{28}$$

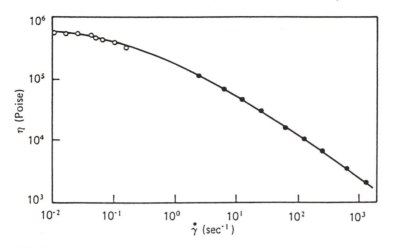

FIGURE 23.   Viscosity versus shear rate for a commercial polystyrene melt. Open circles are data obtained with a cone and plate instrument; filled circles are data obtained with a capillary instrument. (Reproduced with permission from reference 16. Copyright 1975 Syracuse University Press.)

It follows that $\eta(\dot\gamma) = \sigma_w/\dot\gamma_w$. Results obtained for a commercial polystyrene sample (*16*) are shown in Figure 23. Cone and plate measurements cover the low-shear-rate range, and capillary measurements cover the high-shear-rate range. The two instruments provide complementary information on the viscosity behavior.

Die Swell

Figure 22 also illustrates the phenomenon of die swell, resulting from the spontaneous rearrangement of the extrudate to a larger diameter D_e than the capillary diameter D as it emerges from the capillary. The swell ratio, D_e/D, for polymer liquids increases with increasing flow rate in the capillary. It also depends on the length/diameter ratio of the capillary (Figure 24). This length dependence reflects a "memory" of shape that the liquid elements had prior to entering the capillary, but D_e/D still depends on Q even in the long-capillary limit. In the long-capillary limit, the swell ratio mainly reflects the normal stress differences (primarily N_1) generated by shearing flow in the capillary itself. The tension along the lines of flow draws the extrudate back when the confinement of the wall is no longer acting, the action producing a recoil that is analogous to the retraction on release of a stretched rubber band.

Figure 25 shows a plot of D_e/D versus $\dot\gamma_w$ for long capillaries in relation to the plot of η versus $\dot\gamma$ for a commercial polystyrene (*16*). At low shear rates, the viscosity levels off at η_0. Normal stress differences are small in that region, as discussed before, and D_e/D is about 1.1, which is the value expected for slow flows in a highly viscous Newtonian (nonelastic) liquid (*5*). The swell ratio then begins to rise near $\dot\gamma \sim 1/\tau_0$, which marks the onset of shear rate dependence in the viscosity and also locates the range when $N_1 \sim \sigma$, as discussed earlier. The values of D_e/D increase steadily at still higher shear rates, where N_1 exceeds σ by increasing

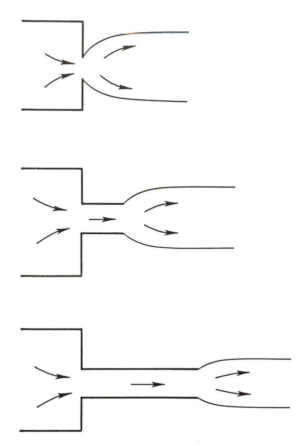

FIGURE 24. Diagram illustrating the effect of viscoelastic memory on die swell for capillaries of different lengths.

amounts, and the viscosity assumes a power law dependence, $\eta \propto \dot{\gamma}^{-a}$, where $a \sim 0.65$ in this particular case. Theories relating die swell and normal stress difference are qualitatively successful; the phenomenon itself is complicated even for Newtonian liquids. The Tanner equation (5) captures the essential features for polymer melts:

$$D_e/D = 0.13 + \left[1 + \frac{1}{8} \left(\frac{N_1}{\sigma} \right)^2_w \right]^{1/6} \tag{29}$$

Temperature Dependence

As with linear response, temperature has a large and systematic effect on nonlinear viscoelastic behavior, and time–temperature superposition can again be very useful. Indeed, the temperature-shift factors are indistinguishable from those obtained from linear viscoelastic measurements on the same material. Stress plays the role of modulus, and shifts along the stress axis with temperature are relatively small. Shear rate plays the role of frequency, and shifts with temperature along the shear rate axis are typically large and governed by a_T. Thus, for example, curves of shear stress versus shear rate are shifted by temperature along the shear rate

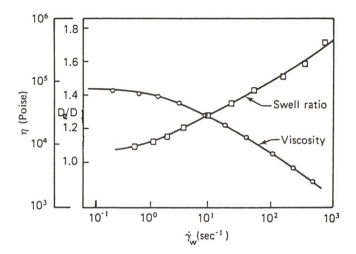

FIGURE 25. Viscosity and swell ratio versus shear rate at the wall for a commercial polystyrene in capillary flow. (Reproduced with permission from reference 16. Copyright 1975 Syracuse University Press.)

axis without change of shape. The same data, plotted as viscosity–shear rate curves, shift by very nearly equal amounts along each axis (*17*), as shown for a nearly monodisperse polystyrene sample at different temperatures in Figure 26. Reduction to master curves can be achieved by normalizing the viscosity values with η_0 and plotting as a function of $\dot{\gamma}\tau_0$, as shown in Figure 27. Because $\tau_0 = J_e^{\circ}\eta_0$ and J_e° is insensitive to temperature in most cases, superposition can frequently be obtained simply by plotting $\eta(\dot{\gamma})\eta_0$ versus $\dot{\gamma}\eta_0$. Another useful feature, the temperature independence of the relationship between σ and N_1, follows from the temperature superposition principle as stated earlier. Both depend on shear rate, and thus, both shift along the shear rate axis with temperature, but when $\dot{\gamma}$ is eliminated by plotting N_1 versus σ, the result is quite insensitive to temperature.

The effect of temperature on die swell provides an interesting application of this principle. As shown in Figure 28 for a commercial polystyrene, at each shear rate the swell ratio decreases with increasing temperature (*16*). However, the data at all temperatures superpose rather well when plotted instead as a function of shear stress (Figure 29). That result turns out to be quite general and of course quite useful for extrapolation purposes. It is a natural consequence of the temperature invariance of the relationship between σ and N_1 and the idea, embodied by equation 29, that D_e/D for long capillaries depends only on N_1/σ in the capillary.

Effects of Molecular Structure

The importance of the terminal zone in viscoelastic response to flow behavior and its strong dependence on chain length, distribution, and branching has already been discussed. In this section, the molecular effects are considered in greater detail. It is important, first of all, to distinguish effects due to the local chain structure, which, aside from

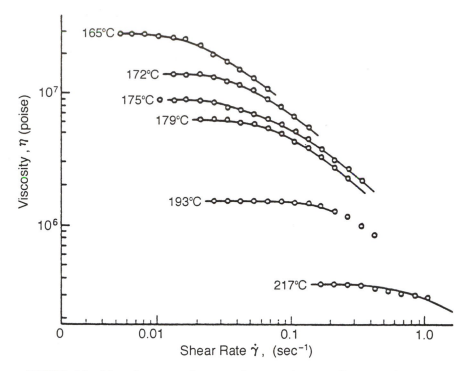

FIGURE 26. Viscosity versus shear rate for a nearly monodisperse polystyrene at several temperatures. (Reproduced with permission from reference 17. Copyright 1974 John Wiley and Sons, Inc.)

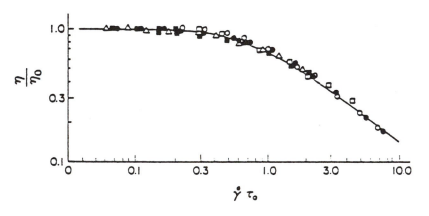

FIGURE 27. Viscosity–shear rate master curve for results shown in Figure 26. The various symbols represent data obtained at different temperatures. (Reproduced with permission from reference 17. Copyright 1974 John Wiley and Sons, Inc.)

nonequilibrium effects induced by incomplete relaxation from some previous flow history, controls the basic physical properties of polymers in the solid state, from those due to the large-scale chain structure, which mainly controls the viscoelastic properties of polymers in the liquid state (Figure 30). Commercial polymers are typically very heterogeneous in large-scale chain structure. Commercial polyolefins, which may contain significant numbers of molecules distributed over 4 or more orders of magnitude in

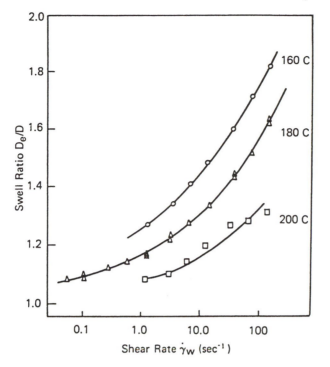

FIGURE 28.   Swell ratio versus shear rate at the wall for capillary flow of a commercial polystyrene at three temperatures. (Reproduced with permission from reference 16. Copyright 1975 Syracuse University Press.)

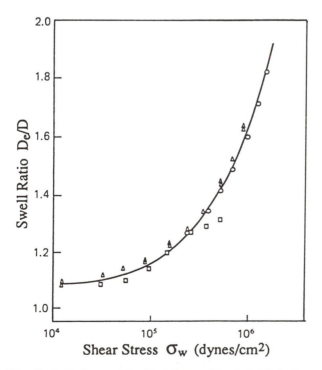

FIGURE 29.   Data in Figure 28 plotted as a function of shear stress at the wall. (Reproduced with permission from reference 16. Copyright 1975 Syracuse University Press.)

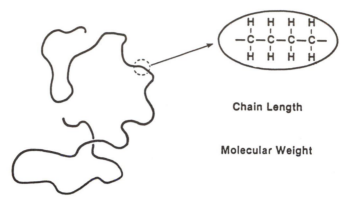

Chain Length

Molecular Weight

FIGURE 30. Schematic illustrating the distinction between chemical microstructure and large-scale chain architecture.

chain length, are rather extreme examples, but in general, the molecular weight distributions of commercial polymers are quite broad and highly variable. Fortunately, the effect of molecular weight and molecular weight distribution appears to be essentially universal: large-scale architecture affects liquid-state viscoelasticity according to general laws. The laws themselves are independent of the local chain structure: aside from proportionality constants, which do depend on the particular species (the local structure), the laws are the same for all flexible-chain polymers.

Molecular Weight Distribution

Figure 31 shows the molecular weight distribution for another commercial polymer, in this case poly(vinyl chloride). The weight distribution function, $W(M)\,dM$, is the fractional weight of the sample contributed by molecules with molecular weights ranging from M to $M + dM$. The distribution is commonly characterized by its averages, $\overline{M}_n$, $\overline{M}_w$, $\overline{M}_z$, and $\overline{M}_{z+1}$ defined as follows by ratios of successively higher moments of the distribution:

$$\overline{M}_n = \frac{\int W(M)\,dM}{\int \frac{1}{M} W(M)\,dM} \qquad \text{(number average)} \qquad (30)$$

$$\overline{M}_w = \frac{\int M W(M)\,dM}{\int W(M)\,dM} \qquad \text{(weight average)} \qquad (31)$$

$$\overline{M}_z = \frac{\int M^2 W(M)\,dM}{\int M W(M)\,dM} \qquad \text{(z average)} \qquad (32)$$

$$\overline{M}_{z+1} = \frac{\int M^3 W(M)\,dM}{\int M^2 W(M)\,dM} \qquad \text{(z + 1 average)} \qquad (33)$$

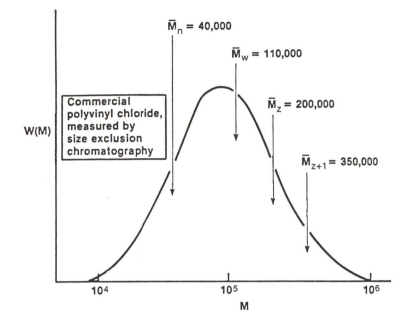

FIGURE 31. Molecular weight distribution and averages for a commercial poly(vinyl chloride).

The values for this sample of poly(vinyl chloride), not particularly broad in distribution compared with most polyolefins, were calculated from measurements with a calibrated size exclusion chromatograph (SEC). The values of $\overline{M}_z$, and especially $\overline{M}_{z+1}$, are only estimates; the results for those averages in particular are exceedingly sensitive to SEC baseline determinations.

Rheological properties are far more sensitive to molecular weight distribution, and particularly to the high-molecular-weight tail of the distribution, than dilute-solution methods such as SEC. It is not unusual to find, for example, two polymer samples with SEC results that are the same within the limits of reproducibility and yet differ significantly in melt flow behavior. Sensitivity to high-molecular-weight components is, in fact, to be expected even from the very simplest of molecular theories. Obviously, it is essential to understand how molecular structure affects rheological behavior and to use samples of well-defined structure to explore those effects experimentally. The discussion begins with the properties of nearly monodisperse linear polymers, then considers polydispersity contributions, and finally some effects due to long-chain branching.

Nearly Monodisperse Linear Polymers

In melts and concentrated solutions the chains are random coils at equilibrium, and the following relation exists for linear polymers:

$$R_G{}^2 = K_\theta M \qquad (34)$$

R_G is the root-mean-square radius of gyration of the coils. The coefficient K_θ is known for many species from measurements in θ solvents (18) or small-angle neutron scattering on the melt (Chapter 7). Deformation distorts the distribution of conformations (i.e., the chains are carried out of equilibrium by frictional interactions with their surroundings), and Brownian motion tends to restore the equilibrium. The competition of these opposing effects determines the average conformational distortion of any moment and thus the stress at that moment.

The chain dynamics depend on the interplay of three types of forces acting on the monomeric units: a frictional force that is proportional to the relative velocity of mer and surrounding medium, a connector force from adjacent mers on the same chain, and a random force due to Brownian motion. The Rouse model (19, 20) describes the effects of these forces on the slow dynamics (terminal response and diffusion) for perfectly flexible chains that move independently and in an unrestricted manner in a viscous medium. Excluded volume and long-range hydrodynamic interactions are omitted.

The diffusion coefficient for Rouse chains (D_r) is given by equation 35:

$$D_r = \frac{kT}{n\zeta_0} = \left(\frac{m_0 RT}{N_a \zeta_0}\right)\frac{1}{M} \tag{35}$$

In this equation, n is the number of mers in the chain, ζ_0 is the monomeric friction coefficient, k is the Boltzmann constant ($k = R/N_a$, where R is the universal gas constant and N_a is Avogadro's number), T is the absolute temperature, and m_0 is the molecular weight per mer ($m_0 = M/n$). Conceptually, the diffusion coefficient of unattached mers is given by $D_0 = kT/\zeta_0$ and $D_r = D_0/n$ because the n mers of the chain must move in a collective fashion.

The stress relaxation modulus, $G(t)$, at long times for a liquid of long, monodisperse Rouse chains is given with sufficient accuracy for our purposes by equation 36:

$$G(t) = \frac{cRT}{M} \sum_{p=1}^{\infty} \exp\left(-\frac{p^2 t}{\tau_r}\right) \tag{36}$$

In this equation, c is the polymer concentration (or the melt density, ρ; ρ is the running index in the summation for an undiluted polymer), and τ_r (Rouse relaxation time) is defined as follows:

$$\tau_r = \frac{n\zeta_0 R_G^2}{\pi^2 kT} = \left(\frac{\zeta_0 N_a K_\theta}{\pi^2 m_0 RT}\right) M^2 \tag{37}$$

Each term in the summation in equation 36 describes the relaxation of a normal mode (1) of the chain motion, τ_r/p^2 being the relaxation time associated with segments of length L_0/p, where L_0 is the chain length. The viscosity and recoverable compliance (obtained by applying equations 7 and 8) are as follows:

$$(\eta_0)_r = \frac{\pi^2}{6}\frac{cRT}{M}\tau_r = \left(\frac{\zeta_0 N_a K_\theta}{6m_0}\right)cM \tag{38}$$

$$(J_e^o)_r = \frac{2}{5}\frac{M}{cRT} \tag{39}$$

The Rouse predictions are consistent with some of the observations described earlier. Thus, the longest relaxation time scales with $\eta_0 J_e^o [\tau_r = 15/\pi^2 \cdot (\eta_0 J_e^o)_r]$ (from equations 38 and 39), and the temperature dependence of recoverable compliance are very weak $[(J_e^o)_r \propto (cT)^{-1}]$. The temperature dependence and local structure specificity of diffusion coefficient, viscosity, and relaxation time reside in the monomeric friction coefficient, ζ_0, which is the only adjustable parameter in the model and which, aside from chain-end effects, should be the same for all molecular weights of a given species. On the other hand, the model predicts no shear rate dependence of viscosity: η_r is independent of $\dot{\gamma}$ even when $\tau_r \dot{\gamma}$ is much greater than unity, contrary to the experimental observations.

How well does the Rouse model work in other respects? Viscosity and recoverable compliance as functions of molecular weight are shown in Figures 32 and 33, respectively, for nearly monodisperse samples of polyisoprene (*21*). The behavior for other species is similar. Below some characteristic molecular weight, M_c, the viscosity prediction is rather good: η_0 is proportional to M (after a chain end correction), and even the magnitudes (with ζ_0 from small-molecule-diffusion data) are about right (*1*). Above M_c, however, the viscosity varies with a much higher power of molecular weight (*22*):

$$\eta_0 = KM^b \tag{40}$$

The value of b is approximately 3.4. The pattern for recoverable compliance is analogous. Below another characteristic molecular weight, M_c', the

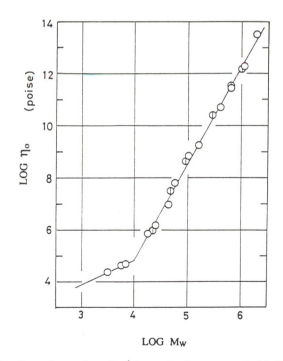

FIGURE 32. Zero shear viscosity (at constant monomeric friction coefficient) versus molecular weight for nearly monodisperse polyisoprene. The two lines have slopes of 1.0 and 3.7. (Reproduced from reference 21. Copyright 1972 American Chemical Society.)

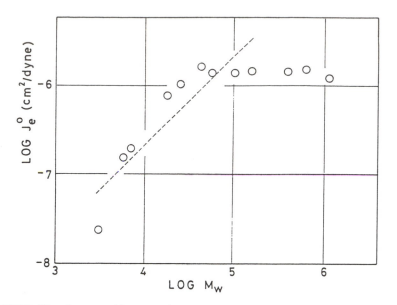

FIGURE 33. Recoverable compliance versus molecular weight for nearly monodisperse polyisoprenes. The dashed line is the Rouse prediction (equation 39). (Reproduced from reference 21. Copyright 1972 American Chemical Society.)

Rouse model prediction, shown by the dashed line, is reasonably good, but then the behavior changes, and J_e^o becomes independent of molecular weight. A crossover is also found for self-diffusion coefficients in the melt state, in this case from $D \propto M^{-1}$ for short chains (after a chain end correction) to $D \propto M^{-2}$ for long chains (23, 24).

All these phenomena appear to be universal for linear polymers and apply to concentrated solutions as well as melts. Rouse-like behavior gives way to a new set of relationships for properties that are controlled by the slow dynamics of the chains. The characteristic molecular weights depend on species and concentration. The crossovers are associated with the onset of entanglement effects, and the new relationships reflect their influence on the slow dynamics (25). In the earlier discussion of the plateau modulus, it was remarked that a liquid of long chains acts, at intermediate times or frequencies, like a network. The theory of rubber elasticity predicts a relationship between shear modulus and the concentration of network strands (Chapter 1). This relationship is used to evaluate M_e, the apparent molecular weight of strands in the entanglement network, called the entanglement molecular weight (1):

$$M_e = cRT/G_N^o \tag{41}$$

Values of M_e are given in Table I for several polymer species in the melt state (c is the melt density, ρ, in this case). The values of M_c and M_c' are greater than, but clearly related to, M_e. Thus, $M_c \sim 2\text{--}3\ M_e$ and $M_c' \sim 5\ M_e$. All three characteristic molecular weights are insensitive to temperature, a fact indicating interactions of a geometrical (topological) character, such as would be anticipated for simple "uncrossability" constraints on the

Table I. Plateau Modulus and Entanglement Molecular Weight for Various Polymer Species in the Undiluted State

Polymer	T (°C)	$G_N^o \times 10^{-6}$ (dyne/cm^2)	M_e
Polyethylene	150	22.	1100
Polypropylene (atactic)	75	8.5	2800
Poly(1-butene) (atactic)	30	1.9	11,600
Poly(1,4-butadiene)	25	11.5	1900
Poly(1,2-butadiene)	50	4.2	5700
Polyisoprene	25	4.4	5100
Polyisobutylene	25	3.2	6900
Poly(dimethylsiloxane)	25	2.4	10,000
Polystyrene	190	2.0	18,700

Note: Selected values were provided by L. J. Fetters (personal communication); *see* Table 13-1 (p 374) in reference 1 for a more extensive compilation.

chain dynamics. Also, the characteristic molecular weights are rather small compared with typical molecular weights for commercial polymers of the same species. Thus, entanglement effects dominate the flow behavior of most commercial polymers.

Small molecule solvents, such as plasticizers, change G_N^o and the characteristic molecular weights in a rather simple manner.

$$G_N^o(\phi) = (G_N^o)_{\text{melt}}\phi^d \qquad (2.1 < d < 2.3) \qquad (42)$$

In equation 42, ϕ is the volume fraction of polymer ($\phi = c/\rho$), and d is the concentration exponent. From equation 41,

$$M_c(\phi) = (M_c)_{\text{melt}}/\phi^{d-1} \qquad (43)$$

The values of M_c and M_c' increase with dilution in the same way as M_e. In no case does the specific nature of the diluent play a role (*26*). The exponent d may depend slightly on the polymer species (thus the range given with equation 42), but it is never far from 2, the value expected for interactions that are proportional to the concentration of pairwise contacts between the chain units.

Viscosity and recoverable compliance also change with dilution. For J_e^o, the dependence is essentially universal (*26*). Thus, for $M > M_c'(\phi)$,

$$J_e^o(\phi) = (J_e^o)_{\text{melt}}/\phi^d \qquad (44)$$

Recoverable compliance, therefore, increases with decreasing concentration, and the product $G_N^o J_e^o$ remains constant. That product appears to be universal for monodisperse linear polymers: $G_N^o J_e^o = 2.0 \pm 0.4$ is found from measurements on nearly monodisperse samples of many polymer species (*25–27*). Two factors, one universal and the other system specific, influence the viscosity (*22*). Thus, for $M > M_c(\phi)$,

$$\eta_0(\phi) = (\eta_0)_{\text{melt}} \frac{\zeta_0(\phi)}{(\zeta_0)_{\text{melt}}}\phi^g \qquad (3.4 < g < 3.9) \qquad (45)$$

where $\zeta_0(\phi)$ is the monomeric friction coefficient for the solution. The ratio $\zeta_0(\phi)/(\zeta_0)_{melt}$ reflects primarily how the glass temperature, and thus the local dynamics, is changed by dilution, whereas the ϕ^g factor accounts for the reduction in both the concentration of chains, c/M, and the number of entanglements per chain, M/M_e. Methods for establishing $\zeta_0(\phi)/(\zeta_0)_{melt}$ are described elsewhere (22, 26). All these equations are applicable to concentrated solutions and eventually fail if the dilution is carried far enough. The limit is perhaps 20% polymer, unless M/M_e for the melt is very large.

Molecular Interpretation

Molecular theories of entangled chain dynamics, based on tube models and reptation, have evolved rapidly in recent years. Figure 34 illustrates the problem of individual chain motion in a liquid filled with long chains. The chains overlap extensively to provide a kind of mutually shared meshwork in which each chain lies along its own tunnel through the mesh. No chain can move sideways very far without crossing through other chains, which is forbidden. As pointed out by de Gennes (28), however, a linear chain can always move freely along its own tunnel and thereby, over time, change its conformation and its location in the liquid. Thus, this snakelike motion, called reptation, provides a mechanism for stress relaxation and diffusion in highly entangled liquids, and it became the basis of a detailed molecular theory by Doi and Edwards (20).

The Doi–Edwards theory assumes that reptation is the dominant mechanism for the relaxation of entangled linear chains. Each molecule has the dynamics of a Rouse chain, but its motions are now restricted spatially by a "tube" of uncrossable constraints (Figure 35). The tube has a diameter corresponding to the mesh size, and the chain diffuses along it at a rate that is governed by its Rouse diffusion coefficient. If the liquid is deformed, as in a step strain experiment (Figure 36), the tubes are dis-

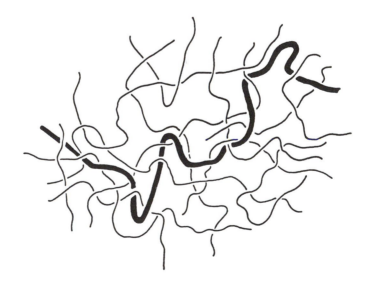

FIGURE 34.   Schematic of an entangled polymer melt. (Reproduced with permission from reference 30. Copyright 1982 Springer-Verlag.)

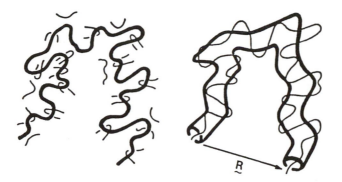

FIGURE 35. Schematic of an entangled polymer chain and representation of uncrossability constraints by a tube. (Reproduced with permission from reference 30. Copyright 1982 Springer-Verlag.)

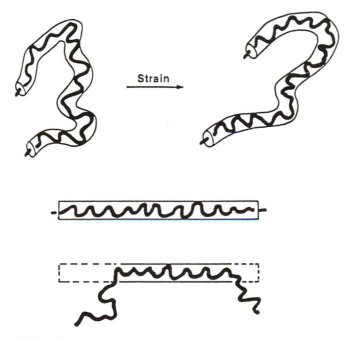

FIGURE 36. Progression of conformational and stress relaxation by reptation.

torted, and the resulting distortion of chain conformations produces a stress. The subsequent relaxation of stress with time corresponds to the progressive movement of chains out of the distorted tubes, and into random conformations, by reptation. The theory contains two experimental parameters, the monomeric friction coefficient, ζ_0 and the tube diameter, a, but once these parameters are independently established, predictions for all properties of the slow dynamics follow directly (*29*).

Despite the simplicity of its basic premises about entanglement interactions, which must be very complex in local detail, the Doi–Edwards theory has been remarkably successful. It encompasses a diverse range of dynamical phenomena within a single molecular framework, and its predictions for nearly monodisperse linear polymers are seldom in gross conflict with observations (*30*). In many cases, its agreement with experiment is essentially quantitative. Thus, self-diffusion coefficients are predictable with some accuracy (ζ_0 and a having been established independently), and the predicted molecular weight independence and magnitude of the recoverable compliance are reasonably consistent with the data ($J_e^\circ G_N^\circ \sim 2$).

$$J_e^\circ = \frac{6}{5}(G_N^\circ)^{-1} \tag{46}$$

The predicted molecular weight dependence of viscosity, $\eta_0 \propto M^3$, is slightly weaker than the observed $M^{3.4}$, however, and the magnitudes of viscosity in the experimental range are too large. The predicted shear rate dependence of viscosity appears to be too strong, but the theory gives a value for the normal stress ratio, N_2/N_1, of the correct sign and with about the right magnitude.

Some deficiencies of the theory, such as its prediction for $\eta_0(M)$, have been attributed to competing mechanisms for relaxation that are not considered in the original theory (*30, 31*). One factor is the contribution of time-dependent fluctuations in the length of tube occupied by the chain. Even if the chain did not reptate, the fluctuations in tube length over time would still relax the stress, albeit much more slowly than reptation. Another factor is the contribution from the finite lifetime of the constraints that define the tube. For any chain, these constraints are the strands of neighboring chains, which are themselves diffusing through the liquid and releasing constraints as they go. In this case, the tubes themselves undergo a random Rouse-like motion over time. The basic elements of these two mechanisms are shown schematically in Figure 37. Fluctuations are thought to be responsible for the discrepancy in the earlier-mentioned prediction for η_0 versus M (*31*) and to be the dominant mechanism for relaxation when long branches suppress reptation (as will be discussed later). Constraint release, although relatively unimportant for nearly monodisperse linear polymers, appears to assume a rather major role in polydisperse systems (*32*).

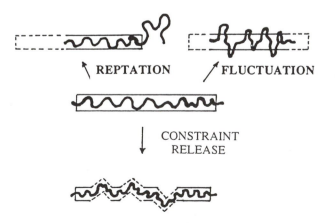

FIGURE 37. Illustration of reptation, fluctuation, and constraint release mechanisms for relaxation of entangled chains.

Effects of Molecular Weight Distribution

How does polydispersity affect the viscoelastic properties of entangled linear polymers? As mentioned previously, the terminal zone broadens, and the recoverable compliance increases with increasing distribution breadth. However, much of what is known quantitatively about such effects is based on studies of model systems, typically mixtures of two nearly monodisperse components with different molecular weights. When both components (S and L for short and long chains, respectively) are well entangled (i.e., M_S and M_L much larger than M_e), the viscosity depends mainly on the weight-average molecular weight. Thus, to a remarkably good approximation (32), equation 40 applies, with M simply replaced by $\overline{M}_w$:

$$\eta_0 = K \cdot \overline{M}_w{}^b \tag{47}$$

where

$$\overline{M}_w = \phi_S M_S + \phi_L M_L \tag{48}$$

and ϕ_S and ϕ_L are either volume or weight fractions. The recoverable compliance is more striking in its behavior. For the pure components, $(J_e^0)_S \sim (J_e^0)_L$, and is larger, sometimes by factors of 30 or more, depending on composition.

Although perhaps surprising at first, this enhancement in J_e^0 for chain-length mixtures is readily understandable as a special kind of dilution effect. Thus, J_e^0 is increased by dilution with small molecules (equation 44), and the effect of adding S chains to an L chain matrix should be similar if S relaxes rapidly enough compared with L. In fact, equation 44 (with $\phi_L = \theta$) works rather well over much of the composition range if M_L is much larger than M_S. The effect of dilution on J_e^0 is best understood physically in terms of its definition in the recovery phase after steady-state creep (equation 6). Dilution reduces the concentration of strands in the entanglement network that supports the stress. Thus, for any given stress, the strands are more deformed (because the stress per strand is higher) so that the recoverable strain, and thus J_e^0 is larger in the diluted system.

The Rouse model for mixtures gives the following:

$$(J_e^0)_r = \left[J_e^0 \left(\overline{M}_w \right) \right]_r \frac{\overline{M}_z \overline{M}_{z+1}}{\overline{M}_w{}^2} \tag{49}$$

$[J_e^0(\overline{M}_w)]_r$ is the monodisperse value at $M = \overline{M}_w$. The Rouse enhancement factor, $\overline{M}_z \overline{M}_{z+1}/\overline{M}_w{}^2$, can thus be much greater than unity, and in fact, it predicts values that are not grossly different from those observed (32) for $J_e^0/(J_e^0)_S$. Other combinations of molecular weight averages have been proposed, but without significant improvement. A simple extension of the Doi–Edwards theory to mixtures is inadequate (32), but a recently proposed law of mixtures for $G(t)$ looks promising (33, 34). The idea is that stress is proportional to the concentration of unrelaxed entanglements, as opposed to the Doi–Edwards supposition that stress is proportional to the fraction of chain length still occupying distorted tubes. The result for binary mixtures is given by equation 50:

$$G(t) = \left[\phi_S G_S^{1/2}(t) + \phi_L G_L^{1/2}(t) \right]^2 \qquad (50)$$

$G_S(t)$ and $G_L(t)$ are the respective stress relaxation moduli for the pure components. The predictions for η_0 and J_e^o (equations 7 and 8) agree rather well with experiment (35). The natural generalization of this form to arbitrary distributions still remains to be tested:

$$G(t) = \left[\int_0^\infty W(M) G^{1/2}(M, t) \, dM \right]^2 \qquad (51)$$

$G(M, t)$ is the stress relaxation modulus for monodisperse samples of the species.

Polydispersity Effects on Nonlinear Response

Nonlinear flow properties and melt-processing behavior are also strongly dependent on polydispersity. The relationship between steady-state viscosity and shear rate is shown schematically in Figure 38. At low shear rates, the viscosity is constant $[\eta(\dot{\gamma}) = \eta_0]$. This is called the Newtonian region and is commonly difficult to reach in very polydisperse commercial polymers. At high shear rates, the viscosity varies inversely with a power of shear rate $[\eta(\dot{\gamma}) \propto \dot{\gamma}^{-p}]$. This is called the power law region, and for melts and concentrated solutions, the exponent p ranges from ~0.85 for nearly monodisperse polymers to ~0.60 for highly polydisperse polymers. (Eventually, at still higher shear rates, the viscosity levels off again, but this second plateau is seldom observable in entangled polymers.) The crossover from Newtonian behavior to power law behavior is located by the characteristic shear rate $\dot{\gamma}_0$, whose value is closely related to the characteristic relaxation time, $\tau_0 = \eta_0 J_e^o$. For example (25),

$$\dot{\gamma}_0 \eta_0 J_e^o = 0.6 \pm 0.2 \qquad (52)$$

if $\dot{\gamma}_0$ is defined for convenience as the shear rate at which the viscosity has decreased to 0.8 η_0. The relationship seems to be universal for entangled polymers (branched or linear, monodisperse or polydisperse, solution or

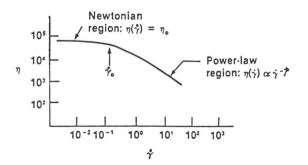

FIGURE 38.   Schematic for viscosity–shear rate relationship.

melts), although the number of tests with commercial polymers is necessarily rather limited.

The shape of the viscosity–shear rate curve for entangled polymers appears to depend only on polydispersity. Thus, for example, the data for nearly monodisperse samples of several polymer species superpose when plotted in reduced form, $\eta(\dot{\gamma})/\eta_0$ versus $\dot{\gamma}/\dot{\gamma}_0$. The same master curve describes results for entangled polymers obtained at different temperatures, concentrations, chain lengths, and even for various branching architectures. Figure 39 illustrates this behavior with data for concentrated solutions of various polystyrenes.

The effect of molecular weight distribution is shown in Figure 40. Viscosity–shear rate data for two polystyrenes are given: a commercial sample ($\overline{M}_w/\overline{M}_n \sim 2.4$) and a nearly monodisperse sample ($\overline{M}_w/\overline{M}_n < 1.1$). Polydispersity broadens the transition from Newtonian behavior to power law behavior. In this example, $\overline{M}_w$ is larger for the polydisperse sample, and so it has a higher value of η_0. However, non-Newtonian behavior appears at a much lower shear rate for that sample: $\dot{\gamma}_0$ is much smaller because both η_0 and J_e° are larger (equation 52), the latter because of polydispersity alone. The two curves cross, and at high shear rates, the polydisperse sample has a lower viscosity than the nearly monodisperse sample. The combination of high viscosity at low shear rates and low viscosity at high shear rates is a desirable feature for certain melt-processing operations (blow molding, for example), and the effect of polydispersity can be used to advantage in those cases.

Some success has been achieved in predicting the shape of viscosity–shear rate curves from information on the molecular weight distribution. The curves drawn in Figure 40 were calculated from SEC data by using a simplified model that attributes the progressive reduction in viscosity with increasing shear rate to a flow-induced disentanglement of the chains (*36*).

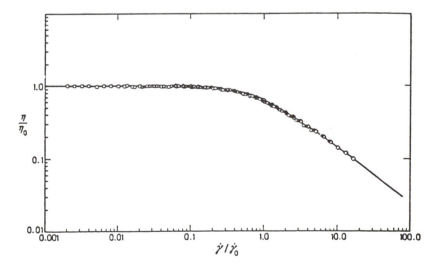

FIGURE 39.   Viscosity–shear rate master curve for concentrated solutions of polystyrene. Data for several concentrations, molecular weights, and temperatures are shown. (Reproduced with permission from reference 25. Copyright 1974 Springer-Verlag.)

Figure 41 demonstrates the extreme importance of distribution breadth in melt elasticity. The die swell ratio, D_e/D_0, begins to increase at much lower capillary shear stress for the broad-distribution sample, and the change is more gradual than that for the narrow-distribution sample. The sensitivity of die swell to distribution breadth follows naturally from the Tanner expression (equation 29), according to which D_e/D_0 is a function of N_1/σ alone. Because $N_1 \sim 2J_e^o\sigma^2$, as noted earlier, and J_e^o increases rapidly with distribution breadth, the ratio N_1/σ, and thus D_e/D_0, at

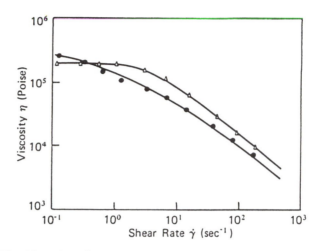

FIGURE 40. Viscosity–shear rate results for polystyrene melts with different molecular weight distributions. The filled symbols represent data for a commercial polystyrene ($M_w = 260\,000$; $\overline{M}_w / \overline{M}_n \sim 2.4$). The open symbols represent data for a nearly monodisperse polystyrene ($\overline{M}_w = 160\,000$; $\overline{M}_w / \overline{M}_n < 1.1$). (Reproduced with permission from reference 16. Copyright 1975 Syracuse University Press.)

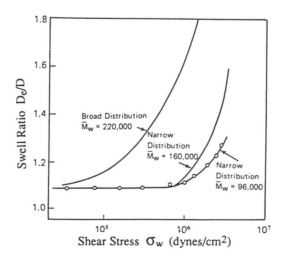

FIGURE 41. Swell ratio versus shear stress at the wall for polystyrene melts with different molecular weights and distribution. (Reproduced with permission from reference 16. Copyright 1975 Syracuse University Press.)

constant shear stress should increase with the distribution breadth of the polymer.

Effects of Branching

Broadly speaking, viscoelastic behavior is only slightly modified if the branches are not too long. However, if the branches are long enough to be well entangled (i.e., if the branch molecular weight is much larger than M_e), then the effects of branching can be profound (37). Nonlinear architectures in commercial polymers are typically generated by some random-branching chemistry, such as polymer transfer, during polymerization. Conventional low-density polyethylene is a prominent example. Random branching invariably introduces a broad distribution of structures, and it becomes extremely difficult to separate the effects due explicitly to branching from those due to polydispersity alone. Most of what is known about branching effects, per se, has come from the study of model systems. Molecular stars, composed of three or more linear strands joined to a common junction, can be made with highly uniform structure by anionic polymerization, followed by an appropriate linking reaction. Branch length and branch point functionality can thus be varied, and their effects can be studied over wide ranges (15, 38, 39). Each molecule contains only one branch point, however, so that some branching effects cannot be studied with stars. A limited amount of data is available for regular combs (40) and randomly branched chains, prepared by fractionation of polydisperse samples (41).

The Rouse theory has been applied to nonlinear polymers (42, 43). It predicts that both η_0 and J_e^o are smaller for branched chains than for linear chains of the same total molecular weight. Qualitatively, the viscosity is less because the coil size is smaller, and the recoverable compliance is less because the coils are less easily deformable. Those predictions hold true experimentally, even in the entanglement region, until the branches themselves are long enough to be significantly entangled ($M_b > 2-4\ M_e$, M_b is branch molecular weight). For longer arms, as shown for stars in Figure 42, the viscosity for branched polymers rises very rapidly relative to that for linear polymers and soon exceeds the latter by factors of 100 or more (44). Viscosity and diffusion coefficient (45) no longer have simple power law dependences on molecular weight. The viscosity increases exponentially with branch length, such that for stars (31)

$$\eta_0 \propto \exp\left(\gamma \frac{M_b}{M_c}\right) \tag{53}$$

where $\gamma \sim 0.6$, irrespective of branch point functionality. Viscosity decreases very rapidly with dilution (27) ($M_e \propto \phi^{1-d}$; *see* equation 43) and will eventually fall below the viscosity for a similarly diluted linear polymer of the same total molecular weight.

The variation of recoverable compliance with molecular weight is also different for linear and nonlinear polymers. In contrast with the behavior of nearly monodisperse linear polymers, for which J_e^o becomes a constant ($\sim 2/G_N^o$) beyond about $5\ M_e$, the values for stars simply continue to increase in direct proportion to M_b, remaining, probably fortuitously, in reasonable agreement with the Rouse prediction for stars (Figure 43).

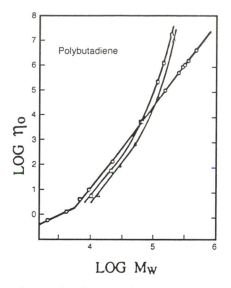

FIGURE 42. Viscosity–molecular weight relationships for linear and star polybutadienes at 107 °C. The symbols represent data for linear samples (○), three-arm stars (□), and four-arm stars (△). (Reproduced with permission from reference 44. Copyright 1965 John Wiley and Sons.)

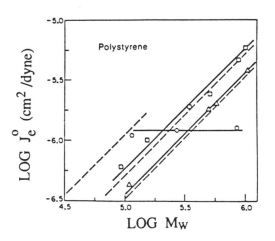

FIGURE 43. Recoverable compliance–molecular weight relationships for linear and star polystyrenes. The symbols represent data for linear samples (○), four-arm stars (□), and six arm stars (△). The dashed lines are Rouse model predictions. (Reproduced from reference 38. Copyright 1979 American Chemical Society.)

Experimentally, the behavior of nearly monodisperse stars is well described by equation 54 irrespective of branch point functionality (*31*).

$$J_e^o = 0.6\frac{M_b}{cRT} \qquad (54)$$

Thus, for branched polymers in the entanglement region, both η_0 and J_e° may be quite large in comparison with the values for linear polymers of the same molecular weight.

The terminal zone is inherently broader for well-entangled branched polymers than for linear polymers of comparable polydispersity (*27*). The dynamic moduli for nearly monodisperse linear and three-arm star polybutadienes (*46*) are compared in Figure 44. The transition from the terminal zone to the plateau zone is much more gradual for the star, and the prominent peak in $G''(\omega)$, which is typical for nearly monodisperse linear polymers, is absent.

All the observed characteristics of viscoelasticity in star polymers are natural consequences of the tube model. The presence of a long branch would be expected to suppress reptation (*28*), as suggested by the sketch in Figure 45. There is no longer any direction for the chain to move freely into new positions and conformations so that relaxation and diffusion must occur by some other, presumably slower process. A theory based on the fluctuation mechanism has been proposed (*47*). It leads directly to a broadened terminal region and to expressions for η_0 and J_e° that agree well with the observed behavior for stars (equations 53 and 54).

The effect of branching on viscosity–shear rate behavior is illustrated in Figure 46. As noted before, in reduced form [$\eta(\dot{\gamma})/\eta_0$ versus $\dot{\gamma}/\dot{\gamma}_0$], the viscosity–shear rate relationship for stars is indistinguishable from that for linear chains (*15*). (Perhaps in polydisperse branched systems it would depend on the molecular size distribution, that is, the distribution of radii of gyration.) Also as noted before, the connection between characteristic

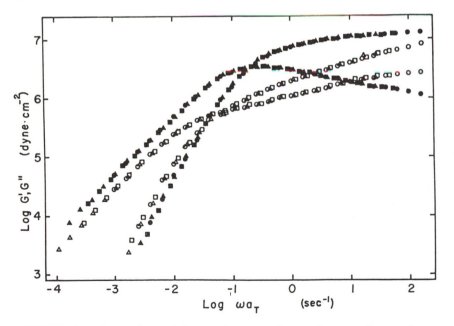

FIGURE 44. Dynamic modulus master curves for nearly monodisperse linear and star polybutadienes at 25 °C. The filled symbols represent data for the linear sample ($\overline{M}_w = 435\,000$), and the open symbols represent data for a three-arm star ($\overline{M}_w = 127\,000$). (Reproduced with permission from reference 46. Copyright 1984 M. J. Struglinski).

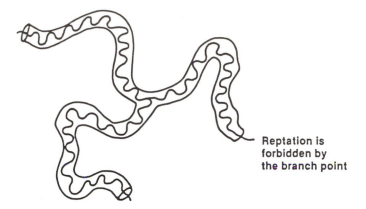

Reptation is
forbidden by
the branch point

FIGURE 45.   Schematic of a star polymer in its tube of entanglements with the surrounding chains.

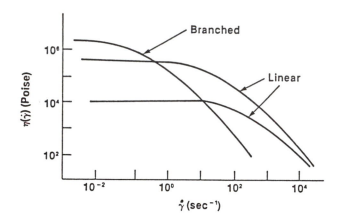

FIGURE 46.   Alteration of viscosity–shear rate behavior due to the presence of long branches.

shear rate, $\dot{\gamma}_0$, and the values of η_0 and J_e^o (equation 52) is not changed by branching (15). Thus, the larger values of J_e^o for long-arm stars make $\dot{\gamma}_0$ smaller than would be obtained for a linear polymer with the same viscosity. As a result, the viscosity at high shear rates is lower for stars, and the viscosity–shear rate curves for linear and branched polymers may cross, as shown in Figure 46 (*see* Figure 3 of reference 15, for example.) Branched systems in general can probably reach domains of viscosity–shear rate behavior that are unattainable by systems of linear chains. Unusual elastic effects would presumably be present as well, but the entire subject of nonlinear response in branched polymers is still relatively unexplored.

Summary

The viscoelastic character and flow behavior of polymer melts and concentrated solutions have been considered from both the macroscopic and microscopic points of view. The universal nature of the behavior, and particularly its dependence on the large-scale molecular architecture for flexible-chain, nonassociating polymers in the homogeneous liquid state, has been emphasized. Various experimental methods for characterizing viscoelastic response have been described, and the main features of current molecular theories about the dynamics of polymer liquids have been outlined. Several important topics were omitted or touched upon only briefly, but some sources in the literature are provided in the *References*.

Acknowledgments

I am grateful to Gary Ver Strate and Carol Gell for their many helpful comments on an earlier version of this chapter and to the Department of Energy, Office of Basic Energy Sciences, for financial support (DEFG02-91ER45452) during the period it was written.

References

In assembling the bibliography, I have used important books and review articles whenever possible, rather than original sources. The interested reader can use these references to trace the history and get a more complete picture of the individual parts of the subject. However, some important topics have been omitted or mentioned only briefly, and so I have included some references at the end that deal extensively with such topics. Thus, the books by Dealy (*48*) and Barnes et al. (*49*) have chapters on extensional flows. Larson (*50*) provides a critical evaluation of the many expressions that have been proposed for relating stress and deformation history. The work of Osaki et al. (*51*) on stress relaxation after finite step strains gives an excellent overview of this simple and important means of investigating nonlinear viscoelastic response. Useful surveys of rheo-optical techniques, principally flow birefringence, are given by Janeschitz-Kriegl (*52*), and Ylitalo et al. (*53*). References to the various methods for correlating the plateau modulus with chemical microstructure are given in the recent article by Zang and Carreau (*54*). Alternative viewpoints about the reality of reptation are well-summarized in the recent review by Lodge et al. (*55*). The use of computer simulation as a tool for investigating rheological questions is summarized in the recent article by Kremer and Grest (*56*). The flow behavior of liquid-crystalline polymers, a topic of growing importance and interest, has been reviewed recently (*57*). The diffusion of polymers in melts is the subject of a review (*58*) and a recent article (*59*) that surveys the literature. Articles on these topics are also available in the *Encyclopedia of Polymer Science and Engineering* (*60*).

1. Ferry, J. D. *Viscoelastic Properties of Polymers*, 3rd ed.; John Wiley and Sons: New York, 1980.
2. Bird, R. B.; Armstrong, R. C.; Hassager, O. *Dynamics of Polymeric Liquids*, 2nd ed.; John Wiley and Sons: New York, 1987; Vol. 1.
3. Astarita, G.; Marrucci, G. *Principles of Non-Newtonian Fluid Mechanics*; McGraw-Hill: Maidenhead, England, 1974.
4. Schowalter, W. R. *Mechanics of Non-Newtonian Fluids*; Pergamon: Oxford, 1978.

5. Tanner, R. I. *Engineering Rheology;* Clarendon: Oxford, 1985.

6. Lodge, A. S. *Elastic Liquids;* Academic: New York, 1964.

7. Treloar, L. R. G. *The Physics of Rubber Elasticity,* 3rd. ed.; Clarendon: Oxford, 1975.

8. See, for example, Graessley, W. W. *Macromolecules* **1982**, *15*, 1164.

9. Markovitz, H. *J. Polym. Sci.* **1975**, *Symp. 50*, 431.

10. Onogi, S.; Masuda, T.; Kitagawa, K. *Macromolecules* **1970**, *3*, 109.

11. Masuda, T.; Kitagawa, K.; Inoue, T.; Onogi, S. *Macromolecules* **1970**, *3*, 116.

12. Coleman, B. D.; Markovitz, H. *J. Appl. Phys.* **1964**, *35*, 1.

13. Ramachandran, S.; Gao, H. W.; Christiansen, E. B. *Macromolecules* **1985**, *18*, 695.

14. Walters, K. *Rheometry;* Chapman and Hall: London, England, 1975.

15. Graessley, W. W.; Masuda, T.; Roovers, J.; Hadjichristidis, N. *Macromolecules* **1976**, *9*, 127.

16. Graessley, W. W. In *Characterization of Materials in Research;* Burke, J. J.; Weiss, V., Eds.; Syracuse University Press: Syracuse, NY, 1975; Chapter 15.

17. Penwell, R. C.; Graessley, W. W.; Kovacs, A. *J. Polym. Sci. Polym. Phys. Ed.* **1974**, *12*, 1771.

18. *Polymer Handbook*, 3rd ed.; Brandrup, J.; Immergut, E. H., Eds.; Interscience: New York, 1989.

19. Bird, R. B.; Curtiss, C. F.; Armstrong, R. C.; Hassager, O. *Dynamics of Polymeric Liquids,* 2nd ed.; John Wiley and Sons: New York, 1987; Vol. 2.

20. Doi, M.; Edwards, S. F. *The Theory of Polymer Dynamics;* Clarendon Press: Oxford, 1986.

21. Odani, H.; Nemoto, N.; Kurata, M. *Macromolecules* **1972**, *5*, 531.

22. Berry, G. C.; Fox, T. G. *Adv. Polym. Sci.* **1968**, *5*, 261.

23. Bartels, C. R.; Crist, B.; Graessley, W. W. *Macromolecules* **1984**, *17*, 2702.

24. Green, P. F.; Kramer, E. J. *Macromolecules* **1986**, *19*, 1108.

25. Graessley, W. W. *Adv. Polym. Sci.* **1974**, *16*, 1.

26. Colby, R. H.; Fetters, L. J.; Funk, W. G.; Graessley, W. W. *Macromolecules* **1991**, *24*, 3873.

27. Raju, V. R.; Menezes, E. V.; Marin, G.; Graessley, W. W. *Macromolecules* **1981**, *14*, 1668.

28. De Gennes, P. G. *Scaling Concepts in Polymer Physics*; Cornell University Press: Ithaca, NY, 1979.

29. Graessley, W. W. *J. Polym. Sci. Polym. Phys. Ed.* **1980**, *18*, 17.

30. Graessley, W. W. *Adv. Polym. Sci.* **1982**, *47*, 68.

31. Pearson, D. S. *Rubber Chem. Technol.* **1987**, *60*, 439.

32. Struglinski, M. J.; Graessley, W. W. *Macromolecules* **1985**, *18*, 2630.

33. Tsenoglou, C. *Prepr. Am. Chem. Soc. Div. Poly. Chem.* **1987**, *28*(2), 185.

34. Des Cloiseaux, J. *J. Europhys. Lett.* **1988**, *5*, 437; **1988**, *6*, 475; *Macromolecules* **1990**, *23*, 4678.

35. Tsenoglou, C. *Macromolecules* **1991**, *24*, 1762.

36. Graessley, W. W. *J. Chem. Phys.* **1967**, *47*, 1942.

37. Graessley, W. W. *Acc. Chem. Res.* **1977**, *10*, 332.

38. Graessley, W. W.; Roovers, J. *Macromolecules* **1979**, *12*, 959.

39. Carella, J. M.; Gotro, J. T.; Graessley, W. W. *Macromolecules* **1986**, *19*, 659.

40. Roovers, J.; Graessley, W. W. *Macromolecules* **1981**, *14*, 766.

41. Mendelson, R. A.; Bowles, W. A.; Finger, F. L. *J. Polym. Sci. Part A2* **1970**, *8*, 105.

42. Ham, J. S. *J. Chem. Phys.* **1957**, *26*, 625.

43. Pearson, D. S.; Raju, V. R. *Macromolecules* **1982**, *15*, 294.

44. Kraus, G.; Gruver, J. T. *J. Polym. Sci. Part A* **1965**, *3*, 105.

45. Bartels, C. R.; Crist, B.; Fetters, L. J.; Graessley, W. W. *Macromolecules* **1986**, *19*, 785.
46. Struglinski, M. J. Doctoral thesis, Chemical Engineering Department, Northwestern University, 1984.
47. Helfand, E.; Pearson, D. S. *J. Chem. Phys.* **1983**, *79*, 2054. Pearson, D. S.; Helfand, E. *Macromolecules* **1984**, *17*, 888.
48. Dealy, J. M. *Rheometers for Molten Plastics*; Van Nostrand Rheinhold: New York, 1982.
49. Barnes, H. A.; Hutton, J. F.; Walters, K.; *An Introduction to Rheology*; Elsevier: New York, 1989.
50. Larson, R. G. *Constitutive Equations for Polymer Melts and Solutions*; Butterworth: Stoneham, MA, 1988.
51. Osaki, K.; Takatori, E.; Tsunashima, Y.; Kurata, M. *Macromolecules* **1987**, *20*, 525, and earlier references therein.
52. Janeschitz-Kriegl, H. *Polymer Melt Rheology and Flow Birefringence*; Springer-Verlag: Heidelberg, Germany, 1983.
53. Ylitalo, C. M.; Kornfeld, J. A.; Fuller, G. G.; Pearson, D. S. *Macromolecules* **1991**, *24*, 749.
54. Zang, Y.-H.; Carreau, P. J. *J. Appl. Polym. Sci.* **1991**, *42*, 1965.
55. Lodge, T. P.; Rotstein, N. A.; Prager, S. *Adv. Chem. Phys.* **1990**, *79*, 1.
56. Kremer, K.; Grest, G. *J. Chem. Phys.* **1990**, *92*, 5057.
57. Berry, G. C. *J. Rheology* **1991**, *35*, 943.
58. Tirrell, M. *Rubber Chem. Technol.* **1984**, *57*, 523.
59. Shull, K. R.; Dai, K. H.; Kramer, E. J.; Fetters, L. J.; Antonietti, M.; Sillescu, H. *Macromolecules* **1991**, *24*, 505.
60. *Encyclopedia of Polymer Science and Engineering*, 2nd ed.; Mark, H. F.; Bikales, N. M.; Overberger, C. S.; Menges, G., Eds.; Interscience: New York, 1986.

Chapter 4

The Crystalline State

Leo Mandelkern

Department of Chemistry and Institute of Molecular Biophysics,
Florida State University, Tallahassee, FL 32306

This chapter will be concerned with the basic principles governing the crystallization behavior of flexible long-chain molecules. The more rigid type polymers will be discussed in Chapter 5. The subject matter is divided into the following topics: thermodynamics of crystallization, crystallization kinetics and mechanisms, structure and morphology, and microscopic and macroscopic properties. Each of these topics will be discussed in terms of fundamental physical and chemical concepts. The interdependence of these aspects of polymer crystallization will become readily apparent. This chapter is not meant to be a review of current research activity in this field. Serious efforts have been made, however, to keep the discussion timely. The primary concern is to develop the basic principles involved. To accomplish this objective, the equivalent of an introductory first course in polymer chemistry or physics is assumed. Knowledge of the basics of molecular constitution and chain structure is essential in understanding the discussion that follows. The level of the chapter is intended to be between that of an introductory polymer science course and a critical review of current research.

The study of crystalline polymers closely parallels the development of polymer science itself. Certain areas of polymer crystallization behavior are well-developed, well-understood, and well-accepted. Other areas are under intensive study. Difficulties in interpretation in areas that are not yet fully understood have been resolved gradually, and a set of unifying concepts are emerging. The guiding principles needed to understand the thermodynamics of fusion and the kinetics of crystallization are firmly established by both theory and experiment. Modern emphasis has, therefore, been directed to the understanding of the structure and morphology of crystalline polymers and their influence on properties. Because crystallization thermodynamics and kinetics are extensively documented in the literature, these areas will be reviewed only briefly here to establish their

2505–2/93/0145 $14.40/1

salient features. The major emphasis will be on understanding structure–property relations. The basic principles that evolve will then be demonstrated with selected examples. Once the principles are understood, they can be applied in the resolution of various problems.

Long-chain molecules can exist in either one of two states. These states are characterized by the conformation of individual molecular chains and their organization relative to one another. The liquid state is the state of molecular disorder. In this state, the individual chains adopt a statistical conformation, commonly called the random coil. The centers of mass of the molecules are also randomly arranged relative to one another in the liquid state. All the thermodynamic and structural properties observed in this state are those that are commonly associated with a liquid, although usually a very viscous one. The liquid exhibits characteristic long-range elasticity. The liquid state in polymers is also commonly called the amorphous state.

The crystalline or ordered state is characterized by three-dimensional order over at least a portion of the chains. The ordered conformation may be fully extended or may represent one of many known helical structures. Irrespective of the details of the ordered chain structure, the molecules are organized into a regular, three-dimensional array. The chain axes are aligned parallel to one another, and the substituent groups are brought into regular register. Such ordered systems diffract X-rays in the conventional manner and display all the properties characteristic of the crystalline state. All chain molecules that have a reasonable structural regularity will crystallize, under suitable conditions.

In contrast to the liquid state, the crystalline state is relatively inelastic and very rigid. For example, the difference between the moduli of elasticity of the two states is about 5 orders of magnitude. Major differences also exist in other properties, including spectral and thermodynamic properties. Moreover, within the crystalline state, properties may be changed by control of structure. The control of properties is of major concern in the application and end use of a polymeric system.

The difference in the molecular conformations in the two states is schematically illustrated in Figure 1. In the crystalline state, the bonds

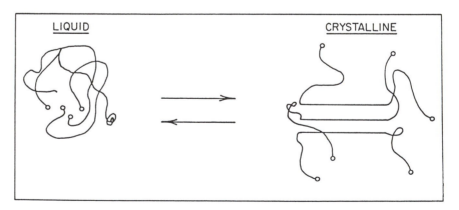

FIGURE 1.   Schematic diagram illustrating differences in the conformations of chain molecules in the liquid and the crystalline state. A straight line represents an ordered conformation. Details of interfacial structure are not being considered at this point.

adopt a set of successive preferred orientations; participation of the complete molecules in the ordering process is not required. In the liquid state, the bond orientations are such that the chain adopts a statistical conformation.

The primary goal of this chapter is to explain how the properties of the crystalline state are influenced by the chemical nature of the repeating unit, the crystallite structure above the level of the unit cell, and the organization of crystallites. Many properties are of interest, including thermodynamic and physical properties, spectroscopic characteristics, complex mechanical behavior, and ultimate strength.

Thermodynamics of Crystallization

Melting Temperatures of Homopolymers

From the point of view of formal thermodynamics, the transformation from the liquid state to the crystalline state can be properly treated as a first-order phase transition in the classical sense. The transformation is very similar to the fusion of low-molecular-weight substances. A typical example of the melting of homopolymers is given in Figure 2 for unfractionated and fractionated samples of linear polyethylene (*1*). When carried

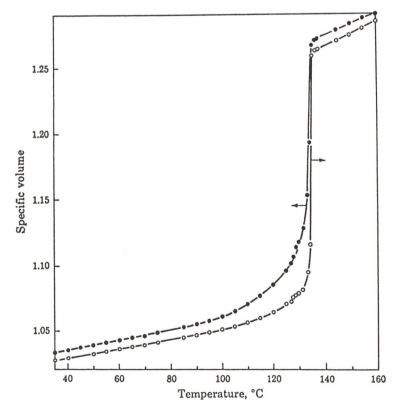

FIGURE 2. Specific volume versus temperature for the melting of linear polyethylenes. Key: ●, unfractionated polymer, and ○, fraction with a molecular weight of 32 000. (Reproduced from reference 1. Copyright 1961 American Chemical Society.)

out carefully, the melting process is relatively sharp, and a well-defined melting temperature is clearly discerned. As easily seen in Figure 2, fusion takes place over a very narrow temperature interval for the molecular weight fraction. In the specific volume–temperature plot, the disappearance of the last traces of crystallinity, which defines the melting temperature, is clear for both examples. The transition, although a diffuse one, can still be classified as a first-order phase transition.

Normal alkanes, as well as oligomers of other type of repeating units, can form molecular crystals at sufficiently low temperatures, because all the molecules are of precisely the same length. For such molecules, the chain ends are paired, one with the other, so that well-defined planes delineating the end groups are developed (Figure 3A). For polymers, however, no matter how well the system is fractionated, the individual molecules will not be exactly of the same length. Consequently, for polymers, the necessary condition for the formation of molecular crystals cannot be fulfilled. The equilibrium state for polymer crystallization has been established by statistical mechanical analysis and by experiment. In the equilibrium state, the end portions of the molecules are disordered or unpeeled, as shown schematically in Figure 3B. A chain of x repeating units is characterized by an equilibrium crystallite length, ζ_e, with $x - \zeta_e$ end-repeating units being disordered. For such system, the dependence of melting temperature on chain length can be expressed as follows (2):

$$1/T_{me} - 1/T_m^0 = (R/\Delta H_u)\{(1/x) + [1/(x - \zeta_e + 1)]\} \qquad (1)$$

$$2\sigma_e = RT_{me}\{[\zeta_e/(x - \zeta_e + 1)] + \ln[(x - \zeta_e + 1)/x]\} \qquad (2)$$

In these equations, T_m^0 is the equilibrium melting temperature for a chain of infinite molecular weight, T_{me} is the corresponding temperature for a fraction containing x repeating units, and R is the gas constant. The effective interfacial free energy associated with the basal plane of an equilibrium crystallite of length ζ_e is σ_e, and ΔH_u is the enthalpy of fusion per repeating unit.

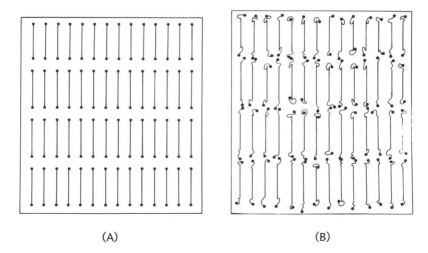

(A)   (B)

FIGURE 3. Schematic representation of crystallites of extended chain crystals. (A) *n*-Alkanes with paired end groups. (B) Polymer fractions with end sequences in disordered conformation.

For a polydisperse system with a most probable chain length distribution, the melting temperature–molecular weight relation can be expressed as follows:

$$1/T_{\mathrm{m}} - 1/T_{\mathrm{m}}^0 = R/\Delta H_{\mathrm{u}} \cdot 2/\bar{x}_{\mathrm{n}} \qquad (3)$$

In equation 3, $\bar{x}_{\mathrm{n}}$ is the number-average degree of polymerization, and T_{m} is the equilibrium melting temperature of the system. For this molecular weight distribution, the quantity $2/\bar{x}_{\mathrm{n}}$ represents the mole fraction of noncrystallizing units. Equation 3 results from the stipulation of the conditions for phase equilibrium. The melting temperature–molecular weight relation for each polydisperse system has to be treated individually.

The application of formal phase equilibrium thermodynamics leads to an expression for the depression of the melting temperature by a low-molecular-weight diluent, when it is excluded from the crystalline phase (*2*):

$$1/T_{\mathrm{m}}^* - 1/T_{\mathrm{m}}^0 = R/\Delta H_{\mathrm{u}} \cdot (V_{\mathrm{u}}/V_1)\left(v_1 - \chi_1 v_1^2\right) \qquad (4)$$

Here T_{m}^0 is the equilibrium melting temperature of the pure system; T_{m}^* is the melting temperature corresponding to a volume fraction of diluent, v_1; V_{u}/V_1 is the ratio of the molar volume of the chain repeating unit (V_{u}) to that of the diluent (V_1); χ_1 is the polymer–diluent thermodynamic interaction parameter; and ΔH_{u} is the enthalpy of fusion per chain repeating unit of the completely crystalline polymer. ΔH_{u} is characteristic of the chain repeating unit and does not depend on the specific nature of the crystalline state. Equation 4 is simply the adaptation to polymers of the classical freezing-point-depression expression when the crystalline phase remains pure. A few exceptions have been noted when the diluent enters the crystal lattice. For this case, equation 4 is obviously no longer valid. Equation 4 has been verified experimentally for many different polymers (*3, 4*). The same value of ΔH_{u} is obtained for a given polymer when studied with a series of different diluents. Thus, by use of equation 4, ΔH_{u} can be obtained for a given polymer. From ΔH_{u} and the equilibrium melting temperature, ΔS_{u}, the entropy of fusion per repeating unit, can be obtained. These thermodynamic parameters for a selected set of polymers are given in Table I. This table is not meant to be exhaustive. However, the examples have been selected to illustrate typical key situations.

The data in Table I illustrate the guiding structural principles that determine the melting temperature. These examples make clear that, for polymers, no correlation between the melting temperature and the enthalpy of fusion exists, as is found in many monomeric systems. The ΔH_{u} values of polymers generally fall into two classes: values of the order of just a few thousand calories per mole and values of about 10 000 cal/mol. For the examples given here, and as is also found more generally, many high-melting-point polymers have low heats of fusion, and conversely, many low-melting-point polymers have high heats of fusion. Consequently, the entropy of fusion (ΔS_{u}) is a key factor affecting the melting temperature. A striking correlation exists between the entropy of fusion and the chain conformation in the completely molten state. Polymers commonly designated as elastomers, such as poly(dimethylsiloxane) and poly(*cis*-1,4-isoprene), have relatively low melting temperatures and high entropies of fusion, which reflect the compact, highly flexible nature of the chain. At the other extreme, the so-called engineering plastics, such as poly(ether

Table I. Thermodynamic Quantities Characterizing the Fusion of Selected Polymers

Polymer	T_m^0 (°C)[a]	ΔH_u (cal/mol)[b]	ΔS_u (cal/deg · mol)
Polyethylene	145.5 ± 1	990	2.36
Polypropylene	208	2100	4.37
Poly(cis-1,4-isoprene)	35.5	1050	3.46
Poly(trans-1,4-isoprene)	87	3040	8.75
Poly(trans-1,4-chloroprene)	107	2000	5.08
Polystyrene (isotactic)	243	2075	4.02
Poly(oxymethylene)	200	1676	3.55
Poly(oxyethylene)	80	2080	5.91
Poly(2,6-dimethoxy-1,4-phenylene oxide)	287	761	1.36
Poly(decamethylene adipate)	79.5	10 200	29
Poly(decamethylene sebacate)	80	12 000	34
Poly(ethylene terephthalate)	282	5600	10.2
Poly(decamethylene terephthalate)	138	11 000	27
Poly(tetramethylene terephthalate)	230	7600	15.1
Poly(hexamethylene adipamide)	269	10 365[c]	45.8
Poly(decamethylene sebacamide)	216	8300	17
Poly(decamethylene azelamide)	214	8800	27
Poly(tetrafluoroethylene)	346	1220[c]	24.4
Poly(dimethylsiloxane)	−38	650	2.76
Poly(tetramethyl-p-silphenylene siloxane)	160	2700	6.20
Poly(ether ether ketone)	338	11 319[c]	18.5
Cellulose trinitrate	>700	900–1500	1.50
Cellulose tributyrate	207	8800	8.1

[a] Best estimate of equilibrium melting temperature.
[b] Determined from the depression of the melting temperature by monomeric diluents, unless otherwise indicated.
[c] Determined by means of the Clapeyron equation.

ether ketone), poly(tetrafluoroethylene), and poly(2,6-dimethoxy-1,4-phenylene oxide), have high melting temperatures and low entropies of fusion, which reflect the more extended chain structures. Cellulose derivatives are characterized by very high melting points and low heats of fusion. The low entropy of fusion must result from the highly extended nature of the chain.

The introduction of ring structures into a linear chain substantially raises the melting temperature relative to that of the aliphatic chain. This effect would be expected because of the decreased conformational entropy of the melt that results. Striking examples of this phenomenon are found in the melting temperatures of aliphatic and aromatic polyesters and polyamides.

Another example of the influence of the entropy of fusion is found in the melting temperatures of aliphatic polyesters and polyamides. For the same type of repeating units, the melting temperatures of the polyamides are substantially higher than those of the corresponding polyesters. Despite the hydrogen bonding capacity of polyamides, there is no significant difference in the enthalpies of fusion of the polyesters and the polyamides. Hence, the 150–200 °C difference in melting temperatures must result from differences in the entropy of fusion.

These few examples indicate that, as a general rule, the chain structure influences the melting temperature through its conformational properties and, thus, the entropy of fusion. In fact, by using rotational isomeric state theory, a quantitative correlation can be made between the entropy of fusion at constant volume and the chain structure of many polymers.

Melting Temperatures of Copolymers

By applying classical phase equilibrium theory, the melting temperatures of copolymers, relative to that of the parent homopolymers, can be derived. From the point of crystallization behavior, different types of chemical repeating units, structural irregularities such as stereoirregularity, branch point head-to-head structures, and geometric irregularities all behave as copolymeric units when incorporated into the chain. The melting point–composition relations for co-polymers are similar to those of binary mixtures of monomers. It has to be decided a priori whether the crystalline state remains pure, that is, whether the counit enters the lattice. If the counit enters the lattice, then it has to be specified further whether this situation represents the equilibrium state or nonequilibrium defects are involved. Detailed calculations have been made for the specific, but very common, situation in which the counits or structural irregularities do not participate in the crystallization; that is, the crystallization phase remains pure. For this case, equation 5 is obtained (*2*, *5*):

$$1/T_m - 1/T_m^0 = -R/\Delta H_u \cdot \ln p \qquad (5)$$

In this equation, p is the sequence propagation probability, that is, the probability that in the copolymer a crystallizable unit is succeeded by another such unit. T_m^0 and ΔH_u are as already defined for equation 1; T_m is the melting temperature of the copolymer. Equation 5 shows the very interesting fact that the melting temperature of a copolymer does not depend directly on its composition but rather depends on the nature of the sequence distribution. This unique result is a consequence of the chainlike character of polymers. For copolymers, therefore, emphasis must be on the nature of the sequence distribution rather than the nominal composition. This requirement also applies when the counits enter the lattice.

Three distinct types of sequence distributions can be differentiated in terms of X_A, the mole fraction of crystallizable units. For an ordered or block copolymer, $p > X_A$, and in many cases, p approaches unity. For most ordered or block copolymers, the melting temperature of the copolymer will, at most, be only slightly lower than that of the corresponding homopolymer. On the other hand, for an alternating copolymer, $p < X_A$, and the melting temperature of the copolymer will be drastically lower than that of the corresponding homopolymer. For a random copolymer, $p = X_A$, and equation 5 is expressed as follows:

$$1/T_m - 1/T_m^0 = -R/\Delta H_u \cdot \ln X_A \qquad (6)$$

From a theoretical point of view, copolymers that have exactly the same composition are expected to have drastically different melting temperatures, depending on the sequence distribution of the counits. This hypothesis has been verified experimentally. Figure 4 gives some typical

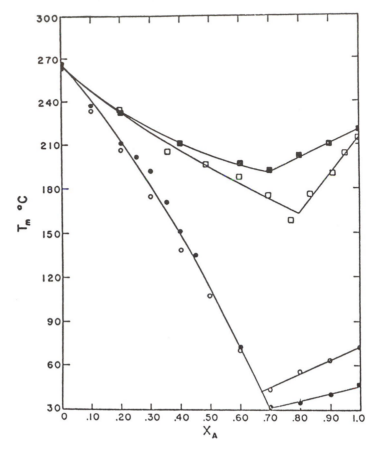

FIGURE 4. Melting temperature versus composition for typical random copolymers and copolyamides. Key: ●, poly(ethylene terephthalate-*co*-adipate); ○, poly(ethylene terephthalate-*co*-sebacate); ■, poly(hexamethylene adipamide-*co*-sebacamide); and □, poly(hexamethylene adipamide-*co*-caproamide).

examples of the melting temperature–composition relations for a set of random copolyesters and copolyamides. As predicted by theory, the melting temperature decreases monotonically with increasing concentration of the noncrystallizing counit. The equivalent of a eutectic temperature is reached at an appropriate composition commensurate with the melting temperature of each component.

In Figure 5 (6), melting temperature–composition relations are given for block poly(ethylene terephthalate) copolymers with different counits. The data for several random copolymers are also included for reference. The difference in the melting temperature–composition relations of the two types of copolymers are quite marked and in agreement with theoretical predictions. As expected for the block copolymers, the melting points remains constant for a wide range of counit contents and are independent of the chemical nature of the counit. Only when the counit content become extremely large does the melting point decrease, an effect consistent with the crystallization of the added species. The results shown in Figure 5 are typical of all types of block copolymers irrespective of their chemical constitution. With the data in Figure 5 as examples, melting

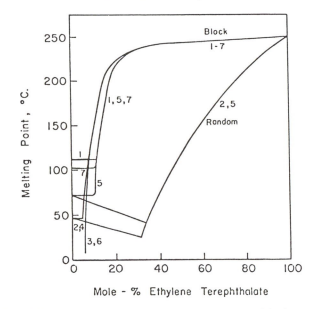

FIGURE 5. Melting temperature versus composition for block copolymers of poly (ethylene terephthalate) with ethylene succinate (1), ethylene adipate (2), diethylene adipate (3), ethylene azelate (4), ethylene sebacate (5), ethylene phthalate (6), and ethylene isophthalate (7). For comparative purposes, data for random copolymers with ethylene adipate and with ethylene sebacate also are given. (Reproduced with permission from reference 6. Copyright 1968 Society of Plastics Engineers.)

point differences of as much as 200 °C can be observed for the same nominal composition depending on whether the two types of units are arranged randomly or in blocks. The melting temperatures of alternating copolymers also follow theoretical expectations.

The large variation in melting temperatures that can be attained for a given copolymer composition gives great versatility in the control of properties. For example, with block copolymers, a crystalline polymer can be modified without a significant reduction of its melting point, modulus, tensile strength, and elongation. Thus, desirable mechanical properties may be maintained while other properties, such as dyeability, water sorption, or elasticity, are enhanced by the proper selection of counits. On the other hand, if a lower melting temperature is required, perhaps for processing purposes, then the random introduction of counits would be appropriate.

A pure crystalline phase is an a priori assumption for the equilibrium analysis of any multicomponent–multiphase system, as is very common with copolymers. The examples described earlier fulfill this condition. However, a pure crystalline phase is not a universal requirement. For some copolymers, the counits or structural irregularities enter the lattice. Melting point–composition relations based solely on the liquidus composition are not adequate to establish this condition. The solidus has to be examined by independent methods. An example of a phase diagram for a copolymer with a crystalline phase that is not pure is illustrated in Figure 6 (7) for random poly(ethylene terephthalate-*co*-ethylene isophthalate) copolymers. Figure 6 resembles a classical phase diagram of a low-

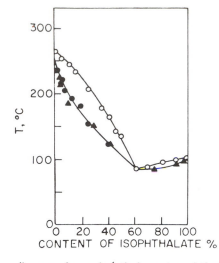

FIGURE 6. Phase diagram for poly (ethylene terephthalate-*co*-ethylene iso-phthalate) copolymers. Key: ○, liquidus, and ▲, ●, solidus. (Reproduced with permission from reference 7. Copyright 1960 Marcel Dekker.)

molecular-weight system. It even shows an azeotropic point. For low-molecular-weight systems, the liquid and the solid must have the same composition at the azeotrope. For copolymers, the comparable require-ment would be that the sequence propagation probability must be the same for both phases. In analyzing a system such as is illustrated in Figure 6, the parameter p must be known for both the crystalline and the liquid states. Analysis based solely on composition would not be correct. Obtain-ing the sequence distribution in the crystalline state is often difficult, and this difficulty presents a serious impediment to analysis.

Although homopolymers and block copolymers have relatively sharp melting points, the fusion of random copolymers usually occurs over a very wide temperature range. Diffuse melting is a consequence of a greatly exaggerated impurity effect caused by sequence length requirements. As fusion progresses, the shorter crystalline sequences melt at lower tempera-tures and shift the equilibrium, and thus, the melting range is broadened. This phenomenon is illustrated in Figures 7 (*8*) and 8 (*9*) for a set of polybutadienes and polypropylenes, respectively. Although the repeating units are chemically identical for each of these polymers, they are properly treated as copolymers because of the structural irregularities of the chain. The polybutadienes have various amounts of the 1,4-*trans*-crystallizing unit. As the content of the crystallizing component decreases, the melting temperature is lowered, and the fusion process occurs over a broader temperature range (Figure 7). The fusion process eventually becomes very difficult to detect, as evidenced by curve C in Figure 7. The existence of even small amounts of crystallinity, nevertheless, must be established because of the influence of crystallinity on mechanical and physical properties. This type of melting behavior typifies the fusion of random co-polymers.

The melting of the stereoirregular polypropylenes (Figure 8) follows a very similar pattern. The melting temperature and the level of crystallinity

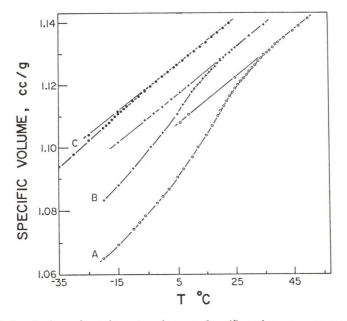

FIGURE 7. Fusion of random copolymers. Specific volume versus temperature for polybutadienes with various mole fractions, X_A, of the 1,4-*trans*-crystallizing unit. Key: A, $X_A = 0.81$; B, $X_A = 0.73$; and C, $X_A = 0.64$. Curves B and C are arbitrarily displaced along the ordinate. (Reproduced with permission from reference 8. Copyright 1956 John Wiley and Sons, Inc.)

decrease as the content of the crystallizing isotactic decreases. Concomitantly, the fusion process occurs over a very much broader temperature range. The crystallinity and the fusion of random copolymers with high contents of structural irregularities are often difficult to recognize.

Another important type of chain irregularity is branching, because the branch points are structurally different from the other chain repeating units. An example of the effect of branching on the crystallization behavior is shown in Figure 9 (*10*) for two polyethylene polymers. Curve A is for the linear polymer; curve B is for the branched polymer. The melting temperature of the branched polymer is significantly less than that of its linear counterpart. In addition, as expected, the melting range of the branched polymer is very much broader. For the linear polymer, most of the melting takes place over a 3–4 °C interval only. The polyethylenes are a striking example of two almost essentially chemically identical polymers that have markedly different crystallization behaviors.

In a general sense, copolymers display some unique aspects of crystallization behavior, which can be explained by the application of phase equilibrium theory. The theory leads to the very important conclusion that the sequence distribution of the units, rather than composition, is the key factor in determining behavior of copolymers.

In summary, formal phase equilibrium thermodynamics can be applied successfully to the fusion of homopolymers, copolymers, and polymer–diluent mixtures. This conclusion has many far-reaching consequences. The same principles of phase equilibrium also can be applied to the analysis of the influence of hydrostatic pressure of deformation to the fusion process (*3*).

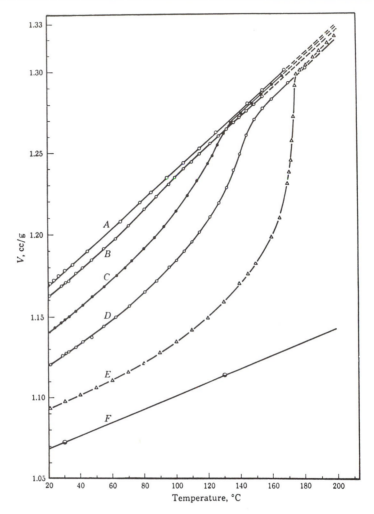

FIGURE 8. Fusion of random copolymers. Specific volume versus temperature for polypropylenes of different stereoregularities. Key: A, ether extract, quenched; B, pentane extract, annealed; C, hexane fraction, annealed; D, trimethylpentane fraction, annealed; E, experimental whole polymer, annealed; and F, calculated curve for pure crystalline polymer. (Reproduced with permission from reference 9. Copyright 1960 John Wiley and Sons, Inc.)

Crystallization Kinetics

An understanding of the mechanisms involved in the crystallization process is very important. These mechanisms are major factors in determining both microscopic and macroscopic properties of actual crystalline systems. Crystallization mechanisms are conventionally deduced from studies of the kinetics of the transformation from the liquid state to the crystalline state. The crystallization kinetics of a polymer from the pure melt can be investigated mainly in two ways. One procedure is to study the overall crystallization rate by such methods as dilatometry, calorimetry, or spectroscopy. The other method is to measure spherulitic growth rate by

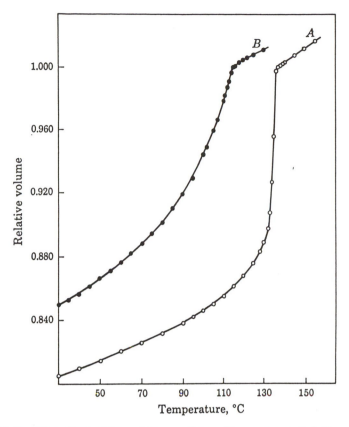

FIGURE 9. The effect of branching on the melting process. Relative volume versus temperature for linear polyethylene (A) and branched polyethylene (B). (Reproduced from reference 10. Copyright 1953 American Chemical Society.)

direct examination with a light microscope. The two methods complement one another and yield essentially the same basic information.

Crystallization from the Melt

A formal understanding of the processes involved in crystallization from the pure melt has been substantially developed. With only minor modification, the general mathematical theory developed many years ago by Avrami (*11*) for the crystallization of metals and other low-molecular-weight substances can be applied to polymer crystallization. Accordingly, the extent of the transformation, $1 - \lambda(t)$, can be expressed as follows:

$$1 - \lambda(t) = 1 - \exp(-kt^n) \qquad (7)$$

Here, t is the time, k is the rate constant, and n is an integer that depends on the specifics of nucleation and growth processes. An essential feature of the Avrami formulation is that, when two growing centers impinge upon one another, their growth ceases. At a small extent of transformation, equation 7 is reduced to the "free growth" approximation:

$$1 - \lambda(t) = kt^n \qquad (8)$$

A typical set of crystallization isotherms for a pure polymer (linear polyethylene with a molecular weight of 284 000) crystallizing from the melt is given in Figure 10 (*12*). In this figure, the extent of the transformation, or degree of crystallinity, is plotted against the logarithm of time for different crystallization temperatures. Some very important features of the crystallization process are illustrated by the curves. The isotherms have the characteristic sigmoidal shape typical of all homopolymers. An initial induction time, which is more apparent than real, is evident and is followed by a period of accelerated crystallization. A retardation of the crystallization process, sometimes called secondary crystallization, then occurs, and a pseudoequilibrium level of crystallinity is reached. After sufficient time, the same limiting value of the extent of transformation is attained at each crystallization temperature for this homopolymer. The time rate of change of the degree of crystallinity (or crystallinity level) is extremely small in this region. Complete crystallinity is rarely, if ever, attained for polymers. The degree of crystallinity that is attained depends on the molecular weight (cf. seq.) and the structural regularity of the chain. These kinetic studies indicate that polymers are best considered as polycrystalline systems.

A very strong and dramatic negative temperature coefficient is also apparent from these plots. As the temperature decreases, the crystallization rate becomes much more rapid. This behavior is quite the opposite of what is found for the usual chemical reaction. The negative temperature coefficient is rather severe. In the example given, the crystallization rate changes by 5 orders of magnitude over a temperature interval of only 7 °C. This behavior clearly indicates a nucleation-controlled crystallization process (*3*). It illustrates an extremely important principle that underlies and controls many aspects of polymer crystallization.

Steady-State Nucleation Theory

The steady-state nucleation rate, $\dot{N}$, can be expressed in its most general form as follows:

$$\dot{N} = N_0 \exp\left\{\left(\frac{E_D(T)}{RT}\right) - \left(\frac{\Delta G^*}{RT}\right)\right\} \tag{9}$$

This simple statement applies to all classes of substances, including polymers. In equation 9, N_0 is a constant that is only slightly temperature dependent, E_D is the energy of activation for the transport of chain units across the crystal–liquid interface, and ΔG^* is the free energy change required to form a nucleus of critical size. Nuclei that are smaller than the critical size are unstable, and those larger than the critical size can develop and grow into mature crystallites. The strong, negative temperature coefficient in the vicinity of the melting temperature arises from the variation of ΔG^* with temperature:

$$\Delta G^* \sim 1/\Delta G_u \tag{10a}$$

$$\Delta G^* \sim (1/\Delta G_u)^2 \tag{10b}$$

Either equation 10a or 10b is valid, depending on whether a two- or a three-dimensional nucleation process is involved. ΔG_u is the free energy of

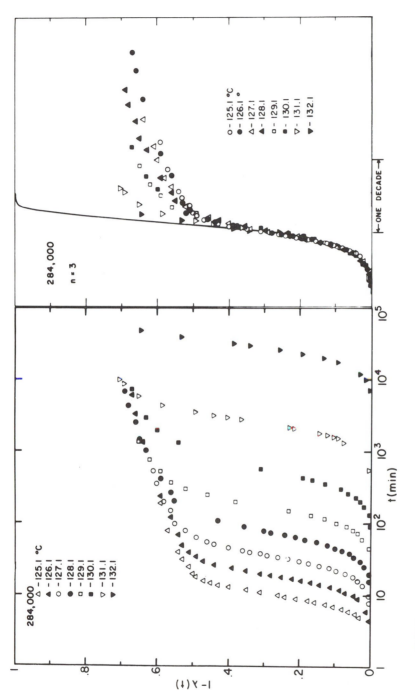

FIGURE 10. Crystallization kinetics from the pure melt. (Left) Degree of crystallinity versus the logarithm of time for the crystallization of linear polyethylene with molecular weight fraction of 2.84×10^5 at the indicated temperatures. (Right) Master isotherm after superimposition of isotherms from the left. (Reproduced from reference 12. Copyright 1972 American Chemical Society.)

fusion per chain repeating unit. In the standard manner, ΔG_u can be expanded in the vicinity of the melting temperature and expressed as follows:

$$\Delta G_u = \Delta S_u (T_m^0 - T) \tag{11}$$

The strong dependence of $\dot{N}$ on temperature in the vicinity of the melting temperature becomes clear from equations 9 and 11, as does the increase in nucleation rate with decreasing temperature.

Polymer crystallization adheres to the general steady-state nucleation theory embodied in equation 9. This conclusion is reached on the most general grounds and is based on the very marked negative temperature coefficient of the crystallization process that is universally observed in the vicinity of the melting temperature. The same temperature coefficient is observed, irrespective of the disposition of the chain within the nucleus. Moreover, different kinds of nucleation processes are characterized by this unique temperature coefficient. Consequently, it is very difficult, if not impossible, to distinguish between the different types of nucleation processes on the basis of kinetic data. For this reason, one cannot assume or deduce, except in an ad hoc manner, that a particular nucleus structure and nucleation type is operative on the basis of crystallization kinetics data only. Despite this constraint, control of the crystallization process by nucleation is an important conclusion that has many ramifications in the structure and properties of crystalline polymers.

The shapes of the isotherms at each temperature in Figure 10 are very similar. They are, in fact, identical and can be superimposed merely by shifting them along the horizontal axis. The superimposition yields one master isotherm, as shown in the right panel of Figure 10. This superimposability of isotherms shows that a single reduced time variable, which depends on temperature, describes the crystallization process. The solid line in Figure 10 represents the Avrami equation, equation 7, with $n = 3$. In this example, the experimental data adhere to the theory until about 50% of the transformation has occurred. Beyond this point, significant deviation from theory occur, and the crystallization rate is significantly retarded as a pseudo-equilibrium level of crystallinity is approached. The deviation from theory and the final level of crystallinity that can be attained are very dependent on molecular weight. They are controlled, in the main, by topological restraints to the crystallization process.

When the crystallization process can be extended over a larger temperature range, well removed from the melting temperature, a well-defined maximum is observed in the rate. This phenomenon is illustrated in Figure 11 (*13*) by the classical crystallization kinetics study of Wood and Bekkedahl (*13*) with natural rubber, poly(*cis*-1,4-isoprene). As the crystallization temperature is lowered, the rate of growth of crystallites becomes more dominant relative to the nucleation rate. Segmental motion and transport, which are essential to growth, are reduced as the glass transition temperature of the polymer is approached. Consequently, the two mechanisms compete in the crystallization process. The nucleation rate increases rapidly as the temperature is lowered, while the rate of transport of chain segment to a growing crystallite is reduced. Because of this competition, a maximum results in the crystallization rate. Such maxima are observed in all homopolymers, as long as the crystallization rate does not become so rapid that it cannot be recorded.

Effects of Molecular Weight

The molecular weight influences not only the level of crystallinity that can be achieved but also the time scale, or the rate of crystallization. Figure 12 gives a summary of crystallization times for linear polyethylene fractions covering a wide range of molecular weights and isothermal crystallization temperatures. In Figure 12, the time for 1% of the absolute amount of crystallinity to develop, $\tau_{0.01}$, on a logarithmic scale, is plotted against the

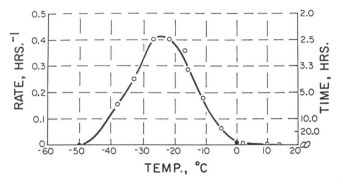

FIGURE 11. Plot of rate of crystallization of natural rubber, poly(1,4-*cis*-isoprene), over an extended temperature range. The rate plotted is the reciprocal of the time required for one-half the total volume change. (Reproduced with permission from reference 13. Copyright 1946 American Institute of Physics.)

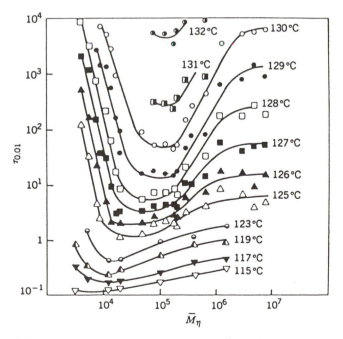

FIGURE 12. Double logarithmic plot of $\tau_{0.01}$ (time for 1% of the transformation) versus molecular weight for the indicated crystallization temperature. (Reproduced from reference 12. Copyright 1972 American Chemical Society.)

molecular weight. Several important features are illustrated in this figure. In the low-molecular-weight range, the crystallization times decrease by several decades as the molecular weight increases. However, a minimum in the time scale (or a maximum in the crystallization rate) is reached. The molecular weights at the extremes depend on the crystallization temperature. For the highest crystallization temperatures, the maximum in the rate occurs in the molecular weight range of 1×10^5 to 2×10^5. The locus defining the maximum rate decreases with decreasing temperature and is in the molecular weight range of 1×10^4 to 2×10^4 for the lowest isothermal crystallization temperatures. Concomitantly, $\tau_{0.01}$ at the maximum rate decreases from about 10^4 min at 132 °C to 1 min at 123 °C. At the left of the maxima, the relation between $\tau_{0.01}$ and molecular weight is qualitatively independent of the crystallization temperature. However, at molecular weights greater than that corresponding to the maximum rate, this relation is dependent on the crystallization temperature. Results, such as are illustrated in Figure 12 are not limited to linear polyethylene but are also observed with various other polymers when the studies are made over an extensive molecular weight range.

Spherulite Growth Rate

Measurement of spherulite growth rate is another convenient method to study crystallization kinetics. Spherulites are morphological forms that are very common, but not universal, modes of polymer crystalization (cf. seq.). An extraordinary number of studies (too many to enumerate) have been made of the spherulite growth rates of virtually all crystalline polymers (*3, 4, 14*). The salient features of spherulite growth rate are common to all polymers. As an example, a plot of the radius of a growing spherulite as a function of time for isotactic polystyrene is given in Figure 13 (*15*). For all homopolymers, the radius (r) increases linearly with time so that the growth rate, $G = dr/dt$, is a constant. G has a strong negative temperature coefficient in the vicinity of the melting temperature. In the data for polystyrene, as well as many other polymers, G reaches a maximum value as the crystallization temperature is decreased. The spherulite growth rate can be expressed as follows:

$$G = G_0 \exp\left\{\left(-\frac{E_D(T)}{RT}\right) - \left(\frac{\Delta G^*}{RT}\right)\right\} \qquad (12)$$

Equation 12 is of the same form as equation 9. Not surprisingly, therefore, experimental results obtained from studies of either the overall rate of crystallization, as illustrated in Figure 10, or the spherulite growth rates are compatible, and essentially the same information is obtained.

The steady-state nucleation rate, as embodied in equation 9, is central to our understanding of the crystallization kinetics of all types of polymeric systems, including copolymers and polymer–diluent mixtures. It also accounts for the effects of chain length, polydispersity, strain-induced crystallization, and applied hydrostatic pressure. Changes in the parameters involved, but not the form, can be effected by adapting specific types of nucleation or chain structures within the nucleus.

The growth rate of a crystal face is reflected in the overall crystallization rate or spherulite growth rate. This growth rate will depend on the relative rates of nucleation and the spreading of the crystallizing units

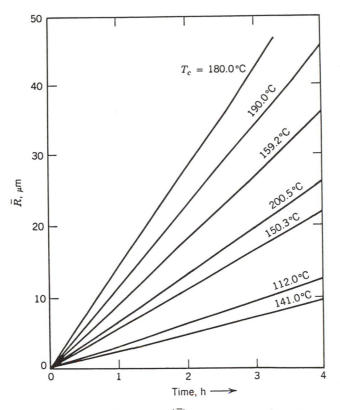

FIGURE 13. Plot of spherulite radius ($\bar{R}$) of isotactic polystyrene as a function of time. (Reproduced with permission from reference 15. Copyright 1970 Society of Polymer Science, Japan.)

along the face. This problem was initially analyzed for low-molecular-weight inorganic systems (*16*) and subsequently extended to polymers (*17*). Several typical cases, termed regimes, can be envisaged. In one extreme (regime I), subsequent to nucleation, the spreading of units along the face is completed before a new layer is initiated (nucleated). In another case (regime II), nuclei are allowed to initiate new layers before the complete coverage of the initial nucleating layer. In another extreme (regime III), very rapid nucleation on a given face does not allow subsequent spreading along the face. In terms of kinetics, the effect of these different physical situations is to alter the proportionality factors in equation 10 and, thus, the specific forms of equations 9 and 12. The expected changes in the temperature coefficient of the different regimes have been observed in many systems (*3, 18*).

Structure and Morphology

General Aspects

Although the basic framework of the subject, fusion thermodynamics and crystallization kinetics, is well-established, many problems still remain to be solved. Consider the crystallization of a normal hydrocarbon with less than 100 carbon atoms per molecule. For such normal hydrocarbons, crystallization will take place very rapidly when the temperature is lowered only infinitesimally below the equilibrium melting temperature. To crystallize the polymeric analogue, linear polyethylene, the temperature must be lowered well below the melting temperature even for low-molecular-weight fractions. For the hydrocarbon, the chains are completely extended, and molecular crystals are found, because each molecule is exactly the same length. Molecular crystals cannot develop in polymers even for the best fractionated samples, because there is a distribution of chain lengths. Thus, the crystallization of long-chain molecules will occur at finite or reasonable rates only at large undercoolings, that is, 20–40° C below the melting temperature. As a consequence, polymers form a polycrystalline system, which is in fact only partially or semicrystalline. For low-molecular-weight fractions containing up to several hundred chain atoms, extended, but not molecular, crystals may be formed. Higher molecular weight polymers form folded structures (*cf. seq.*). (Normal hydrocarbons with more than 150 carbon atoms per molecule also can form folded structures under appropriate conditions.) The crystallite structure, as well as the associated morphology, is complex, and these structures and morphological features determine the actual polymer properties. The fact that polymers can crystallize only at a finite rate under conditions well removed from equilibrium presents the basic problem. Therefore, to describe and understand properties, we have to deal with a very morphologically complex nonequilibrium system. These considerations bring us to the more modern aspects of the problems involving the crystalline state in polymers, that is, the relation between structure and properties.

To put the problems involved into perspective, the chart in Figure 14 (*19*) illustrates the interrelations among the various pertinent subjects. Essentially, all properties are controlled by molecular morphology. Molecular morphology is, in turn, determined by crystallization mechanisms. Such mechanisms are deduced from detailed studies of crystallization kinetics. The equilibrium requirements are necessary to analyze properly the kinetics. Obviously, the different aspects of the problem are very closely interrelated. Very few, if any, of the problems concerned with crystallization behavior, or properties, of the crystalline state can be examined in isolation. Keeping the different facets of the whole problem in

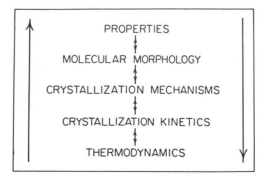

FIGURE 14. Representation of problem areas in the study of crystalline polymers. (Reproduced with permission from reference 19. Copyright 1979 Royal Society of Chemistry.)

mind, we focus our attention on the relationship between the molecular morphology and the properties of homopolymers crystallized from the pure melt. The principles that will be established can and have been extended to include polymer–diluent mixtures and the various types of copolymers. Crystallization under an applied stress, or oriented crystallization, presents another distinct area but will not be discussed here.

The problem of understanding structure and morphology can be simplified by examining different hierarchies: the unit cell, the crystallite and associated structures, and the supermolecular structures. The unit cell structures are essentially the same as those in the conventional crystallography of low-molecular-weight substances. The description of the crystallite structure, resulting from the polycrystalline nature of the system, involves a description of the structure of the actual crystallite; its associated interfacial region or zone; and the interconnections, if they exist, between crystallites. The supermolecular structure is concerned with organization of crystallites into larger structures.

The unit cell structure is determined by classical X-ray crystallography. The problem, initially thought to be very complicated, became relatively simple when it was recognized that the complete long-chain molecule need not be in the unit cell. The deduction of the unit cell has not presented any real interpretive problems. In contrast, elucidation of the crystallite structure has been a very controversial and, unfortunately, divisive problem for the past 20–25 years. However, a rational analysis and resolution of this problem is finally at hand. Systematic work on the supermolecular structure also has evolved, particularly the specification of the different kinds of superstructures than can be developed under different conditions and their influence on properties.

Molecular morphology differs in a very important and significant way from what might be called gross morphology. Both of these concepts, however, are important. The gross morphology is observed directly by microscopic examination; it specifies the form and shape of the structures of interest. The molecular morphology is a description of the arrangement and disposition of the chain units that is consistent with the gross morphology. Obviously, the molecular morphology cannot be observed directly. Both morphological descriptions must be consistent with one another.

Crystallite Structure

A lamellarlike crystallite habit is the characteristic gross morphological form developed by homopolymers during crystallization from the pure melt. Such lamellar structures were initially observed for crystallites formed from dilute solution. The characteristic, thin lamellar habit for solution-formed crystals is shown in Figure 15 for linear polyethylene. Such structures have now been observed for all homopolymers studied and can be taken to be a universal mode of homopolymer crystallization. These crystallites possess some very characteristic features. The lamellar thickness for crystals formed from dilute solutions is of the order of 100–200 Å, depending on the crystallizing solvent and temperature. The chain axes are preferentially oriented perpendicular to the basal planes of the lamellae. Such crystal habits are found for polymers of very high molecular weight. Because the thickness of the crystallites in the chain direction is only of the order of 100–200 Å, a single chain must traverse

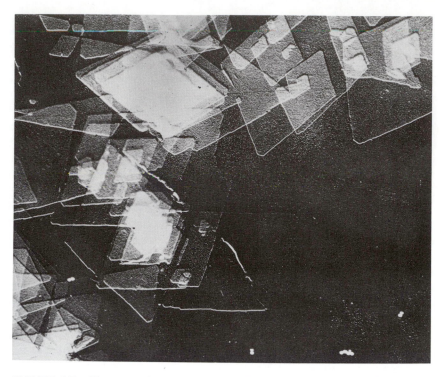

FIGURE 15. Electron micrograph of lamellae formed by linear polyethylene crystallized from dilute solution.

the crystallite from which it originates many times. The detailed nature of the interface that develops is quite important and unique. The interfacial structure is not obvious and cannot be deduced solely from microscopic studies. Despite the aesthetic pleasantness of the crystallites shown in Figure 15, the interfacial structure is not at all apparent from this and similar kinds of microscopy.

Electron microscopic observations do not yield a description of the interfacial structure on a molecular level but establish quite well the gross morphological form and the orientational features. The molecular interfacial structure is consistent with several extremes, as schematically indicated in Figure 16 (20). In one extreme, called the regularly folded adjacent reentry structure (Figure 16a), the molecular chains are accordionlike and make precise hairpin turns to yield optimum crystallinity. Equally consistent with the gross morphological features is the "switchboard" model (Figure 16b), which has a distinct, disordered, amorphous overlayer. Either of these interfacial structures is consistent with the electron micrographs. The reason for introducing these concepts here is that a lamellar-type crystallite is also the universal mode of homopolymer crystallization in the bulk.

The first lamellae observed in bulk-crystallized systems were obtained by surface replica electron microscopy. A typical example of such lamellae crystallites in a bulk-crystallized polymer is shown in Figure 17. Unfortunately, the thicknesses of the lamellae are only of the order of 100–200 Å. These dimensions were originally thought to be typical of and unique to the crystallites formed during bulk crystallization. We know now that lamellar thicknesses, depending on molecular weight and crystallization

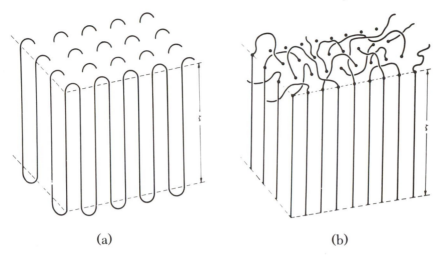

(a)   (b)

FIGURE 16. Schematic diagrams of possible chain structures within lamellar crystallite. (a) Regularly folded array and (b) nonregularly folded chains. Loop lengths are variable. (Reproduced from reference 20. Copyright 1962 American Chemical Society.)

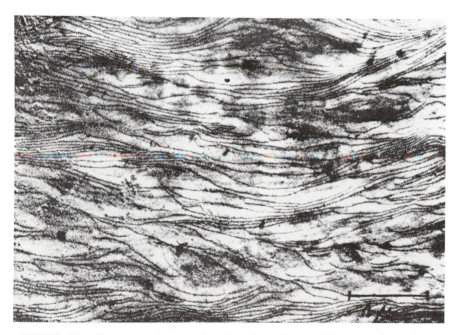

FIGURE 17. Electron micrograph of surface films of linear polyethylene showing the lamellar structures of bulk-crystallized polymers. (Reproduced with permission from reference 21. Copyright 1959 John Wiley and Sons, Inc.)

temperatures, can reach 1000 Å or more, even when crystallization is at atmospheric pressure. Even thicker lamellae can be obtained after crystallization at higher temperatures and pressures. Visual examination shows a regularity in the gross morphological structures shown in Figure 17. Because the crystallite thickness of bulk-crystallized polymer in this early work is about the same as that of solution crystals, a connection between the two situations was immediately made. It was proposed that the lamellar crystallites observed in bulk-crystallized polymers were made up of regularly folded chains that formed a smooth interface; that is, the model in Figure 16a was followed. Moreover, it was also proposed that crystallites had no molecular interconnections; that is, on the basis principally of gross morphological observations, the existence of noncrystalline regions was ruled out.

Lamellae-type crystallites are widely recognized and universally accepted as the characteristic mode for bulk crystallization of homopolymers. Surprisingly, copolymers with a relatively high counit content also form lamellar crystallites. The visual observation of lamellae or even the occasional viewing of defined sectors within lamellae is not sufficient to describe the interfacial structure, the presence or absence and the nature of connecting regions, or even the type and concentration of internal defects. The apparent geometric regularity, as perceived by the electron microscope, is obviously a very important gross morphological observation. However, it cannot be taken by itself as evidence for any detailed structure on a molecular level. This fact, unfortunately, has not always been recognized. The appearance of lamellae per se gives no detailed information about the disposition and arrangement of the chains within the crystallites. Selected area diffraction studies tell us only about the chain orientation.

For bulk-crystallized samples, major deviations exist in thermodynamic and other properties from those of the perfect crystal. These other properties include, among others, halos in the wide-angle X-ray scattering patterns, the nature of the infrared and Raman spectra, and the proton and carbon-13 NMR spectra. These deviations were thought to be due to contributions from the smooth interface, because small crystals are involved, and to a major contribution from defects believed to exist within the interior of the crystallite. A crystalline polymer was viewed as consisting of disordered material, or defects, imbedded within a crystalline matrix. Chain units in nonordered conformations, which would connect crystallites, did not exist in this view. The establishment of the structure of the crystallite on a molecular level is very crucial. It goes to the heart of the relationships between structure, morphology, and properties.

In analyzing the crystallite and related structures, a set of independent structural variables, related to properties, is important (*22*). These variables may be classified into two types. One is the molecular constitution concerned with molecular weight, polydispersity, and structural regularity of the chain. The other includes the degree of crystallinity, the structure of the residual or liquidlike isotropic regions (i.e., the region between lamellae), the distribution of crystallite thickness, the extent and structure of the interfacial region, and the internal structure of the lamellar crystallites. These variables serve as the basis for relating structure to properties. The independent structural variables act in synergy with the molecular constitution of the chain and the crystallization conditions.

Degree of Crystallinity

Various experimental methods have conclusively demonstrated that the degree of crystallinity is a quantitative concept. These methods include measurement of density and enthalpy of fusion, infrared and Raman spectroscopy, wide- and small-angle X-ray scattering, and proton and carbon-13 NMR spectrometry, among others. The basic principle involved is the assignment of a specific value of the quantity of interest to each element of the phase structure. In general, the different methods are in qualitative agreement. Small but significant differences, however, are observed between some of the methods. These differences can be attributed to the sensitivity of the elements of the phase structure that are being probed.

The level of crystallinity that can be attained at a given crystallization temperature depends on the molecular weight and the structural regularity of the chain. Figure 18 (*22*) illustrates this point for several different homopolymers that were crystallized under isothermal crystallization conditions. The level of crystallinity is relatively high at the lower molecular weights. However, as the molecular weight increases, the level of crystallinity decreases monotonically until a limiting value of about 25–30% is reached. The wide range of crystallinity levels for homopolymers rules out the possibility that the crystallites are made up of a regularly folded chain structure, with perhaps just minor perturbations. These results cannot be attributed to the influence of end groups, cilia, or similar structures, whose concentrations decrease with increasing molecular weight. As noted in the discussion of crystallization kinetics, molecular-weight-dependent topological restraints, such as chain entanglements, restrain the crystallization process. These structures are molecular-weight-dependent and are reflected in the level of crystallinity that can be attained.

The level of crystallinity is further reduced by the introduction of noncrystallizing structural units into the chain. Random copolymers are one example of structurally irregular chains. An example of the influence of counit content on the level of crystallinity at ambient temperature is illustrated in Figure 19 (*23*) for random ethylene copolymers. In this example, the molecular weights are restricted to the range $5 \times 10^4 - 1 \times 10^5$. The introduction of the noncrystallizing counits into the chain leads to a rapid and continuous decrease in the level of crystallinity with increasing side-group content. The levels of crystallinity vary from about 48% for 0.5 mol % branches to about 7% for 6 mol % branches. The level

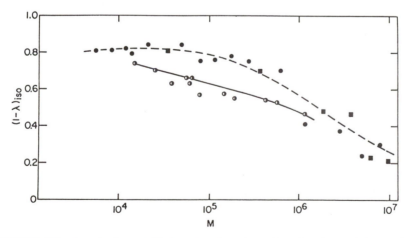

FIGURE 18. Level of crystallinity as a function of molecular weight under isothermal crystallization conditions. Key: ●, linear polyethylene; ■, poly(ethylene oxide); and ⬚, poly(tetramethyl-*p*-silphenylene siloxane.) (Reproduced from reference 22. Copyright 1990 American Chemical Society.)

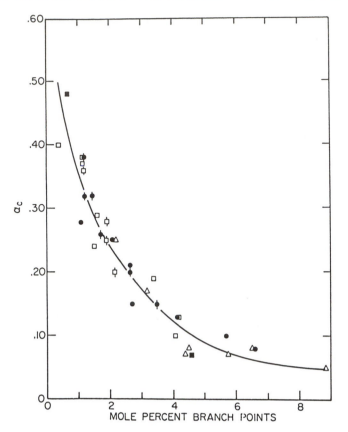

FIGURE 19. Degree of crystallinity calculated from Raman internal modes, α_c, versus mole percent branch points. Key: △, hydrogenated polybutadiene; ●, ethylene–vinyl acetate; ⊡ and □, ethylene–butene; ■ and ▪, ethylene–octene; and ●, ethylene–hexene. (Reproduced from reference 23. Copyright 1989 American Chemical Society.)

of crystallinity may be assumed to decrease even further at higher counit contents. The chemical nature of the branches on counits has virtually no influence on the level of crystallinity for a given counit content. The result is to be expected for random-type copolymers when the crystalline phase remains pure.

The level of crystallinity obtained after cooling a sample from the isothermal crystallization temperature is also of interest. The results of such a study, in terms of the density, for linear polyethylene are shown in Figure 20 for two different modes of crystallization. Here, the densities obtained after isothermal crystallization are compared with those observed after very rapid crystallization. The densities, and the levels of crystallinity derived therefrom, systematically depend on both the molecular weight and the crystallization conditions. For example, the densities, measured at room temperature, range from 0.99, which is very close to that of the unit cell, to 0.94 after crystallization at 130 °C and subsequent cooling. For linear polyethylene, a density as low as 0.92 can be observed after rapid crystallization of a high-molecular-weight fraction. After high-temperature, isothermal crystallization and subsequent cooling, the densities of the lower molecular weights fractions approach that expected for the unit cell.

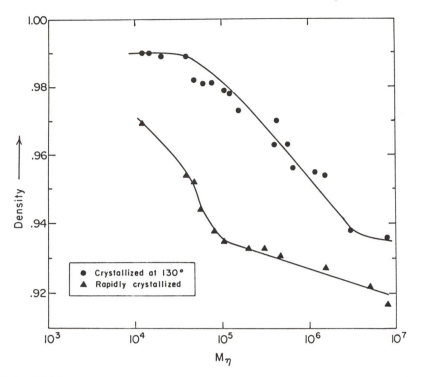

FIGURE 20. Density, measured at room temperature, versus molecular weight for linear polyethylene fractions crystallized under the indicated conditions. (Reproduced from reference 24. Copyright 1971 American Chemical Society.)

The monotonic decrease in density now starts at a slightly lower molecular weight compared with that for isothermal measurements. A constant value is reached in the very high molecular weight range. More rapid, non-isothermal crystallization results in much lower densities at comparable molecular weights. The molecular-weight dependence of the density is no longer as severe. The main changes now occur at molecular weights less than about 10^5. For molecular weights greater than 10^5, only a small decrease in density is observed with increasing chain length.

Properties measured by other experimental methods change similarly with molecular weight and crystallization conditions. The very large range in the level of crystallinity that can be attained by control of molecular weight, chain structure, and crystallization conditions is very striking. Because many structural, physical, and mechanical properties depend directly or indirectly on the level of crystallinity, these results portend large changes in many of the other properties of crystalline polymers. The quantitative nature of the degree of crystallinity is quite general and not limited to polyethylene. The classical work with natural rubber, for which much lower levels of crystallinity are obtained, substantiates this conclusion. It has also been established for other polyolefins, polyamides, polyesters, and poly(tetrafluoroethylene), to cite a few examples. All crystalline polymers behave similarly.

Although the degree of crystallinity is a quantitative concept, some of the experimental methods give small but significantly different values. These differences reflect sensitivity to different aspects of the phase structures. A striking example is given in Figure 21, which shows that the degree of crystallinity obtained from density measurements, $(1 - \lambda)_d$, for linear polyethylene is always greater than the degree of crystallinity obtained from measurement of the enthalpy of fusion, $(1 - \lambda)_{\Delta H}(25)$. Similar results are found with the structurally irregular polyethylenes (i.e., long-chain branched polymers and copolymers). $(1 - \lambda)_{\Delta H}$ corresponds to the core crystallinity (only unit cell contributions), which also can be obtained from an analysis of the Raman internal modes. Poly(tetramethyl-p-silphenylene siloxane) and poly(hexamethylene adipamide) show similar differences in the degrees of crystallinity determined by the two methods. Other flexible chains can be expected to behave similarly. These differences can be attributed quantitatively to the proportion of the interfacial region. The degree of crystallinity obtained from density measurements includes contributions for both the core crystallinity and the interfacial region.

In addition to establishing the quantitative nature of the degree of crystallinity, another important feature emerges from this analysis. The deviations in values from that expected for the unit cell (the perfect crystal) occur systematically with changes in the molecular weight and the mode of crystallization. These deviations are far from trivial and are, in fact, quite significant. To develop a complete and meaningful picture of the crystalline state, the wide range in values must be explained. A single piece of information, for example, an isolated density value, can be interpreted in virtually any arbitrary manner. Focusing attention on an isolated piece of information can thus be very misleading. Examining the complete set of

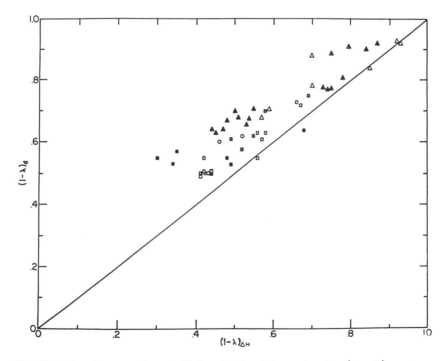

FIGURE 21. Degree of crystallinity obtained from density, $(1 - \lambda)_d$, versus degree of crystallinity obtained from enthalpy of fusion, $(1 - \lambda)_{\Delta H}$, for linear polyethylene. (Reproduced with permission from reference 25. Copyright 1985 Society of Polymer Science, Japan.)

data imposes rather extensive, rigorous demands that must be satisfied before any structural analysis can be given to the crystalline state.

Internal Structure

Lamellarlike crystallites are formed by all homopolymers over the complete range of molecular weights and crystallization temperatures. This crystal habit is also observed in random copolymers, to a surprisingly high counit content, even when the counits are excluded from the crystal lattice. Despite this very important generalization about lamellar structures, the detailed internal crystallite structure varies systematically with the molecular weight and the undercooling at which the crystallization is conducted. For low-molecular-weight fractions, the lamellae have very large lateral dimensions and are geometrically well developed at all crystallization temperatures. However, with increasing molecular weight, the lamellae become curved and segmented, and the lateral dimensions become smaller. This effect is accentuated at very high molecular weights ($\geq 10^6$). The random introduction into the chain of noncrystallizing counits causes a gradual deterioration of the lamellar structure. Eventually, at crystallinity levels of about 10–15%, lamellae are no longer observed, and a micellarlike habit develops.

An important feature of the internal structure of the crystallites is the tilt angle, the angle of inclination between the chain axis and the normal to the basal plane of the lamella. For linear polyethylene at high crystallization temperatures (i.e., low undercoolings), the tilt angle is about 19–20°. The tilt angle gradually decreases with decreasing crystallization temperature, and at low temperatures, it is approximately 45°. Other polymers for which tilt angles have been determined show a qualitatively similar behavior. The tilt angle is an important factor that needs to be considered in developing a detailed crystallization mechanism.

The concentration of defects within the crystallite interior may influence macroscopic properties. As already noted, a wide range in crystallinity levels can be obtained for many different polymers. The macroscopic density of linear polyethylene ranges from about 0.92 to 0.99 as the molecular weight decreases from 10^6 to 10^4. This change in the density raises the question of the integrity of the crystal structure. Do the changes in density reflect alterations within the crystallite interior or structural features that are exterior to the crystallite itself? However, as illustrated in Figure 22 (*26*), as the macroscopic density of linear polyethylene varies over the range of 0.92–0.99, the actual lattice parameters, as reflected in the unit cell density, remain constant. Therefore, the observed deviations in density from that of the ideal crystal cannot be attributed to a concentration of imperfections within the lattice. These deviations, therefore, must be due to specific structures located outside of the crystalline region, that is, structures external to the crystallite itself. This conclusion is consistent with the quantitative concept of the degree of crystallinity and involves analysis of interfacial and interlamellar structures.

Distribution of Crystallite Thickness

Different methods can be used to determine the distribution of crystallite thickness. These include thin-section electron microscopy, analysis of the

Raman longitudinal acoustical mode (LAM), and long-period, small-angle X-ray scattering. These methods give concordant results for narrow distributions of crystallite thickness. Agreement is not usually observed when broad size distributions are involved. However, when this variation is recognized, a rational interpretation of size distribution can be made. After rapid, nonisothermal crystallization, a narrow size distribution is obtained, as illustrated in Figure 23 (*22*) for linear polyethylene. The crystallite thicknesses range from 120 to 150 Å and are independent of molecular weight. Thickness depends only slightly on the crystallization temperature. In contrast, after isothermal crystallization, a broad size distribution results, because of crystallite thickening at the isothermal crystallization temperature. The rate of thickening depends on the molecular weight and the temperature. Consequently, by control of these variables, a wide range in sizes and distributions can be obtained. Crystallite thick-

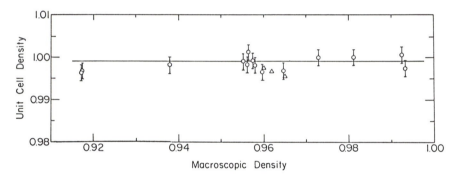

FIGURE 22.   Linear polyethylene lattice parameters: unit cell density versus macroscopic density for linear polyethylene fractions. (Reproduced with permission from reference 26. Copyright 1970 John Wiley and Sons, Inc.)

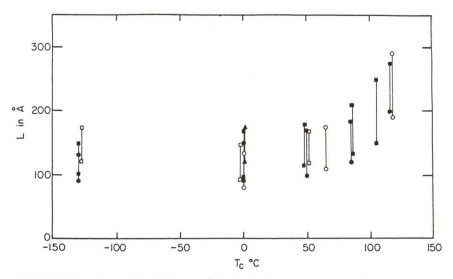

FIGURE 23.   Crystallite thickness (L) distribution versus quenching temperature (T_c) for molecular weight fractions of linear polyethylene. Key: $\bigcirc$, 1.97×10^4; $\bullet$, 4.6×10^4; $\square$, 2.26×10^5; $\blacksquare$, 4.28×10^5; and $\blacktriangle$, 1.62×10^6. (Reproduced from reference 22. Copyright 1990 American Chemical Society.)

nesses of several thousand angstroms have been obtained after long-time, high-temperature, isothermal crystallization at atmospheric pressure. However, to obtain thick crystallites, the narrow distribution is sacrificed. Copolymers and branched polymers, with counit exclusion, give thicknesses in the range of about 40–100 Å. Therefore, depending on molecular weight, chain structure, and crystallization conditions, crystallite thicknesses can vary from about 40 Å to several thousand angstroms, with a variety of size distributions.

Interlamellar Structure

Measurements of the level of crystallinity have shown that the interlamellar region can be a significant portion of the total system. A detailed analysis of the structure of this region has been elusive, because of the complex structure of the initial melt. In the pure melt, polymer chains assume randomly coiled configurations with dimensions identical with Θ conditions. For these dimensions to be maintained in a very dense system, the chains must be intertwined. This requirement leads to the formation of chain entanglements, loops, knots, interlinks, and other structures that cannot be reversed or dissipated during the time course of crystallization. Consequently, these structures will be rejected from the growing crystallite and concentrated within the noncrystalline regions. The level of crystallinity and the structure of the residual noncrystalline portion of the system are governed by these factors. The noncrystalline, interlamellar region influences many macroscopic properties.

A large body of experiments gives strong evidence that the chain units in the interlamellar region are in nonordered conformations without any preferential orientation; that is, the region is isotropic. In this region, the disordered chain units and their properties are similar to those in the completely molten or random state. This requirement is a natural consequence of the quantitative nature of the concept of the degree of crystallinity. Certain aspects of vibrational spectroscopy support this concept of isotropy. In addition, semicrystalline polymers display well-defined glass transition temperatures. Although the mechanism of glass formation is not yet completely understood, it is universally accepted that it is a property of the liquid state. The glass transition temperature of a semicrystalline polymer is either identical, or very close, to that of the corresponding completely amorphous polymer. Glass formation then lends further support to the presence of random structures. The sequences of chain-unit-connected crystallites are not complete molecules. The term "tie molecules," which has often been applied, is a misnomer. It implies that the connections are extended or straight and represent complete molecules. These connections represent only portions of molecules and are clearly in random conformation. A chain can adopt various trajectories after it leaves a crystallite. Therefore, many different structures result that are consistent with the condition of isotropy. A highly schematic representation of the complex structure of the interlamellar region is given in Figure 24. A portion of the chain can traverse unimpededly the space between lamellae. Some chains, however, will become entangled and knotted one with the other. Others will form long loops contained within the domain of a given crystallite, and loops from two adjacent lamellae can interlink with one another and connect the crystallites.

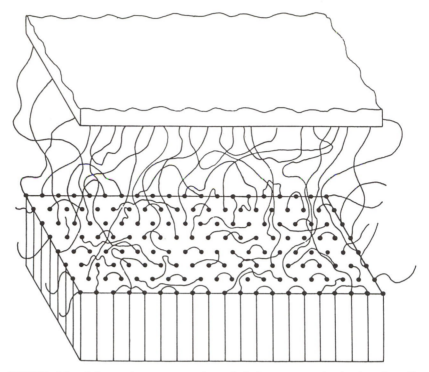

FIGURE 24. Schematic representation of chain structure in the interlamellar region.

Analysis of the small-angle neutron-scattering patterns of mixtures of hydrogenated and deuterated chains in the semicrystalline state has been very informative. (For more details, *see* Chapter 7 by G. Wignall.) The radius of gyration of a chain in the pure melt is the same as that of a chain crystallized from the melt. The virtual identity of the radii of gyration in the two states indicates that there is not much readjustment in chain conformation as the crystallizing growth front advances. Moreover, these results indicate that the lamellar crystallites do not contain any significant concentration of regularly folded chains. If they did, the radius of gyration would be quite different from that observed.

The interlamellar region can be structurally quite complex, although isotropy is maintained. The detailed quantitative description of the structures in this region remains a major problem in the area of crystalline polymers.

Interfacial Structure

The nature of the boundary between a lamellar crystallite and the disordered interlamellar region has been a matter of intensive study and discussion for over three decades. Several divergent views have emerged, as shown schematically in Figure 16. Recent experimental and theoretical developments have led to a resolution of the problem.

Examination of transmission electron micrographs suggests that the basal planes of the lamellae are molecularly smooth. This property was immediately identified with regular chain folding. However, detailed analysis of the electron micrographs, including their overall characteristics, decoration, sectorization, and interfacial dislocation networks for solution-

and bulk-crystallized samples, shows that there is no a priori need to identify the lamellarlike crystallites with regularly folded chains to satisfy morphological features. All of these key morphological characteristics are found in nonfolded *n*-alkanes and low-molecular-weight polymers and are thus are not unique to folded chain structures. As already pointed out in the analysis of the temperature coefficient of crystallization kinetics, polymer crystallization is a nucleation-controlled process. This conclusion was reached on the most general grounds, irrespective of the structure of the chains within the nucleus and the type of nucleation process that is involved. An assumption has been made that the nuclei involved are composed of regularly folded chains that grow into mature crystallites of the same structure. This assumption was then incorporated into a theory for polymer crystallization. Because the observed temperature coefficient is typical of nucleation, it was concluded that the chains in both the nucleus and the mature crystallite are regularly folded. This argument is a circular one. By itself, it does not have any bearing on the structure of either the nucleus or the crystallite.

Other structures can satisfy the well-established morphological and kinetic characteristics of polymer crystallization. Some of the chains could traverse the crystallite only once and then join a nearby crystallite. Others could return to the crystallite of origin but not necessarily in juxtaposition after traversing the interfacial and interlamellar regions. Some chains could return in adjacent reentry positions. Small-angle neutron-scattering studies (*see* Chapter 7), the analysis of chain statistics, and many different properties do not allow for regularly folded chain structures. For example, the observed variation of the crystallinity level with the molecular weight is not compatible with a regularly folded structure. Other experimental studies involving ^{1}H and ^{13}C NMR spectrometry, small-angle neutron scattering, specific heat measurements, dielectric relaxation, analysis of Raman internal modes, and electron microscopy clearly demonstrate the presence of an appreciable interfacial region that is characterized by partial ordering of the chain units (*3, 22*). In fact, there is no substantive experimental or theoretical basis for the view that lamellar crystallites, formed either in bulk or in dilute solution, are made up of regularly folded chains. This conclusion does not preclude some type of chain folding taking place. As will be discussed later, some amount of adjacent reentry can be expected, but not on the basis of nucleation theory. This conclusion, however, is not incompatible with nucleation-controlled kinetics, and this point is quite important. Two basic questions remain: Why is a lamellar crystallite characteristic of polymer crystallization, and what, in fact, is the true structure of the interphase?

Interfacial Free Energy

Flory pointed out in 1949 (*2*) that the boundary between the crystalline and liquidlike regions in polymers is not sharply defined. This behavior is fundamentally different from the behavior of low-molecular-weight systems. The continuity of a long-chain molecule imposes severe constraints on the transition between the two regions. The major differences of chain conformations in the two states require a boundary or interface that allows the crystalline order to be dissipated. The problem is that the flux of chains (the number of chains per unit area) emanating from the basal plane of the crystallite cannot usually be accommodated in the isotropic, liquidlike region. (Exceptions to this generalization are crystal structures, such as the α-helical polypeptides, with chains that are sufficiently far apart in the crystal so that their flux is reduced and can thus be accommodated in random conformation. This problem is not pertinent to the formation of a critical-sized nucleus, because not enough ordered sequences are involved.) One obvious solution to this problem is the return of the chain to the crystallite of origin. These returns, however, do not have to be in juxtaposition. For a crystallite to grow laterally, a significant amount of chain bending or folding must, therefore, occur. An expenditure of free energy, that is, a gain in free energy, will be involved. This free energy gain can be compensated by the crystallization of long sequences resulting in lateral crystallite growth. Thus, a straightforward mechanism exists by which well-developed lamellae will form at the expense of folded structures without the need to invoke or to manipulate monomeric nucleation theory.

Several detailed theoretical analyses have pursued this idea quantitatively with essential agreement in major conclusions (27–32). Several principal factors have to be taken into account. One factor is the density of the chains at the crystal surface. This quantity is determined by the tilt angle and the ratio of the cross-sectional area of a chain segment in the crystalline and the liquidlike regions. When the cross-sectional area in the crystalline state exceeds the corresponding area in the liquidlike region, the problem of flux dissipation is severely reduced. Another very important factor is the free energy increase necessary to make a bend or a tight fold. All of these concerns make clear that the structure of the interfacial zone will be specific for a given polymer and that generalizations will be difficult to make.

As a first example, consider the hypothetical polyethylene chain for which no free energy is expended in forming a fold and no conditions are placed on the chain incidence probability. For this case, about 70–75% of the sequences would return to the lamellae of origin in tight, adjacent folds. This result is not surprising, because with no free energy cost in making a bend, the regularly folded structure is clearly the easiest way in which the flux of chains can be dissipated. In reality, however, the incidence of a tight fold or an immediate adjacent reentry for polyethylene is only 30–40% of the sequences because of the free energy increase necessary to make a fold. Consideration of the surface chain density (tilt angle of 45°) reduces the possibility of adjacent reentry to about 20%. Adjacent reentry will, thus, not make a major contribution to the interfacial structure for the real polyethylene chain, even for these idealized calculations. The adjacent folds that are formed will be randomly distributed along the lamellar surface.

The chemical nature of the chain, as reflected in the crystal structure and in the disordered-chain conformation, will determine the interfacial structure. Except for the basic principles, generalizations cannot be made about structure. For chains not requiring a free energy expenditure in making a bend, adjacent reentry will predominate. For chains with axes positioned far from one another in the unit cell, as in α-helical polypeptides, or with extended conformations in the disordered liquid state, as in cellulose and its derivatives, folding of any type, including adjacent reentry, will be minimal.

These concepts have been verified experimentally for specific polymers. For example, the excess interfacial free energy associated with the basal plane of a mature polyethylene crystallite has been calculated to be in the range of 50–65 ergs/cm^2. This value is in good agreement with that determined for low-molecular-weight fractions having a folded-type lamellar morphology, for which topological restraints are minimal. The interfacial thickness, L_b, for bulk-crystallized linear polyethylene has been theoretically estimated to be in the range of 10–30 Å. The thickness can be determined from either Raman spectroscopy or small-angle, long-period X-ray scattering. The experimental results show that the interfacial thickness varies from about 14 Å for low-molecular-weight fractions (10^4) to about 25 Å for fractions with molecular weights of 10^5–10^6 (*33*). Good agreement is, thus, found between theory and experiment. The interfacial thickness clearly increases with chain length. The theoretical calculations are for an idealized chain with infinite molecular weight and do not take into account the role of entanglements and other topological restraints and the limitations on chain mobility. These factors are molecular-weight-dependent, and their effects become more pronounced with increasing chain length. For example, the L_b for crystallites formed in dilute solution is about 10 Å, independent of molecular weight (*33*). Under these crystallization conditions, chain entanglements are minimal, and no significant chain mobility restrains the crystallization process. In addition to the interface, a substantial disordered overlayer also is associated with crystals formed in solution. The theoretical calculations, therefore, although highly idealized, give a very satisfactory explanation of the interfacial structure and the molecular factors that are involved.

The fraction of the system in the interfacial region, α_b, depends on the molecular weight. Changes in α_b parallel changes in L_b. As illustrated in Figure 25, α_b for linear polyethylene is about 5% in the low-molecular-weight range, monotonically increases to 15–17% with increasing molecular weight, and then apparently levels off. Because the core level of crystallinity is only about 40% for high-molecular-weight polyethylene, the interfacial region makes up a significant portion of the total system. The

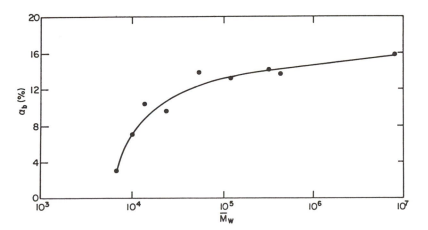

FIGURE 25. Interfacial content (α_b) versus weight-average molecular weight (M_w) for rapidly crystallized fraction of linear polyethylene. (Reproduced with permission from reference 34. Copyright 1988 Elsevier.)

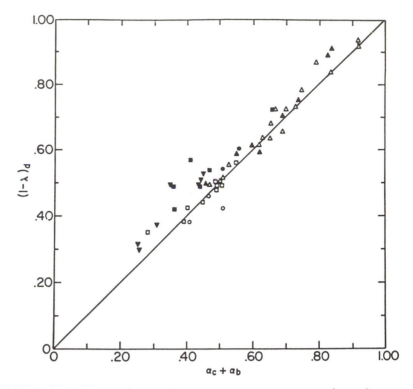

FIGURE 26.   Degree of crystallinity obtained from density, $(1 - \lambda)_d$, versus the sum $(\alpha_c + \alpha_b)$ for linear and branched polyethylenes and ethylene copolymers. (Reproduced with permission from reference 25. Copyright 1985 Society of Polymer Science, Japan.)

analyses of solid-state carbon-13 NMR data gives a comparable value for the interfacial content of bulk-crystallized polyethylene. Similar values are found for polypropylene and poly(tetramethylene oxide).

The interfacial free energy associated with the basal plane of the crystallite can be obtained from the dependence of the melting temperature on the crystallite thickness. This interfacial free energy increases monotonically with molecular weight, the curve paralleling the plot in Figure 25 (*34*) until a molecular weight of 10^5. At this point, the interfacial free energy becomes constant at a relatively high value (295 ergs/cm^2). This quantity is characteristic of the mature crystallite and not of the stable nucleus. It reflects the effect of the initial melt structure on the crystallization behavior and on the resulting crystallite structure. The value of α_b is slightly higher for random copolymers relative to the value for homopolymers. For copolymers, however, α_b constitutes a significantly higher proportion of the core crystallite thickness. Although α_b for copolymers depends on counit content, it does not vary with molecular weight. The accumulation of noncrystalline counits at the crystallite surface is the dominant factor in determining the interfacial structure of copolymers. It surpasses the molecular weight effect of entanglements and similar structures, which is a major factor in determining the interfacial structure of homopolymers.

As previously noted, $(1 - \lambda)_d$ is always greater than $(1 - \lambda)_{\Delta H}$ for polyethylene and other polymers. This discrepancy can be attributed to α_b and the existence of an interfacial region, as illustrated in Figure 26 (*25*).

Here, $(1 - \lambda)_d$ for linear polyethylene is plotted against the sum $(\alpha_c + \alpha_b)$, in which α_c is $(1 - \lambda)_{\Delta H}$, as noted earlier. The data fall quite well on the 45° straight line, a result indicating a one-to-one correspondence between the two quantities. Because α_c and $(1 - \lambda)_{\Delta H}$ measure the core crystallinity and α_b measures the interfacial content, a self-consistent and physically satisfying interpretation of these results is that both the core crystallinity and the partially ordered anisotropic interfacial region contribute to the measured density. The enthalpy of fusion, on the other hand, measures only core crystallinity. Theory and experiment thus lead to the conclusion that a significant interfacial region exists that is characterized by a partial order of the chain units. Although a large number of chains return to the crystallite of origin, the number of chains returning in an adjacent position is generally small.

Summary of Crystallite Structure

The study of the crystallite and related structures point to three major regions with different basic chain conformations traversed by many individual chains: the crystalline region, the interfacial region, and the interlamellar or liquidlike region. The crystalline region is the three-dimensional ordered structure with the typical lamellarlike habit. The imperfection levels within the crystallites are no different from those found in crystals formed by similar low-molecular-weight compounds. The crystallite, or core, thickness is controlled by the nucleation requirements. However, the nucleation control of the crystallization process does not mean regular folded chains. Neither does it require crystallite thicknesses that are identical, or very close, to those of the critical-sized nuclei. The interfacial region is diffuse; it is not the sharp, clearly defined boundary usually associated with the interface of crystals of low-molecular-weight substances. This characteristic is unique to chain molecules. This boundary is characterized by a very high interfacial free energy. The interlamellar, liquidlike isotropic region is the main portion of the noncrystalline region. Although often neglected and not discerned by many kinds of gross morphological observations, this region plays a crucial role in governing many properties.

Two major points have emerged from many studies: (1) the importance of the initial melt structure and (2) the role of the molecular weight in influencing the properties that describe and define the crystallite and related structures. An extremely wide range of values can be attained for a given structural parameter by control of molecular weight and crystallization conditions. Interest in explaining the properties of crystalline polymers in terms of structure continues.

Supermolecular Structures

The supermolecular structure is concerned with the arrangement of the individual lamellar crystallites into a larger scale of organization. This aspect of structure has been extensively studied. The existence of such higher orders of organization is well-established, as evidenced by the common observation of spherulites in semicrystalline polymers. To establish the conditions under which different kinds of supermolecular structures are formed and their influence, if any, on properties, a powerful technique that can be used is small-angle light scattering. Light-scattering studies are usually complemented by light and electron microscopic observations.

The most useful light-scattering method for describing supermolecular structures is the H_v mode, which depends on orientation fluctuations. In this mode, the incident light is polarized in the vertical direction, and the observed scattered light is polarized in the horizontal direction. Although only the results for polyethylene will be discussed in detail, very similar results have been found for poly(ethylene oxide) and polypropylene. Thus, the results described for polyethylene can be taken to be quite general.

The polyethylenes display five distinctly different types of light-scattering patterns, which are illustrated in Figure 27 and which are designated by letters. These patterns range from that of the classical cloverleaf (pattern a) to one that is circularly symmetric (pattern h). The light-scattering patterns can be related by theory to different supermolecular structures, which are listed in Table II. In Table II, patterns a, b, and c represent spherulites of decreasing order; that is, the spherulitic structure is deteriorating. Pattern a, the classical cloverleaf with zero intensity in the center, represents the ideal, best developed spherulite. Pattern d, which

a   b   c

d   h

FIGURE 27. Types of light-scattering patterns observed with polyethylenes. (Reproduced with permission from reference 19. Copyright 1979 Royal Society of Chemistry.)

Table II. Light Scattering and Supermolecular Structure

Pattern Produced by Small-Angle Light Scattering	Supermolecular Structure
a, b, and c (cloverleaf)	a-, b-, c-type spherulites
d (some azimuthal dependence)	d-type thin rods or rodlike aggregates
h (no angular dependence)	g-type sheetlike rods (breadth comparable with length)
	h-type randomly oriented lamellae

has an azimuthally dependent light-scattering pattern, represents lamellae that are organized into thin rods or rodlike aggregates. The circularly symmetric pattern, pattern h, does not represent a unique morphological situation. It can represent rods or sheets with breadths comparable with widths. In this case, the structure is designated as having a g-type morphology. Pattern h also can represent a random collection of uncorrelated lamellae. In this case, the structure is designated as having an h-type morphology. Pattern h, therefore, can represent either of two supermolecular structures that can be differentiated only by the application of some complementary microscopic method.

The formation of supermolecular structures depends on the molecular weight, the crystallization conditions (such as the isothermal crystallization temperature or the cooling rate), the molecular constitution, and the molecular weight polydispersity. Supermolecular structures are uniquely sensitive to polydispersity.

A morphological map that depicts the dependence of supermolecular structure on molecular weight and crystallization conditions is given in Figure 28 (*35*). In this diagram, the almost vertical dashed line represents the boundary for isothermal crystallization. The temperatures below this demarcation are those of the quenching bath to which the sample is rapidly transferred from the melt. Although this experiment is subjective, it is reproducible, and it accomplishes the main purpose of varying the supermolecular structure formed. The nonisothermal portion of the dia-

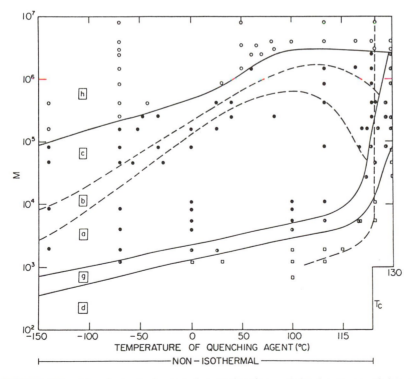

FIGURE 28. Morphological map for molecular weight fractions of linear polyethylene: molecular weight versus quenching temperature or isothermal crystallization temperature. (Reproduced from reference 35. Copyright 1981 American Chemical Society.)

gram merges continuously with the isothermal region. The regions where the different supermolecular structures are formed are given the letter designation of Table II. An important point highlighted by this map is the fact that spherulitic structures are not always observed. Spherulitic structures clearly are not the universal mode of homopolymer crystallization, although they are very common to polydisperse whole polymers. In fact, as the map clearly indicates, supermolecular structures do not always develop. An h-type morphology is found under isothermal and nonisothermal crystallization conditions for molecular weights greater than 10^5. Thus, although no organized supermolecular structures are observed at the highest molecular weights, the crystallinity level is still about 0.50–0.60 for these samples. Detailed examination of the map in Figure 28 shows that the low-molecular-weight polymers form thin rodlike structures. As the molecular weight increases, a g-type morphology is observed at higher crystallization temperatures. Here, the length and breadth of the rodlike structures are comparable so that sheetlike structures are observed. If the crystallization temperature is lowered for this molecular weight range, spherulites will form. The spherulitic structure deteriorates as the chain length increases. Well-developed a-type spherulites also are generated at low temperatures for very low molecular weight fractions.

The map in Figure 28 shows that it is possible to prepare different supermolecular structures from fractions of the same molecular weight by choosing the appropriate crystallization conditions. In certain situations the different supermolecular structures can be formed from fractions with the same molecular weight at the same level of crystallinity. Therefore, another well-defined independent variable must be taken into account in discussing the properties and behavior of crystalline polymers.

Thin-section transmission electron microscopy, when properly carried out, can be very useful in studying the structure of crystalline polymers. It is of interest to compare the morphological structures obtained from small-angle light-scattering studies with those obtained from electron microscopy. A typical set of electron micrographs is given in Figures 29–31 (*36, 37*). In Figure 29, the sheetlike structures, or g-type morphology, obtained from a fraction with a molecular weight of 1.89×10^5 that had been isothermally crystallized at 131.2 °C are shown. The same sample quenched at 100 °C produces well-developed spherulites, as shown in Figure 30. Random lamellae, or h-type structures, are shown in Figure 31 for a fraction with a molecular weight of 6×10^6 and crystallized at 130 °C. Other structures also can be demonstrated by electron microscopy. Comparison of results from light scattering and electron microscopy shows a one-to-one correspondence between the morphological map deduced from small-angle light scattering and the direct electron microscopic observations.

It is of interest to examine the influence of supermolecular structure on properties. The relationships among the supermolecular structure, the density, and the degree of crystallinity are shown in Figure 32. Here densities are plotted for different crystallization temperatures for a set of molecular weight fractions (*19*). The different supermolecular structures that are formed are also indicated. No morphological changes are observed for fractions with a molecular weight of 10^4, and the density changes smoothly with crystallization temperature. In contrast, for fractions with molecular weights of 10^5 and 10^6, major changes in the

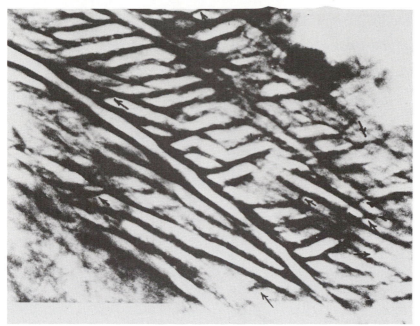

FIGURE 29. Transmission electron micrograph of linear polyethylene fraction with molecular weight of 1.89×10^5 crystallized at 131.2 °C. (Reproduced with permission from reference 36. Copyright 1981 John Wiley and Sons, Inc.)

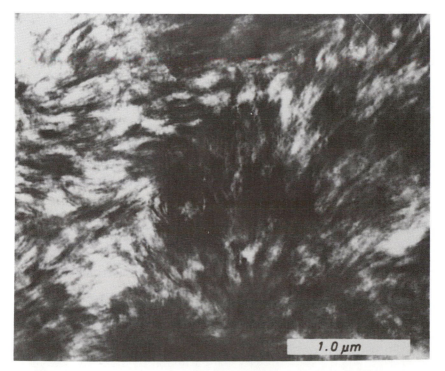

FIGURE 30. Transmission electron micrograph of linear polyethylene fraction with molecular weight of 1.89×10^5 quenched at 100 °C. (Reproduced with permission from reference 36. Copyright 1981 John Wiley and Sons, Inc.)

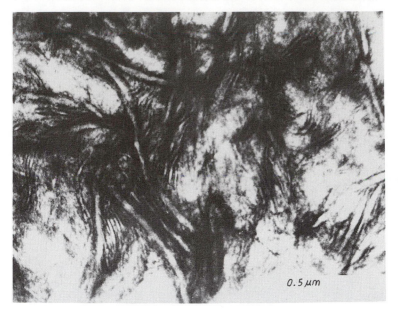

FIGURE 31. Transmission electron micrograph of linear polyethylene fraction with molecular weight of 6.0×10^6 crystallized at 130.0 °C. (Reproduced with permission from reference 37. Copyright 1980 John Wiley and Sons, Inc.)

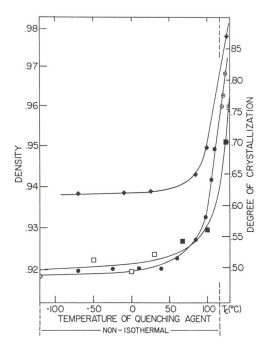

FIGURE 32. Density and degree of crystallization versus isothermal crystallization temperature for three linear polyethylene fractions. Key for fraction with molecular weight of 2.78×10^4: ◆, spherulites, and ◇, rods (d). Key for fraction with molecular weight of 1.61×10^5: ○, random lamellae; ●, spherulites; and ⊙, rods (g). Key for fraction with molecular weight of 1.50×10^6: □, random lamellae; ■, spherulites; and ⊡, rods (g). (Reproduced with permission from reference 19. Copyright 1979 Royal Society of Chemistry.)

supermolecular structure are evident. These structural changes, however, are not reflected in any changes in the density under comparable crystallization conditions, because the density changes smoothly with properties. The enthalpies of fusion and the measured melting temperatures show a similar insensitivity to the supermolecular structure. The supermolecular structure, thus, has very little influence on thermodynamic properties. Different spectroscopic studies also show that supermolecular structure has very little, if any, effect on spectroscopic properties. Many physical and mechanical properties are similarly unaffected. The type and size of the supermolecular structures, however, must obviously influence optical properties. Besides the effect on optical properties, no definite influence of superstructure on properties has been established. The primary contributions to properties come from the lamellar crystallite and associated structures.

The influence of the chain constitution on the supermolecular structure is also important. The incorporation of branched (side) groups, copolymeric units, or other irregularities into the chain alter the major characteristics of the morphological map. In general, as far as the supermolecular structure is concerned, the behavior of a branched polymer is similar to that of a high-molecular-weight linear polymer. Analysis of isothermally crystallized samples of branched polymers is complicated by the small amount of transformation under these conditions and the substantial amount that develops on cooling. Such complexities are avoided in reproducible, nonisothermally crystallized systems.

A typical morphological map obtained in the standard way is given in Figure 33 (*35*) for molecular weight fractions of branched polyethylene. Each of these fractions contains about 1.5 mol % branch points. For a given branching content and molecular weight, the temperature range within which spherulites of different degrees of order can form is very limited. Low-molecular-weight fractions are conducive to the formation of more highly ordered spherulites. When the supermolecular structures that are formed are examined as a function of molecular weight, a dome-shaped curve develops that forms the boundary for spherulite formation. For both higher and lower temperatures outside the dome boundary, the h-type morphology of random lamellae is usually observed. Within the dome, spherulites are formed. These morphological conclusions are confirmed by thin-section, transmission electron micrographs. For branched polymers, lamellae can be formed over a very wide range of molecular weights.

On the basis of experiments with fractions of various branching contents and molecular weight, a schematic representation of the changes that take place in the boundary for spherulite formation is given in Figure 34 (*35*). For a given molecular weight, as the branching concentration decreases, the temperature range over which spherulites can be formed becomes wider. The height of the dome, which is the boundary for spherulite formation, decreases with increasing branching content. Thus, both increased molecular weight and branching content reduce spherulite formation and favor randomly arranged lamellae.

Properties

The basic thermodynamic, kinetic, and structural principles that govern the crystallization behavior of polymers can now be used to explain

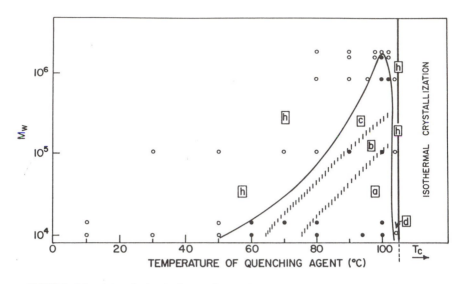

FIGURE 33. Morphological map for molecular weight fractions of branched polyethylenes with 1.5 mol % branched groups: molecular weight versus quenching temperature. Solid line delineates region of spherulite formation. (Reproduced from reference 35. Copyright 1981 American Chemical Society.)

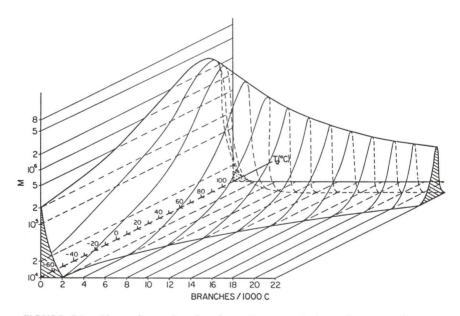

FIGURE 34. Three-dimensional schematic morphological map of non-isothermal crystallization of polyethylenes. The curved, dome-shaped regions define the volume within which spherulitic structures are formed; outside this volume, no defined supermolecular structure exists. (Reproduced from reference 35. Copyright 1981 American Chemical Society.)

various properties of semicrystalline polymers. Interest in explaining the properties of crystalline polymers in terms of structure continues. Because of the nonequilibrium character of the crystalline state in polymers and the attendant morphological complexities, the influence of structure on properties is overriding.

The different elements of chain structure, as well as the crystallization conditions, specify the independent structural variables that, in turn, affect properties. A strategy can be developed where, by control of molecular weight and crystallization conditions, a specific variable can be isolated and its influence on a given property can be assessed. This procedure has been demonstrated successfully for certain mechanical and spectral properties. The independent variables are varied over the widest extent possible by control of molecular constitution and crystallization conditions. The concomitant changes in the property of interest are observed. By following this strategy, problems can be reduced to their important structural features. This procedure applies to virtually all properties of crystalline polymers, and many complex problems can be resolved in this manner. These principles will now be applied to selected examples: analysis of low-frequency, dynamic mechanical properties and tensile behavior of polyethylenes.

Dynamic Mechanical Properties

Dynamic mechanical measurements of crystalline polymers yield a set of relaxation transitions, in addition to melting. A typical, low-frequency dynamic mechanical spectrum for a branched polyethylene (short- and long-chain branching) is illustrated in Figure 35 (*38*). Such a spectrum, with only occasional minor variations, is characteristic of crystalline polymers. In the order of decreasing temperature below the melting temperature, these transitions, or relaxations, for polyethylenes have been designated alpha, beta, and gamma, respectively. (For other polymers, the designation may be different. If labeled in the same order, the transitions could reflect different structural phenomena. Therefore, each polymer needs to be analyzed individually to avoid confusion of the molecular basis of the relaxation.) The transitions are usually between -150 and -120 °C for the gamma transition, between -30 and $+10$ °C for the beta transition, and between 30 and 120 °C for the alpha transition. By using the strategy outlined earlier, the molecular and structural bases for these relaxations can be analyzed.

The Alpha Transition

The alpha transition is observed in all polyethylenes (i.e., linear, copolymer, and long-chain branched polymers). The change in the intensity of the transition when the level of crystallinity is changed indicates that the relaxation involved results from the motion of chain units located within the crystalline portion of the polymers. Two questions need to be addressed next: What are the structural and molecular factors that govern this transition, and why can it be observed over such a wide temperature

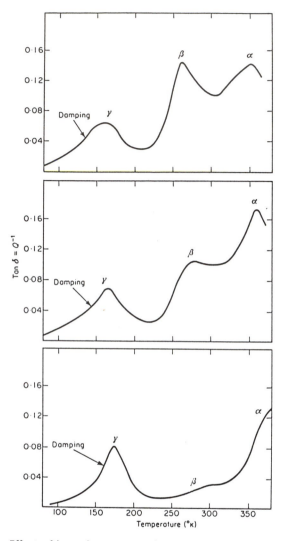

FIGURE 35. Effect of branch concentration on magnitude of beta relaxation in polyethylene for samples containing 32 (Top), 16 (Middle), and 1 (Bottom) branch per 1000 carbon atoms. (Reproduced with permission from reference 38. Copyright 1956 John Wiley and Sons, Inc.)

interval? The influence of the independent variables and their wide range of values indicate that the temperature of the alpha transition, T_α, depends on the crystallite thickness, irrespective of molecular weight, branching type and concentration, and level of crystallinity. This point is illustrated by the plot in Figure 36 (*39*). Here T_α ranges from about -20 °C to $+120$ °C for crystallite thicknesses of 60–300 Å. Careful examination of the data in the figure shows quite clearly that the supermolecular structure does not play a role in determining T_α. Many examples can be found of polymers that have different types of supermolecular structures and have the same value for T_α. The controlling factor for the different types of

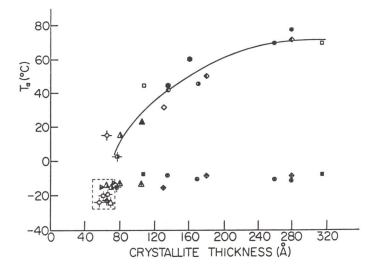

FIGURE 36. Plot of alpha and beta transition temperatures at a frequency of 3.5 Hz for various linear and branched polymers representing the complete range of supermolecular structures. Key: Linear polyethylene: ◇, unfractionated $M_w = 8 \times 10^6$; □, h-type morphology. Linear polyethylene fraction $M_w = 3.6 \times 10^5$: ●, a-type morphology; ◐, b-type morphology; ◑, c-type morphology. Linear polyethylene fraction $M_w = 4.28 \times 10^5$: ◆, c-type morphology; ◇, h-type morphology. Linear polyethylene unfractionated $M_w = 1.5 \times 10^5$: ◆, a-type morphology. Branched, high-pressure polyethylene: ▲, a-type morphology; ▲, b-type morphology; ▲, c-type morphology. Hydrogenated poly(butadiene) $M_w = 1.08 \times 10^5$: ⊡, b-type morphology; ⊡, h-type morphology. Ethylene–vinyl acetate copolymer: ▲, c-type morphology; ⊖, h-type morphology; △, h-type morphology. (Reproduced with permission from reference 39. Copyright 1984 John Wiley and Sons, Inc.)

polyethylene supermolecular structures and the morphologies is the crystallite thickness. A closely related phenomenon, the carbon-13 crystalline spin-lattice relaxation time, T_1, at ambient temperatures increases directly with the crystallite thickness (*40*). A correlation can be made between the NMR results and the location of the alpha transitions. However, when the interfacial structure is drastically altered, by selective oxidation or by the use of extended-chain crystals, a major increase in T_1 is observed. This result indicates that the interfacial structure influences and is coupled with the motion within the crystallite interior. The coupling of the motion of these two regions is also expected in dynamic mechanical and dielectric relaxations.

The Beta Transition

An intense beta transition is universally observed in all branched polyethylenes (short and long chains). However, this transition is found only in very high molecular weight linear polyethylenes. The data plotted in Figure 36 make quite clear that, in contrast to the alpha transition, the temperature of the beta transition does not depend on the crystallite thickness. The temperature of the beta transition, T_β, depends on the chemical nature and concentration of the counit. Thus, each copolymer has its own beta transition.

The universal observation of the beta transition in copolymers, in long-chain branched polymers, and in high-molecular-weight linear polymers suggests a possible relation between the interfacial content and the intensity of the beta transition (39). As noted earlier, these polymers possess the highest fraction of chain units located in the interfacial region. Analysis of experimental data indicate that just the presence of a high noncrystalline content is not sufficient for the observation of a beta transition. Furthermore, at a constant level of crystallinity, the intensity of the beta transition substantially increases with the interfacial content. A compilation of the structures and conditions under which a beta transition is observed and the corresponding interfacial content, α_b, is given in Table III. This summary emphasizes the relation between the interfacial content and the existence and intensity of the beta transition for various situations. When the interfacial content is low, less than about 5–7%, the beta transition is not observed, as indicated for the solution-formed crystals and for low- and medium-molecular-weight, bulk-crystallized linear polyethylene. When the interfacial content is greater than about 10%, well-defined beta transitions are observed in high-molecular-weight, bulk-crystallized linear polyethylene and in both solution- and bulk-crystallized branched polyethylenes. This analysis shows why the beta transition in linear polyethylene has been elusive and controversial. For the linear polymers, a sufficiently high value of α_b, necessary to ensure the observation of this transition, is associated only with high-molecular-weight fractions.

Although a detailed analysis has been given only for the polyethylenes, the extensive amount of experimental data indicates that a similar basis for the beta transition also exists for other crystalline polymers. For example, the dynamic mechanical behavior of poly(oxymethylene) is, in fact, very similar to that of polyethylene. Poly(oxymethylene) displays a crystalline relaxation and two others, which are usually referred to as the beta and gamma relaxations. The introduction of small amounts of ethyl-

Table III. Summary of Beta Transitions in Polyethylenes

Polymer Type	Observation of Beta Transition	Interfacial Content (%)
Solution crystals of linear polyethylene	Not observed	< 5
Solution crystals of branched polyethylene	Transition observed[a]	11–17
Bulk-crystallized linear polyethylene	Not observed for low-molecular-weight fractions ($< 2 \times 10^5$)	< 7
	Observed for high-molecular-weight fractions ($> 2 \times 10^5$)	> 10
Bulk-crystallized branched polyethylene	Strong relaxation always observed	11–21

[a] No report of dynamic mechanical studies on such systems. Transition is observed by indirect measurement of thermal expansion coefficient.

ene oxide counits into the chain greatly enhances the intensity of the originally weak beta transition. These results parallel those for copolymers of ethylene and point to a common origin. Because the ethylene oxide counits are effectively excluded from the crystal lattice, an enhanced interfacial structure would be expected.

Analysis of the experimental data indicates that the beta transition results from the motion of disordered chain units that are associated with the interfacial regions of semicrystalline polymers and copolymers. The presence of crystalline and noncrystalline material is a necessary requirement. Although it might be convenient to consider this transition as some type of pseudo-glass transition temperature, the correlation time for segmental relaxation is many orders of magnitude too large. The transition, or relaxation, is unique to the partially ordered interfacial region. The assignment of the beta transition to the interfacial region also explains the observed unique dependence on counit composition.

The Gamma Transition

The gamma transition can be assigned to segmental motions within the interlamellar or liquidlike regions. The basis for this assignment is the fact that the intensity of the transition parallels the change in crystallinity level. Specific heats measured in the temperature range of the gamma transition have all the characteristics of glass formation. The assignment of the gamma transition to glass formation is also consistent with carbon-13 NMR relaxation measurements.

The analysis of the dynamic mechanical transition in the polyethylenes demonstrates how a complex problem can be resolved in molecular and structural terms by applying the strategy described earlier and by recognizing the wide range in values that can be attained by a given structural parameter. Other types of physical and mechanical properties can be analyzed similarly.

Tensile Behavior

A very important property of a crystalline polymer is its stress–strain behavior. Although this property is not completely understood as yet in molecular terms, some progress has been made (*41, 42*). A highly schematic illustration of ductile deformation, in tension, of crystalline polymers is given in Figure 37. The initial portion of the deformation, about 2 or 3% strain, is usually reversible. The dashed line in Figure 37 accentuates this initial deformation. The initial modulus can be calculated from the slope of this straight line. As the deformation proceeds, a yield point is reached, which is followed by a decrease in the force or stress. Subsequently, the deformation becomes inhomogeneous, or "necking" is said to occur. In this region, the force is invariant with length. The final upsweep in the force–length curve, called "strain hardening," terminates in the fracture or rupture of the sample. Figure 37 represents a specimen undergoing a ductile-type deformation. Brittle fracture also can occur in certain type of samples. Two main types of brittle failure can occur. In one case, fracture occurs just past the yield point. In the other, the specimen does not reach the yield point. The deformation process is time-dependent in that the quantitative character of the force–length curve depends on the

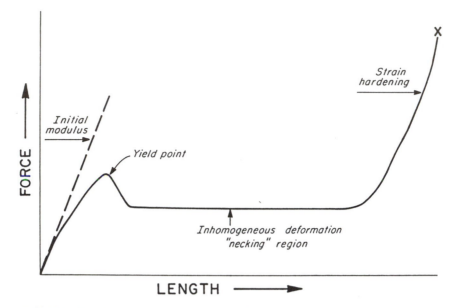

FIGURE 37. Schematic representation of force–length relation for semicrystalline polymers.

quantitative character of the force–length curve depends on the rate of strain. A major challenge is to explain the major characteristic of the deformation, as illustrated in Figure 37, in terms of molecular and structural factors.

Figure 37 gives an overall idealized view of deformation. In real situations, major variations can be observed, depending on the molecular weight, the structural regularity of the chain, and the values of the independent structural parameters (*41, 42*). As an example, Figure 38 shows the force–length curves for linear polyethylene samples with different molecular weights (*34*). For the lowest molecular weight shown, 1.75×10^5, the yielding process is very sharp. The character of the force–length curve generally resembles that shown in Figure 37, although the strain-hardening region is not as well defined. As the molecular weight increases, the yield region becomes more diffuse, and strain hardening becomes more clearly defined. Finally, for the highest molecular weight studied, 8×10^6, yielding is barely discernible, and strain hardening dominates the deformation. As will be demonstrated shortly, the degree of crystallinity establishes the yield stress and, thus, the stress level for the remainder of the deformation. The force–length curves of random copolymers and other structurally irregular chains show further differences from the curves in Figure 38 (*42*).

Among the characteristic features of the stress–strain curves, the initial modulus does not depend on a specific identifiable parameter (*41*). Rather, it depends on various factors, such as molecular weight, crystallinity level, and interlamellar thickness. In contrast to the initial modulus, the yield stress depends specifically on the crystallinity level. The following question needs to be addressed: What type of crystallinity is required, or put in another way, on the basis of our understanding of the phase structure, what is the role of the interfacial region? In Figure 39 (*42*) the yield stress is plotted against the density-derived level of crys-

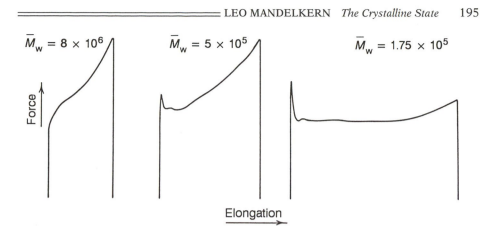

FIGURE 38. Plot of observed force–length relations for unfractionated linear polyethylenes. (Reproduced from reference 34. Copyright 1988 Elsevier.)

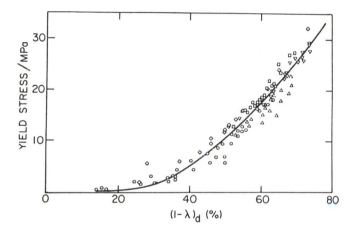

FIGURE 39. Compilation of literature data for yield stress as a function degree of crystallinity calculated on a density basis, $(1 - \lambda)_d$. (Reproduced with permission from reference 42. Copyright 1990 John Wiley and Sons, Inc.)

tallinity for an extensive sampling of data for random ethylene copolymers and long-chain branched polyethylenes. Despite the many different sources, the data fall within a narrow range. However, when the yield stress falls to zero, a significant amount of crystallinity (15–20%) still remains. A plot of the yield stress against the core crystallinity determined from an analysis of the Raman internal modes, for those samples for which data is available, is given in Figure 40 (*42*). In this case, the data give a single straight line that extrapolates very smoothly to the origin. This result is rather significant by itself. However, it becomes more significant when it is recognized that different copolymer types are involved. These include copolymers having 1-butene, 1-hexene, 1-octene, vinyl acetate, and methacrylic acid as counits. In addition, a wide range of branch contents, crystallinity levels, and supermolecular structures (from well-developed spherulites to random lamellar structures) are represented by the data plotted in this figure. Yet the yield stresses for all these structures fall on the same straight line.

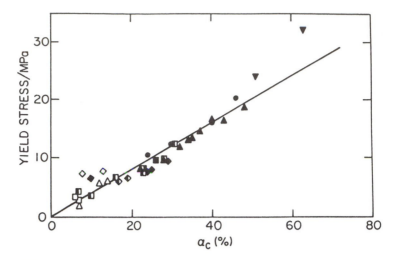

FIGURE 40. Yield stress versus core crystallinity, α_c, for random ethylene copolymers. Key: Ethylene–butene: ● a-type morphology; ethylene–hexene: ▼ a-type morphology; ethylene–octene: ▲ a-type morphology, △ h-type morphology; ethylene–vinyl acetate: ■ a-type morphology, ▣ b-type morphology, ▨ c-type morphology, □ h-type morphology; ethylene–methacrylic acid: ◆ a-type morphology, ◈ b-type morphology, ◇ c-type morphology, ◇ h-type morphology. (Reproduced with permission from reference 42. Copyright 1990 John Wiley and Sons, Inc.)

The data in Figures 39 and 40 demonstrate the strong dependence of the yield stress on the core crystallinity and suggest that yielding is governed by some process or processes involving the deformation or transformation of part or all of the crystalline region. Two distinctly different mechanisms have been suggested as being involved in yielding. One mechanism assumes that partial melting–recrystallization is involved in the deformation process. Partial melting involves the fusion of less perfect crystallites, which then recrystallize in the draw direction with a concomitant reduction in stress. The energy for partial melting comes from the overall stress being concentrated on the less perfect crystallites. A large temperature increase does not have to take place for partial melting to occur. The other mechanism assumes that yielding in crystalline polymers in general, and in polyethylene in particular, involves the thermal activation of screw dislocations with Burger's vectors that are parallel to the chain direction. Without going into details, this theory requires that the reduced yield stress (the yield stress divided by the core crystallinity) increases with the crystallite thickness. Figure 40 makes clear that the reduced yield stress is independent of crystallinity level. It can also be shown that the reduced yield stress also is independent of the crystallite thickness. Thus, by focusing on specific structural variables, it is possible to address and resolve an important problem. Small-angle neutron-scattering studies of mixtures of protonated and deuterated linear polyethylenes have shown that a significant amount of partial melting and recrystallization occurs during the complete deformation process and specifically during yielding. (For a more detailed discussion of this subject, *see* Chapter 7.)

Ultimate Properties

Finally, we consider the ultimate properties of crystalline polymers and focus attention on the draw ratio at break after a ductile deformation. We are interested in assessing the influence of molecular weight, supermolecular structure, degree of crystallinity, crystallite thickness, and structural irregularities on this property. The draw ratio at break, λ_B, for a given sample, will depend on the rate of deformation and the temperature. A plot of λ_B against the weight-average molecular weight, M_w, is given in Figure 41 (*34*) for a set of unfractionated linear polyethylenes at ambient temperature and a draw rate of 1 in./min. Chain length definitely affects λ_B. Under the conditions cited, λ_B varies from about 16 for the sample with the lowest molecular weight ($M_w = 5.3 \times 10^4$) to 2–3 for the sample with the highest molecular weight ($M_w = 8 \times 10^6$). The value of λ_B decreases continuously with molecular weight. Extrapolation of the present data indicates the virtual absence of deformation at much higher molecular weights, despite the fact that the crystallinity level would be extremely low. Results with molecular weight fractions of linear polyethylene fall on the same curve shown in Figure 41, as do results for binary mixtures of fractions. The values of λ_B for random copolymers lie below the curve for fractions of comparable molecular weights.

Sufficient data are now available to examine in detail the influence of independent structural variables on λ_B. For any given molecular weight, λ_B is independent of the supermolecular structure, including the possible

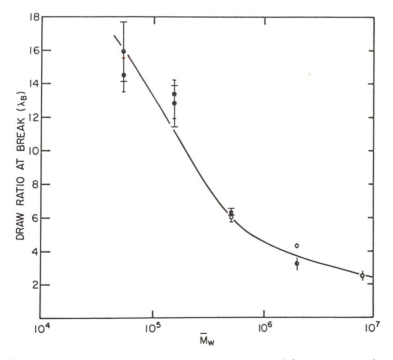

FIGURE 41. Draw ratio after break, λ_B, versus weight-average molecular weight, M_w, for unfractionated linear polyethylenes. (Reproduced with permission from reference 34. Copyright 1988 Elsevier.)

extremes. In the ductile region, λ_B is also independent of the level of crystallinity and the crystallite thickness of the initial isotropic sample. From the major structural factors that have been identified, only the weight-average molecular weight influences the ultimate draw ratio. The true ultimate tensile stress shows a very similar behavior.

The studies of ultimate properties indicate that the deformation of crystalline polymers cannot be understood in terms of crystallographic changes, no matter how attractive the analogy to low-molecular-weight solids may be. The fact that ultimate properties depend solely on M_w focuses attention on the interlamellar, isotropic region as being of prime importance for this phenomenon. The same structures that retard the development of crystallinity (*see* Figure 18) also influence the deformation process. Increasing the concentration of entanglements and other topological structures within the interlamellar region retards the rubber elastic type of deformation. Regions are developed where the stress can be concentrated so that failure will take place before large draw ratios can be attained; as the molecular weight decreases, the entanglement problem becomes less severe, and larger draw ratios can be attained.

Many aspects of the deformation of crystalline polymers have yet to be understood in terms of structure. Although a great deal of work remains to complete our comprehension of this subject, substantial progress has been made by focusing on independent structural variables and molecular constitution and by following the strategy outlined earlier.

Conclusions

The basic thermodynamic, kinetic, and structural principles that govern the crystallization behavior of polymers were developed by starting with an analysis of the conformation of a polymer chain in the liquid (amorphous) and crystalline (ordered) states. The quantitative description of crystallization kinetics and the thermodynamic analysis of the melting–crystallization process are generalized manifestations of the classical processes applicable to low-molecular-weight systems. Consequently, these two areas have reached a high level of maturity and are well-understood.

Because of the nonequilibrium character of the crystalline state in polymers and as a consequence of kinetic restraints to the crystallization process, both microscopic and macroscopic properties depend on specific structural and morphological features that are present. A set of independent structural variables at different hierarchies have been identified that, with the molecular constitution of the chain, determine properties. These structural variables can be determined experimentally. A striking feature of the structure of a crystalline polymer is the wide range of values that a given variable can attain by control of molecular constitution and crystallization conditions. By taking advantage of this feature, specific variables can be isolated, and their influence, singly or in conjunction with others, on a specific property can be assessed. By following this strategy, the spectroscopic, physical, and mechanical properties can and are being explained on the molecular level. This approach applies to virtually all properties of crystalline polymers.

References

1. Quinn, F. A., Jr.; Mandelkern, L. *J. Am. Chem. Soc.* **1961,** *83,* 2857.
2. Flory, P. J. *J. Chem. Phys.* **1949,** *17,* 223.
3. Mandelkern, L. "Crystallization and Melting of Polymers." In *Comprehensive Polymer Science;* Booth, C.; Price, C., Eds.; Pergamon: Oxford, England, 1989; Vol. 2.
4. Mandelkern, L. *Crystallization of Polymers;* McGraw-Hill: New York, 1964.
5. Flory, P. J. *Trans. Faraday Soc.* **1955,** *51,* 848.
6. Kenney, J. F. *Polym. Eng. Sci.* **1968,** *8,* 216.
7. Hachiboshi, M.; Fukuda, T.; Kobayashi, S. *J. Macromol. Sci. Phys.* **1960,** *35,* 94.
8. Mandelkern, L.; Tryon, M.; Quinn, F. A., Jr. *J. Polym. Sci.* **1956,** *19,* 77.
9. Newman, S. *J. Polym. Sci.* **1960,** *47,* 111.
10. Mandelkern, L.; Hellman, M.; Brown, D. W.; Roberts, E.; Quinn, F. A., Jr. *J. Am. Chem. Soc.* **1953,** *75,* 4093.
11. Avrami, M. *J. Chem. Phys.* **1939,** *7,* 1103; **1940,** *8,* 212; **1941,** *9,* 177.
12. Ergoz, E.; Fatou, J. G.; Mandelkern, L. *Macromolecules* **1972,** *5,* 147.
13. Wood, L. A.; Bekkedahl, N. *J. Appl. Phys.* **1946,** *17,* 362.
14. Wunderlich, B. *Macromolecular Physics;* Academic: New York, 1980; Vol. 3.
15. Suzuki, T.; Kovacs, A. J. *Polym. J.* **1970,** *1,* 82.
16. Hillig, W. B. *Acta Metall.* **1966,** *14,* 1868.
17. Lauritzen, J. I., Jr. *J. Appl. Phys.* **1973,** *44,* 4353.
18. Fatou, J. G.; Marco, C.; Mandelkern, L. *Polymer* **1990,** *31,* 1685.
19. Mandelkern, L. *Faraday Discuss. Chem. Soc.* **1979,** *68,* 310.
20. Flory, P. J. *J. Am. Chem. Soc.* **1962,** *84,* 2857.
21. Eppe, R.; Fischer, E. W.; Stuart, H. A. *J. Polym. Sci.* **1959,** *34,* 721.
22. Mandelkern, L. *Acc. Chem. Res.* **1990,** *23,* 380.
23. Alamo, R. G.; Mandelkern, L. *Macromolecules* **1989,** *22,* 1273.
24. Mandelkern, L. *J. Phys. Chem.* **1971,** *75,* 3920.
25. Mandelkern, L. *Polym. J.* **1985,** *17,* 337.
26. Kitamaru, R.; Mandelkern, L. *J. Polym. Sci. Part A-2* **1970,** *8,* 2079.
27. Mansfield, M. L. *Macromolecules* **1983,** *16,* 914.
28. Flory, P. J.; Yoon, D. Y.; Dill, K. A. *Macromolecules* **1984,** *17,* 862.
29. Marqusee, J. A.; Dill, K. A. *Macromolecules* **1986,** *19,* 2420.
30. Marqusee, J. A. *Macromolecules* **1989,** *22,* 472.
31. Kumar, S. K.; Yoon, D. Y. *Macromolecules* **1989,** *22,* 3458.
32. Zuniga, I.; Rodrigues, K.; Mattice, W. L. *Macromolecules* **1990,** *23,* 4108.
33. Mandelkern, L.; Alamo, R. G.; Kennedy, M. A. *Macromolecules* **1990,** *23,* 4721.
34. Mandelkern, L.; Peacock, A. J. *Studies in Physical and Theoretical Chemistry;* Lacher, R. C., Ed.; Elsevier: London, 1988; Vol. 54, p 201.
35. Mandelkern, L.; Glotin, M.; Benson, R. A. *Macromolecules* **1981,** *14,* 22.
36. Voigt-Martin, I. G.; Mandelkern, L. *J. Polym. Sci. Polym. Phys. Ed.* **1981,** *19,* 1769.
37. Voigt-Martin, I. G.; Fischer, E. W.; Mandelkern, L. *J. Polym. Sci., Polym. Phys. Ed.* **1980,** *18,* 2347.
38. Kline, D. E.; Sauer, J. A.; Woodward, A. E. *J. Polym. Sci.* **1956,** *22,* 455.
39. Popli, R.; Glotin, M.; Mandelkern, L.; Benson, R. S. *J. Polym. Sci. Polym. Phys. Ed.* **1984,** *22,* 407.
40. Axelson, D. E.; Mandelkern, L.; Popli, R.; Mathieu, P. *J. Polym. Sci. Polym. Phys. Ed.* **1983,** *21,* 2319.
41. Popli, R.; Mandelkern, L. *J. Polym. Sci. Polym. Phys. Ed.* **1987,** *25,* 441.
42. Peacock, A. J.; Mandelkern, L. *J. Polym. Sci. Polym. Phys. Ed.* **1990,** *28,* 1917.

Supplementary Reading

Mandelkern, L. *Crystallization of Polymers*; McGraw-Hill: New York, 1964.

Wunderlich, B. *Macromolecular Physics*; Academic: New York, 1980.

"Organization of Macromolecules in the Condensed Phase." *Faraday Discuss. Chem. Soc.* **1979,** *68.*

Magill, J. H. In *Treatise on Materials Science and Technology*; J. M. Schultz, Ed.; Academic: New York, 1977; Vol. 10, p 3.

Keller, A. *Rep. Prog. Phys.* **1968,** *31,* 623.

Mandelkern, L.; Jenkin, A. D., Eds. In *Progress in Polymer Science;* Pergamon: Oxford, England, 1970; Vol. 2, p 165.

Mandelkern, L. "Morphology of Semi-crystalline Polymers." In *Characterization of Materials in Research: Ceramics and Polymers;* Syracuse University Press: Syracuse, NY, 1975.

Flory, P. J. *J. Chem. Phys.* **1949,** *17,* 223.

Flory, P. J. *Trans. Faraday Soc.* **1955,** *51,* 848.

Flory, P. J. *J. Am. Chem. Soc.* **1962,** *84,* 2857.

Flory, P. J.; Vrij, A. *J. Am. Chem. Soc.* **1963,** *85,* 3548.

Flory, P. J.; Yoon, D. Y. *Nature (London)* **1978,** *272,* 226.

Maxfield, J.; Mandelkern, L. *Macromolecules* **1977,** *10,* 1141.

Mandelkern, L. *Acc. Chem. Res.* **1990,** *23,* 380.

Mandelkern, L. *Comprehensive Polymer Science;* Booth, R. C.; Price, C., Eds.; Pergamon: Oxford, England, 1989; Vol. 2.

McCrum, N. G.; Read, B. E.; Williams, G. *Anelastic and Dielectric Effects in Polymeric Solids;* John Wiley: New York, 1967.

Selected Works of Paul J. Flory; Mandelkern, L.; Mark, J. E.; Suter, U.; Yoon, D. Y., Eds.; Stanford University Press: Stanford, CA, 1985; Vol. 3.

Fatou, J. G. "Crystallization Kinetics." In *Encyclopedia of Polymer Science and Engineering, suppl.,* 2nd ed.; John Wiley: New York, 1989.

Chapter 5

The Mesomorphic State

Edward T. Samulski

Department of Chemistry, University of North Carolina, Chapel Hill, NC 27514–3290

The term "mesomorphism" (exhibiting an intermediate form) is generally reserved for spontaneously ordered fluids known as liquid crystals. Discovered in 1888 and extensively studied in the early 1900s, liquid crystals essentially remained a laboratory curiosity until the 1960s, when prototypes of the now-commonplace liquid crystal display (LCD) were first demonstrated and other electrooptic applications for these unusual melts were initiated. During this same period of renewed interest in liquid crystals, the correlation between mechanical properties of polymers and mesomorphism was recognized in ultrahigh-strength synthetic poly(arylamide) fibers that were spun from liquid-crystalline polymer solutions (e.g., DuPont's Kevlar and Akzo's Twaron). Now macromolecular-design strategies for synthesizing new high-performance polymers and paradigms for modeling polymer processing must consider the potential role of this relatively recently discovered mesomorphic state in polymers. This chapter will try to develop a qualitative yet comprehensive understanding of this state and thereby provide the context for appreciating how mesomorphism impacts polymer chemistry and physics.

The essential features of the mesomorphic state may be realized in the two ways that ordinary fluid phases are formed from solids: dissolution and fusion. The two corresponding categories of mesomorphism are lyotropism (liquid-crystalline solutions) and thermotropism (melts), respectively. Thermotropic materials consist of single-component substances (simple organic molecules) and include the large variety of low-molar-mass mesogens used in LCDs. More recently, thermotropic specialty polymers have been commercialized; for the most part, these specialty polymers are polyesters (e.g., Hoechst Celanese's Vectra and Amoco's Xydar). Lyotropic

2505–2/93/0201 $15.00/1

polymers, on the other hand, are multicomponent mixtures (solute plus solvent). For low-molar-mass liquid crystals, lyotropism requires specific solute–solvent interactions [e.g., hydrophobic–hydrophilic interactions in amphiphile (soap)–water mixtures] to assemble solute aggregates (micelles). At high solute concentrations, anisometric aggregates with high aspects ratios, L/d (length/diameter), will in turn organize in the excess solvent medium to give fluid, orientationally ordered, supra-aggregate arrangements (cubic, hexagonal, lamellar, etc.). For lyotropic polymer mesophases, specific solute–solvent interactions are not necessary (other than those needed to solubilize the polymer). The most significant class of lyotropic polymers are those with a rodlike secondary structure. Solutions of rigid and high-aspect-ratio macromolecules will spontaneously order above some critical polymer concentration, ϕ, that merely depends on geometry (L/d); excluded-volume interactions among the rodlike polymers simply force the adoption of long-range, quasiparallel organization of discrete, mobile rods in the fluid solvent continuum.

The phenomenon of spontaneously ordered macromolecular solutions was first observed in the 1930s: Solutions of the rodlike virus particles tobacco mosaic virus (TMV) exhibited spontaneous birefringence above some critical volume fraction of TMV. And curiously, despite the dominance of thermotropic systems in both experimental and theoretical liquid-crystal research activity prior to 1950, the first valid theoretical molecular modeling of the phenomenon of liquid-crystal formation, that is, modeling the disorder–order transformation in a fluid phase, was developed in the late 1940s by Onsager (*see Onsager's Virial Expansion*) to describe this rather esoteric, lyotropic TMV solution. Subsequent theories of this disorder–order transition, as well as current extensions of the Onsager model, will be reviewed briefly. The implications of mesomorphism on the viscoelastic behavior of fluid phases of polymers, as evident from theory and experiment, also will be considered. This aspect of polymer mesomorphism is especially important, because the rheological behavior of polymer fluids is intimately related to the morphology that forms in the solid state and in turn determines the ultimate bulk properties of the polymer.

In certain high-molar-mass materials (e.g., deformed elastomers, amorphous regions in semicrystalline polymers, and phase-separated block copolymers) some characteristics of the mesomorphic state are observed. In many instances, researchers have tried to describe the deviation from isotropy in these materials with the vocabulary of mesomorphism. Indiscriminate applications of terminology is confusing, however, and leads to misconceptions about both the nature of liquid crystallinity and the nature of order in amorphous polymers. A consistent manner of describing these superficially related materials is lacking; when encountered here, the related characteristics of these nonmesogenic materials will be put into a proper perspective.

References 1–20 represent a reverse chronology of books and substantive reviews intended to expedite searching the literature for more specialized topics and in-depth treatments of the mesomorphic state in general and especially as it relates to polymers.

General Concepts

Definitions and Terminology

In this chapter, the descriptors "mesogen" and "liquid crystal" (LC) may be interchanged when referring to a molecule that exhibits mesomorphism. The abbreviation mLC represents both low-molar-mass liquid crystal and monomer liquid crystal; PLC and LCP will differentiate between meso-genic polymers synthesized from mLCs and those prepared from conventional (commercial) monomers, respectively.

Initially the qualitative features of low-molar-mass liquid crystals will be considered in a manner that facilitates the transfer of the underlying physics and characteristics of these materials to macromolecular systems. Generally, the focus will not be on differences between low-molar-mass and macromolecular mesogens that derive from idiosyncratic chemical origins, that is, differences stemming from the primary atomic constitution of the mesogen. Rather, the focus throughout will be only on the general features of mesomorphism. In keeping with this goal, for both polymerized liquid crystals (PLCs) and liquid-crystalline polymers (LCPs) derived from conventional (e.g., aromatic ester and amide) monomers, only two classes of primary structural topologies need to be examined: linear (main-chain) polymers and side-chain (comblike) polymers.

To appreciate the mesomorphic state in polymers the subtleties of long-range molecular organization in fluid phases must be understood. Mesomorphism in low-molar-mass materials will be examined first. This examination, in turn, necessitates the identification of variables that quantify translational and orientational order in fluid states. To introduce these variables in a systematic way from a familiar frame of reference, the molecular structure of liquid crystals will be reviewed briefly.

Figure 1 illustrates molecular primary structures, associated schematic secondary structures, and idealized shapes for representative thermotropic mLCs, that is, organic molecules that melt into ordered fluid phases. The mesogenic core is that primitive (central) segment of the mesogen (usually composed of aromatic rings) possessing the requisite excluded-volume interactions (anisometric shape) for liquid crystallinity. The flexible tails (generally hydrocarbon chains) that terminate the rigid mesogenic core facilitate the transformation from the solid state to the fluid LC phase; the flexible tails lower the melting temperature of the crystal by weakening (diluting) attractive intermolecular interactions between rigid mesogenic cores in the solid state. For pedagogical purposes, the secondary structure of mLCs may be idealized to prolate or oblate ellipsoids of revolution ("calamitic" and "discotic" mesogens, respectively) with symmetry axes denoted by **l** (Figure 1; for prolate mesogens, **l** is usually referred to as the molecular long axis). With this low-resolution depiction of the mesogen shape, the common types of supramolecular organizations found in mesophases composed of mLCs can be described.

Figure 2 exaggerates the kinds of organizations found in fluid mesophases of calamitic mLCs. The nematic phase is the most common mesophase. It is a fluid with cylindrical symmetry and having the mesogen

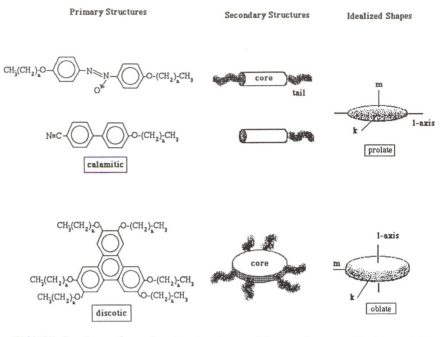

FIGURE 1. Example molecular structures of low-molar-mass liquid crystals. From left to right: primary chemical constitution, low-resolution secondary shapes, and idealized shapes for calamitic (prolate, lathelike, or rodlike) and discotic (oblate or disklike) mesogens.

centers of mass randomly located (translational disorder), but the mesogen long axis **l** is, on average, parallel to the fluid symmetry or optical axis, the director **n**. Local, uniaxial molecular orientational order also is found in the common smectic phases, but in addition, smectic phases exhibit some translational ordering; that is, the mesogens show a tendency to stratify into layers. (This layering also is exaggerated in Figure 2; the stratification in many smectics is detectable only via diffraction techniques that are sensitive to subtle periodic changes in the electron density.) Despite translational segregation, fluidity exists; that is, molecules may move more readily within layers, and to varying degrees, translation between layers also is possible. A large variety (almost a continuum) of smectic organizations exists. The simplest, illustrated in Figure 2b, is a smectic A (S_A) phase, with the molecular **l** axes, on average, normal to the layer plane; that is, the local director **n** is perpendicular to the layers. Tilted smectics also are common; the smectic C (S_C) phase (Figure 2c) has the molecular **l** axes, on average, tilted with respect to the layer normal; hence, the local director **n** makes an angle, α, with respect to the layer normal. The direction of the tilt (the orientation of the projection, $\sin\alpha$, of **n** in the smectic plane) is generally conserved from layer to layer within a macroscopic volume element in the mesophase. There are corresponding normal and tilted smectic phases with antiferroelectric organization (canceling polar orientations in neighboring layers; S_{A_2} and S_{C_2}), smectics with pronounced interdigitation (layers composed of dimers of associated mesogen pairs; S_{A_d} and S_{C_d}), and smectics with mesogens showing both

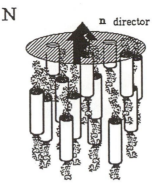

(a) **Nematic**

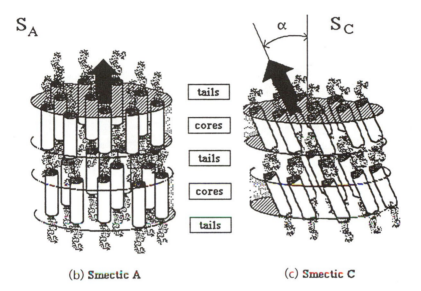

(b) **Smectic A**   (c) **Smectic C**

FIGURE 2. Schematic idealized pictures of calamitic mesogen supramolecular organization: (a) uniaxial nematic (N), (b) uniaxial smectic A (S_A), and (c) biaxial smectic C (S_C with average tilt α).

long-range layer definition plus lateral packing preferences within the layer (S_B) (*21*).

The related supramolecular arrangements of discotic mesogens are as follows: the uniaxial nematic phase (D_N) having the molecular symmetry **l** axes aligned parallel to **n** (Figure 3a); the uniaxial, hexagonally arranged columns with ordered (D_{ho}) and disordered (D_{hd}) stacking of the disklike mesogens in the columns (Figures 3b and 3c, respectively); and analogous to the tilted calamitic smectics, the biaxial arrangements (e.g., D_{obd}) with a translationally disordered stack of disklike molecules making an oblique angle with the column axis.

In the calamitic mesophases, the molecular arrangements with cubic symmetry (S_D) also exist between more conventional phases (i.e., S_C and S_A phases) wherein the S_D bulk properties obviously would appear to be isotropic, just as expected for an ordinary liquid. In the S_B phase, the

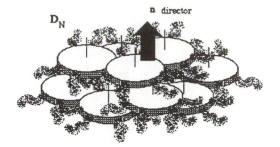

(a) **Discotic Nematic**

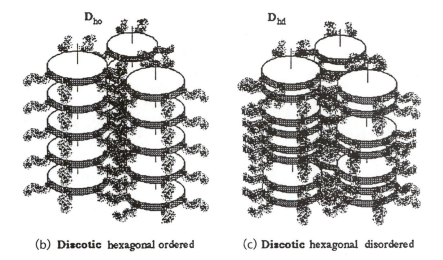

(b) **Discotic hexagonal ordered**   (c) **Discotic hexagonal disordered**

FIGURE 3. Schematic pictures of discotic mesogen uniaxial supramolecular organizations: (a) nematic (D_N), (b) hexagonal ordered (D_{ho}), and (c) hexagonal disordered (D_{hd}).

molecular organization really begs the question: How is a liquid crystal distinguished from a crystal? The focus of this chapter remains on truly fluid phases. Properties that confirm the existence of long-range molecular orientational order (and less than three-dimensional translational ordering) in fluids, the mesomorphic state, are considered next.

Birefringent Fluids

The transmission of visible light by substances placed between crossed polars (a linear polarizer–analyzer pair with principal dichroic axes at right angles) is observed in crystals having direction-dependent refractive indices, that is, birefringence, and is a convenient indicator of long-range molecular orientational order. Birefringence in a homogeneous fluid melt (or solution) composed of dispersed molecules (or aggregates) is one key signature of thermotropic (or lyotropic) mesomorphism. This unique property of thermotropic mesophases might best be understood by recalling the properties and the organization (structure) in the solid state of a molecular crystal. Certain space group symmetries excepted (e.g., crystals with cubic symmetry), most molecular crystals exhibit anisotropic (i.e., direction dependent, relative to the crystallographic axes) physical properties, such as thermal expansivity, refraction of visible light, dichroism (visible, UV, and IR), and magnetic and dielectric susceptibilities. This macroscopic anisotropy has its origins in the arrangement (structure) of the molecules in the crystal, which, in turn, is built on particular molecular juxtapositioning (i.e., relative placements and orientations of molecules in the unit cell) that may amplify intrinsic molecular anisotropy. For example, the refractive index of a crystal is ultimately related to (anisotropic) molecular electronic polarizability. Roughly speaking, the molecular polarizability is additive and collectively anisotropic if all molecules have the same orientation in the crystal. Rotation of plane-polarized light by birefringent crystals is one dramatic macroscopic indication of molecular-based anisotropy. Birefringence is conspicuously absent in ordinary liquids or other isotropic media (glasses), because there is no long-range structure (molecular organization extending over distances comparable with the wavelength of visible light, $\sim 10^{-6}$ m) to manifest the anisotropy of the molecules in these materials. How nematic order manifests molecular anisotropies will be considered in more detail later.

Thermodynamic Properties

On heating a molecular crystal composed of mLCs, very rapid, restricted (low-amplitude) thermal motions gradually increase in amplitude up to the melting point (T_m). Discrete reorganizations of the solid may take place before T_m is reached, for example, a conformational change in the mesogenic core or the population of a *gauche* rotational isomer in an alkyl tail or a shift in the mesogen-packing motif; these reorganizations are called solid-state transitions. At T_m, the long-range translational and orientational order in the crystal collapses abruptly, and in the resulting fluid phase, the molecules interact with one another via motionally averaged intermolecular forces. In fluids composed of anisometric calamitic molecules with aspect ratios, L/d of ≥ 3 (d/L of ≥ 3 for oblate discotic mesogens), the average of dispersion forces over (anisotropic) intermolecular excluded-volume interactions results in effective, anisotropic, attractive forces (*see Theories of Mesomorphism*). In this situation, a delicate balance is established between the residual anisotropy of these forces in the fluid and the thermal energy. Sometimes this balance yields a range of temperature in which long-range orientational order persists in the fluid (i.e., mesomorphism).

At the T_m, a first-order phase transition from the crystal to the mesophase takes place with the usual discontinuities in the extensive properties (volume, entropy, etc.). Figure 4 shows a hypothetical differential scanning calorimetry (DSC) trace and the plot of sample volume versus temperature for an ideal nematic. The values for the enthalpy change (ΔH, ~ 45 kJ·mol^{-1}) and volume change (ΔV) at T_m are typical of those changes in extensive properties on melting of ordinary organic molecular

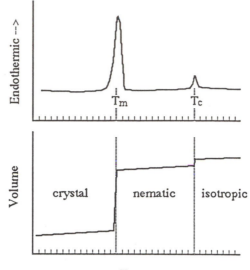

FIGURE 4. Hypothetical endothermic transitions (DSC trace) and volume changes that occur on melting a molecular crystal into a nematic mesophase (T_m) and on melting the nematic phase into the isotropic liquid at the clearing temperature, T_c.

crystals. However, if the melt is heated further, a second transition to the transparent isotropic state occurs at T_c, the "clearing" temperature; nematic melts appear milky and opalescent, because thermal energy excites fluctuations of the refractive index that result in light scattering (*see Anisotropic Properties*). At T_c, the magnitudes of ΔH and ΔV associated with the nematic-to-isotropic transition (N ↔ I) are much less than those observed at T_m but, nevertheless, indicative of a first-order phase transition. The very small value of ΔH and the discontinuity in ΔV imply that the differences between the fluid phases on either side of T_c are only very subtle; that is, despite the apparently dramatic changes in macroscopic properties at the N ↔ I transition (light scattering, birefringence, etc.), the very small thermodynamic changes observed at T_c suggest that a nematic mesophase is a homogeneous fluid with molecular motion and molecular organization very close to that in an ordinary isotropic liquid. Hence, contrary to the exaggerated features of the nematic organization depicted in Figure 2a, a more realistic sketch stressing the subtle differences between mesophases and the isotropic states should be considered. In actuality, the differences between the isotropic liquid and the nematic are very difficult to discern at the molecular level (the sketch on page 209 exaggerates the local packing differences). Likewise, the stratification characteristic of the S_A phase is very exaggerated in Figure 2; it is more aptly represented by the sketches in Figure 5, which emphasize the similarity between the uniaxial N and S_A phases.

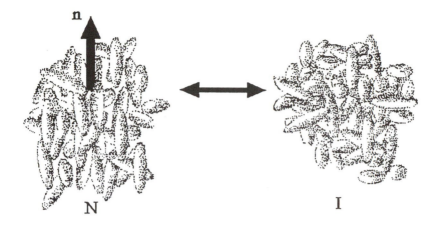

For some mesogens, more than one type of liquid-crystalline molecular organization is exhibited as the temperature is changed; multimesophase formation is called polymorphism. When the transitions are reversible, they are called enantiotropic transitions. Enantiotropic polymorphism is observed for bis(*p*-heptyloxyphenyl) terephthalate (**1**).

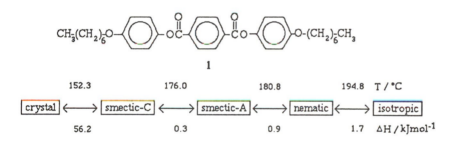

The experimentally observed transition temperatures (°C) and enthalpy changes (kJ · mol^{-1}) for **1** are indicated on the transition map shown below the molecular formula. Mesogen **1** exhibits two smectics and a nematic phase; the more ordered S_C phase occurs at the lower temperature followed by increasingly less ordered mesophases at higher temperatures (S_A and N). On heating of **1**, the crystal melts to an S_C phase, which in turn melts into an S_A phase; the delicate organizational differences between these two smectic phases, which differ only by the average molecular tilt (Figures 2b and 2c), is reflected in the very small enthalpy change (0.3 kJ · mol^{-1}) associated with the $S_C \leftrightarrow S_A$ transition. At a higher temperature, the nematic phase forms, and eventually, it melts into the isotropic liquid.

Monotropic transitions (mesophases encountered on heating only or on cooling only) also are frequently encountered. This phenomenon is exhibited by bis(*p*-heptyloxyphenyl) 2,5-thiophenedicarboxylate (**2**).

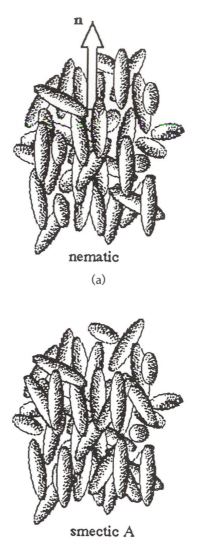

nematic

(a)

smectic A

(b)

FIGURE 5. Molecular organization: (a) in the nematic phase, a translation-ally uniform distribution of molecular centers of mass is found along the director and (b) in a smectic phase, a Fourier analysis of the centers of mass distribution along n exhibits a fundamental frequency component with a wavelength approximately equal to the molecular length. This subtle tendency toward stratification is frequently exaggerated in diagrams (e.g., Figure 2) to give the false impression of discrete, delineated layers in the fluid smectic phases (S_A and S_C).

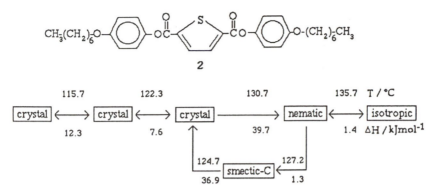

2

	115.7		122.3		130.7		135.7 T / °C

crystal ⟷ crystal ⟷ crystal ⟶ nematic ⟷ isotropic

12.3 7.6   39.7   1.4 ΔH / kJmol⁻¹

124.7 ⟶ smectic-C ⟵ 127.2

36.9   1.3

Mesogen **2** shows an enantiotropic N ↔ I transition at 135.7 °C and a monotropic N ↔ S_C transition on cooling at 127.2 °C. Monotropism occurs, for example, when particular intermolecular interactions are present in the crystal (hydrogen bonds, dipole–dipole interactions, etc.) that make T_m greater than any transition temperatures for the mesophase, because such interactions need to be disrupted before the crystal melts. On cooling, however, a mesophase might appear and remain stable at temperatures lower than T_m (before crystallization occurs). In this instance, the monotropic mesophase is supercooled 6 °C below the crystal melting transition (130.7 °C). In polymers, supercooling and the intervention of a glass transition are common (*see* Chapter 4). (In fact, glass formation is one way to get a solid replica of the molecular organization in the mesophase and may have practical technological implications, especially in optical applications where the mLC is a host matrix for a guest molecule with special orientation-dependent optical characteristics.) Mesogen **2** also shows two solid-state transitions between three distinct crystal phases at 115.7 and 122.3 °C with relatively large enthalpy changes (12.3 and 7.6 kJ · mol⁻¹, respectively.)

Mesophase Textures

The various types of molecular organization in mesophases may be identified by the texture, that is, the pattern of light and dark morphological features superposed on the (colored) birefringent field when the mesophase is observed with a polarizing microscope. For example, Figure 6 shows a typical *schlieren* texture adopted by the nematic phase of mesogen **2**; the two- and four-armed brushes, intersecting pattern of dark bands, arise from disclinations (the analogue of dislocations in a crystal) when the director changes orientation abruptly in the mesophase (*see Disclinations*). The texture of **2** changes dramatically at the N ↔ S_C transition (Figure 7; in this example, the S_C phase was identified by examining its miscibility with a known S_C-forming mesogen; *22*). Figure 8 shows the S_A focal conic texture of mesogen **1**; the dark region at the top of the figure is an example of the *homeotropic alignment* mode of this uniaxial phase having the director anchored normal to the glass microscope slide (and cover slip) and coincident with the viewing direction. In the homeotropic texture, the optically uniaxial mesophase does not rotate the polarization of the incident light, and consequently, this region appears dark, because no light passes through the orthogonal analyzer. (In a *planar alignment* mode, the director would be anchored parallel to the slide.) However, when the mesophase temperature is lowered into the biaxial S_C phase, the director tilts relative to the viewing direction, and a birefringent field with a broken focal conic texture (Figure 9) appears in the formerly dark regions of Figure 8. With some practice and the aid of published photomicrographs of mesophase textures, it is possible to recognize features that are specific to nematic and smectic phases and, moreover, to differentiate among various smectics.

Mesogen Molecular Structure

Significant differences in mesogen supramolecular organization (nematic, smectic, etc.) and stability (temperature range in which mesomorphism is exhibited) may be caused by apparently small chemical changes in the mesogen primary structure (e.g., by substituting a halogen for a hydrogen atom). However, such a high-resolution chemical description of mesomorphism is beyond the scope of this chapter (readers should refer to the numerous tabulations of correlations between mesogen chemical structure and LC phase types; *23*). For a lower resolution description of the phenomenon of liquid crystallinity, the focus is on generalized secondary structure and its influence in promoting specific kinds of organizations. In fact, as will be demonstrated, many of the physical attributes of liquid crystals can be understood in terms of the idealized shapes of mesogens given in Figure 1, without reference to the primary chemical composition of the mesogen at all.

In actuality, molecular structural features cannot be avoided totally. A particularly instructive case in point arises when attempts are made to answer the following question: How far can the calamitic mesogenic core be distorted from the shape of a prolate ellipsoid? Mesogen **1** is transformed to mesogen **2** by replacing the 1,4-phenylene ring in mesogen **1** with the 2,5-thiophene ring, which introduces a bend into the mesogenic core. The bend derives from the geometry of the thiophene ring: An ~150° angle is subtended by the 2,5-exocylic bonds in **2**. Correspondingly, the mesophase stability (temperature range) of compound **2** is less than that of compound **1**; the nematic range is only 5 °C with $T_c = 135.7$ °C for **2**, whereas for **1**, the nematic range is 14 °C and is stable up to a higher temperature, $T_c = 194.8$ °C. Molecules in which the 1,4-phenylene of **1** is replaced by 1,3-phenylene are not mesomorphic at all. The 120° bend in the core of the 1,3-derivative precludes the kinds of molecular packing required for a stable mesophase.

        1   **2**

Another example of the relevance of molecular structure is encountered when the following question is asked: Why do smectic phases form? No single answer exists for this deceptively simple question. A contempo-

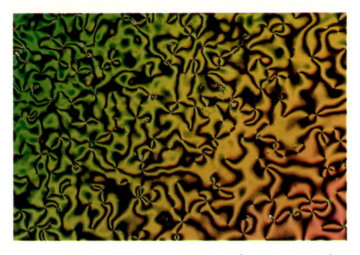

FIGURE 6.   A nematic *schlieren* texture of bis(heptyloxyphenyl) 2,5-thiophenedicarboxylate at 135 °C (crossed polars at 200×). (Reproduced with permission from reference 22. Copyright 1991 Taylor and Francis.)

FIGURE 7.   A focal conic texture of the monotropic smectic C phase of bis(heptyloxyphenyl) 2,5-thiophenedicarboxylate at 128 °C (crossed polars at 100×). (Reproduced with permission from reference 22. Copyright 1991 Taylor and Francis.)

rary reply based on excluded-volume considerations alone (the idealized ellipsoidal shapes in Figure 1) would stress a mechanism of smectic-phase stabilization that is based on increased translational freedom (translational entropy) when the mesogens condense into layers. That is, relative to the layered smectic, lateral diffusion is impeded in the nematic because of the randomly disposed mesogen centers of mass. More traditional replies implicate secondary molecular structural features (Figures 1 and 2). Consider the bis(*p*-alkyloxyphenyl) terephthalate homologous series, that is, a fixed mesogenic core with a systematic progression of the

FIGURE 8. A focal conic texture and homeotropic aligned regions (dark) of the smectic A phase of bis(heptyloxyphenyl) terephthalate at 179 °C (crossed polars at 100×). (Reproduced with permission from reference 22. Copyright 1991 Taylor and Francis.)

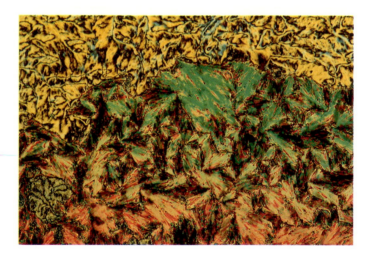

FIGURE 9. Broken focal conic fan texture of the smectic C phase of bis(heptyloxyphenyl) terephthalate at 172 °C (same field of view as Figure 8; crossed polars at 100×). (Reproduced with permission from reference 22. Copyright 1991 Taylor and Francis.)

terminal alkyl chain length y, where $y = 7$ corresponds to the mesogen **1** considered earlier. The experimental transition temperatures (°C) for the members of the series are plotted versus chain length y on the phase diagram in Figure 10. Within the homologous series, the more ordered S_A phase begins to displace the nematic phase at increased chain length ($y \geq 5$). At longer chain lengths ($y \geq 6$), the slighty more ordered (tilted) S_C phase grows in. It seems that nanophase separation tends to occur with increasing chain length so that molecular cores reside preferentially next to neighboring cores and separated from a chain-rich stratum to give the exaggerated alternating pattern of ... tails–cores–tails–cores... shown in

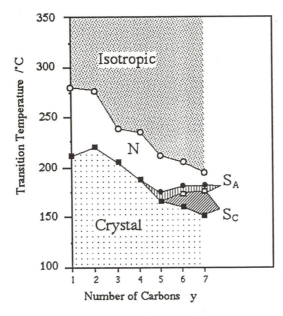

FIGURE 10. Phase diagram showing the transition temperatures (°C) versus the alkyl tail length (number of carbons, *y*) for the homologous series of bis(*p*-alkyloxyphenyl) terephthalates.

Figures 2b and 2c. This hypothesis of smectic phase formation based on secondary molecular structure is reinforced by thermodynamic considerations. Chemically similar parts of the mesogens associate in the smectic organization, and furthermore, such segregation may be enhanced by the increased entropy that would accompany the disordered pendant chains when they are relegated to an aliphatic milieu.

Mechanisms of smectic phase formation based on primary molecular structure also exist. Putative correlations exist between smectic phase formation and the location of permanent electric dipoles on the mesogenic core. Additionally in the case of a PLC, mesogenic cores covalently incorporating into polymers may restrict intercore juxtapositioning and thereby stabilize smectic phases. Although mesogen primary and secondary molecular structures potentially have an important role to play, the remainder of this chapter, however, will focus on aspects of mesomorphism that for the most part can be described in terms of the idealized prolate (or oblate) shape of calamitic (or discotic) mesogens (Figure 1). The next section considers the nature and the implications of the local molecular order in the nematic state after a brief review of the nature of molecular crystals and isotropic liquids.

Monomer Liquid Crystals (mLCs)

Molecular Crystals

The perfect order in a molecular crystal allows probing of its structural features (i.e., molecular organization) with X-ray diffraction. The regular, periodic variation in electron density diffracts X-rays and enables the reconstruction of the relative positions and orientations of molecules in the crystal unit cell. Consider the packing of the idealized shapes of calamitic low-molar-mass mesogens. The relation of the coarse features of the crystal structure to the scattered X-ray intensity is schematically illustrated in Figure 11. In the figure, a microscopic fragment of a crystal composed of prolate molecules having the **l** axes parallel to the crystal **c** axis is shown. Incident X-rays (along the **b** axis) will be diffracted when the Bragg condition, $n\lambda = 2d \sin \theta$ is met (*see* Chapter 7). In the idealized diffraction pattern shown in the figure, two sets of diffraction spots are indicated. Those along the meridian (vertical direction, parallel to the **c** axis) correspond to multiple-order diffraction ($n = 1, 2, 3 \ldots$) with a spacing that is reciprocally related to the molecular length, L. The larger spaced diffraction maxima on the equator (horizontal direction) correspond to the smaller, regular lateral spacing, D, between molecules (along the **a** axis). The absence of azimuthal diffraction intensity (well-defined spots rather than arcs of intensity along χ) indicates perfect orientational order within the crystal, that is, **l**∥**c**. This schematic diffraction pattern may be used as a benchmark for characterizing structure (molecular translational and orientational order) in mesophases. However, to place the molecular organization present in the mesomorphic state into a more general context, it is instructive to review the nature of structure in the state of complete disorder at the other extreme of condensed matter, the ordinary molecular liquid state.

Molecular Liquids

The relative positions of molecules in a liquid may be characterized by the pair distribution function $g(R)$, where $g(R)dR$ is the probability of finding the center of mass of a second molecule within the range dR at a

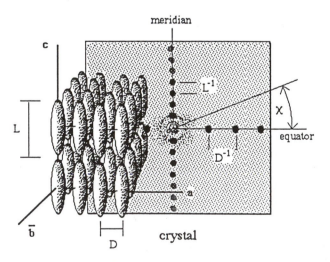

FIGURE 11. Schematic molecular crystal of calamitic molecules and the associated idealized diffraction pattern for incident X-rays parallel to **b̄**. The intermolecular distances, L (approximate length) and D (lateral spacing), show diffraction spots on the meridian of the molecule and equator, respectively; the diffraction intensity along χ shows no appreciable azimuthal spread.

distance R from a given molecule (independent of the direction of R). The pair distribution function can be measured experimentally by X-ray diffraction. Analogous to the way in which the precise and regular periodic electron density associated with a molecular crystal diffracts X-rays according to the Bragg condition, specific diffraction phenomena are characteristic of the "structure" in liquids. For ordinary liquids, it is common to refer to "liquid structure," the short-range biasing of relative (average) orientations and separations of neighboring molecules originating primarily from excluded-volume (local-packing) conditions (*24*). This local-packing anisotropy is more pronounced in liquids composed of anisometric molecular shapes (rods or disks). However, the persistence of such packing correlations (i.e., structure in the radial distribution function) is limited to a few molecular diameters (~ 1 nm). On larger distance scales, molecular orientations and positions are random; the dynamically or ensemble-averaged properties of a molecular liquid are isotropic. Consequently, the diffracted X-ray intensity is very diffuse and exhibits a broad intensity maximum located radially at $\theta = \sin^{-1}(n\lambda/2\langle d\rangle)$, where $\langle d\rangle$ corresponds to an average intermolecular distance in the liquid. The uniform azimuthal intensity distribution, $I(\chi)$ (Figure 12a; the circular diffraction pattern), indicates that the liquid has no preferred orientational order; the figure also depicts an instantaneous "snapshot" of a microscopic volume element of a calamitic fluid taken with an extremely fast "shutter speed" ($<10^{-10}$ s) to freeze molecular reorientation and translation. The half-width of the radial intensity distribution is inversely related to the distance over which the molecules are positionally ordered. Absence of significant higher order diffraction intensity ($n > 1$) is evidence of the very short range nature of the local structure of liquids. The preceding brief characterizations of molecular crystals and isotropic liquids may now be used as a benchmark to contrast their diffraction features with those of monomer liquid crystals.

Nematic Liquid Crystal

Figure 12b shows the diffraction pattern from an idealized mLC, an aligned nematic phase ("aligned nematic" implies that the director **n** has the same orientation throughout the diffracting volume element, ~ 1 mm^3). By referring to the crystal and liquid diffraction patterns (Figures 11 and 12a), the "structure" in the liquid crystal can be inferred. Analysis shows that significant long-range positional order is absent [no high-order ($n > 1$) diffraction spots]; in fact, the first-order ($n = 1$) reflection along the meridian is just visible. Its reciprocal spacing corresponds to the approximate length of the mesogen, whereas the diffraction maximum along the equator indicates the nominal lateral intermolecular distance. These features of the diffraction pattern are reminiscent of a liquid, with one very significant difference: the azimuthal intensity is not evenly distributed over χ. Closer examination of the distribution, $I(\chi)$, leads to the conclusion that, although the translational "structure" is liquidlike, on average, the molecules are aligned along a direction parallel to the meridian (i.e., parallel to **c**). That is, this otherwise normal fluid exhibits mesogen orientational order about a preferred direction in the fluid, called the director and symbolized by **n**, an apolar vector. Moreover this orientational ordering is long-range,

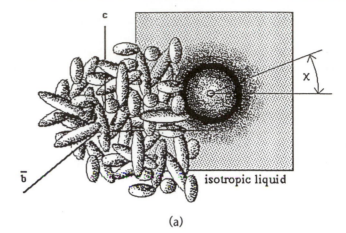

(a)

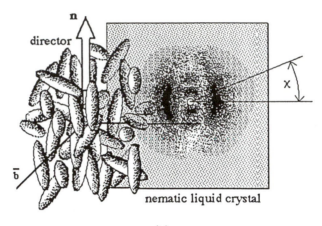

(b)

FIGURE 12. Diffraction patterns from fluid phases: (a) the isotropic liquid shows a uniform azimuthal X-ray diffraction intensity distribution, and (b) the nematic aligned with the director **n** along the vertical direction shows remnants of local intermolecular order in the form of diffraction intensity maxima corresponding to the molecular length (along the meridian) and the lateral intermolecular spacing (along the equator).

uniform, and coherent over the entire diffracting volume element of the nematic fluid. If the volume element is of the order of 1 mm^3 and the mesogen volume is of the order of 100 Å^3, then this orientational order extends over ~10^{19} molecules! What kinds of intermolecular interactions are responsible for this long-range orientational order in an otherwise pure liquid state? To answer this question, the static features of the nematic apparent in the diffraction pattern shown in Figure 12b must be analyzed first.

Order Parameter

The origins of the diffracted azimuthal X-ray intensity distribution [the arcs at fixed Bragg angle in $I(\chi)$; Figure 12b] will now be considered in detail. The scattering intensity, $I(\omega)$, from a single prolate mesogen of length L is a thin line (for large L) with a negligible intrinsic angular width, $\lambda/(L \sin \omega)$, where λ is the wavelength of the X-rays and ω is the angle between $\mathbf{l}$ and the incident-beam direction (25). The observed $I(\chi)$ comes from a superposition of the scattering from many mesogens having a continuous orientation distribution of $\mathbf{l}$ about the director $\mathbf{n}, W(\beta)$, where β is the angle between $\mathbf{l}$ and $\mathbf{n}$. The distribution $I(\chi)$ is related to $I(\omega)$ and

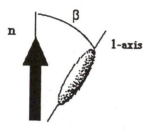

$W(\beta)$ via an integral equation (which must be solved numerically):

$$I(\chi) \cong \int W(\beta) \cdot I(\omega) \sin \omega \, d\omega \qquad (1)$$

The molecular quantity of interest, the average orientation of $\mathbf{l}$ relative to $\mathbf{n}$ (the nematic order parameter S) is defined in terms of $W(\beta)$:

$$S \equiv \int_0^{\pi/2} P_2(\cos \beta) \cdot W(\beta) \sin \beta \, d\beta \qquad (2)$$

$W(\beta) \sin \beta \, d\beta$ is the (normalized) probability of finding $\mathbf{l}$ in the range $d\beta$ about the direction β with respect to the director; $W(\beta)$ is independent of β in a normal isotropic liquid. In equation 2, the order parameter, S [the average of the second Legendre polynomial, $P_2(\cos \beta) = (3 \cos^2 \beta - 1)/2$] assumes the value unity in a perfectly ordered system (when $\mathbf{l} \| \mathbf{n}$, as in the molecular crystal idealized in Figure 11), and the value zero when $\mathbf{l}$ is isotropically distributed (Figure 12a). To extract S from the experimentally observable $I(\chi)$, equation 1 can be used (by recognizing that $\cos \beta = \cos \chi \sin \omega$) with an assumed form for $W(\beta)$ to fit (numerically) $I(\chi)$ (26). If a Gaussian distribution is assumed, then equation 3 follows:

$$W(\beta) = A \cdot e^{-[\beta^2/2\beta_0^2]} \qquad (3)$$

$I(\chi)$ may be fit by adjusting the amplitude, A, and the distribution width, β_0, to obtain $W(\beta)$ and thereby get S with equation 2.

In nematic phases of mLCs, S typically ranges from ~ 0.25 to 0.75. The meaning of these magnitudes for S must be well-understood to get a better feel for the nature of orientational order in nematics. First, however, it must be recognized that nematic order is described with a single number (the scalar S), because assumptions about the molecular symmetry in this uniaxial phase have been made. Namely, an idealized prolate-shaped molecule having cylindrical symmetry is assumed. If this assumption is removed, the orientation of the klm Cartesian frame fixed to the molecule (*see* Figure 1) is described by a second-rank tensor, $\underline{\mathbf{S}}$, the order tensor with five independent elements, $S_{ij} = (3\langle \cos \beta_i \cos \beta_j \rangle - \delta_{ij})/2$, where β_i specifies the orientation of the i axis relative to $\mathbf{n}$ and δ_{ij} is the delta function ($\delta_{ij} = 0$ for $i \neq j$; $\delta_{ij} = 1$ for $i = j$). $\underline{S}$ is a traceless tensor (the sum of its diagonal elements, ΣS_{ii}, is 0) and gives the average orientation of any molecule-fixed frame relative to the director. If the klm axis system is the principal axis system (PAS), $\underline{S}$ is diagonal in the klm frame ($S_{ij} = 0$ for $i \neq j$). If the mesogen shape deviates from cylindrical symmetry (e.g., a biphenyl mesogenic core approximated as a parallelepiped), the average

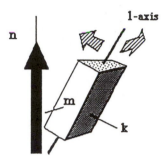

orientation of the mesogen must be specified with two order parameters, S_{ll} and ($S_{kk} - S_{mm}$); the latter is referred to as the molecular biaxiality, the preference for having the l–m, rather than the l–k, plane of the paral-lellepiped remain tangent to $\mathbf{n}$ while the molecular long axis goes through angular librations. When the molecule has cylindrical symmetry, the orientation of the klm PAS is specified by a single element, $S \equiv S_{ll} (= -2S_{kk} = -2S_{mm})$, the average orientation of the molecular symmetry axis l; S is the nematic order parameter referred to in equation 2.

To return to a discussion of the meaning of the magnitude of S, Figure 13a shows the probability distribution, $W(\beta) \sin \beta$, for different Gaussian widths β_0, centered about 0° (and, equivalently, 180° in the apolar nematic phase); the value of S obtained from equation 2 is plotted versus β_0 in Figure 13b. When, for example, an angular spread (β_0) of 60° is used in the Gaussian distribution (equation 3), the mesogen order parameter, S, is 0.5; the average inclination, $\langle \beta \rangle$, of l relative to the nematic director associated with this spread (or for that matter, any value of β_0) is 0°. [It makes no physical sense to invert the expression $S = \langle P_2(\cos \beta) \rangle$ and find an aver-age β ($= 35°$ for the example $\beta_0 = 60°$). Although this manipulation is widely practiced in the literature, such an inversion is valid only if the distribution, $W(\beta)$, is a delta function.] In summary, the diffraction pattern in Figure 12b tells us that, locally, the nematic fluid has a common symmetry axis (the director) defined by the preferred direction in which the molecular axes l spontaneously align in the liquid crystal, and more-over, the average degree of order of l relative to the director may be computed from $I(\chi)$. This method of extracting S from diffraction data applies to both low-molar-mass and polymer liquid crystals (25–29).

Anisotropic Properties

In simple liquids, it is straightforward to relate a bulk macroscopic prop-erty to its microscopic origins. Consider, for example, how the molecular

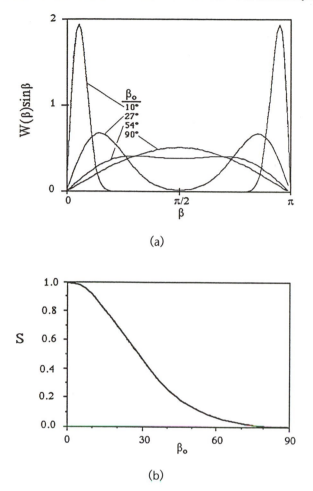

FIGURE 13. (a) Plots of a Gaussian distributed probability density, $W(\beta)\sin\beta$ versus β for different widths, β_0 (equation 3). (b) The computed (equation 2) order parameter, S, versus the width, β_0, of the Gaussian distribution.

electronic polarizability, α, manifests itself in the refractive index, n_r, of a simple liquid. The relative permittivity (the dielectric constant), ε_r, is a simple function of α and the number density, N, of molecules (the number per unit volume) in the liquid:

$$(\varepsilon_r - 1)/(\varepsilon_r + 2) = N\alpha/3\varepsilon_0 \qquad (4)$$

In this equation, the Clausius–Mossotti equation, ε_0 is the vacuum permittivity. At optical frequencies ($\sim 10^{15}$ Hz), n_r and ε_r are related by a quadratic expression:

$$\left(n_r^2 - 1\right)/\left(n_r^2 + 2\right) = N\alpha/3\varepsilon_0 \qquad (5)$$

Thus the relationship between the macroscopic refractive index and the microscopic polarizability is achieved without reference to the fact that the latter molecular property is described by a second-rank tensor, $\underline{\alpha}$; that is, all directions in the molecule do not exhibit the same electronic response to an applied electric field. The simplicity of equations 4 and 5 comes about because the average projection of $\underline{\alpha}$ along the direction of observation, α_{zz}, is independent of the orientation of z in the liquid. This projection is a simple scalar and is related to the mean value of the diagonal elements of $\underline{\alpha}$: $\alpha \equiv \alpha_{zz} = \text{trace}(\underline{\alpha})/3 = (\alpha_{kk} + \alpha_{ll} + \alpha_{mm})/3$. By contrast in the nematic liquid, a non-zero-order parameter has macroscopic implications; the directional dependence of the molecular polarizability tensor manifests itself when one relates macroscopic properties to microscopic properties. That is, the anisotropic part of the polarizability tensor, $\underline{\alpha}' \equiv [\underline{\alpha} - \text{trace}(\underline{\alpha})/3]$, must be accounted for explicitly when α_{zz} is calculated by recognizing that the orientation of the z axis relative to the director plays a key role in the observed macroscopic properties. The orientation of z is important, because the (directional) molecular properties are averaged in a unique way in the nematic.

Consider a small region of a nematic fluid having a uniform director field (a monodomain wherein the orientation of **n** is unchanged throughout). In this uniaxial fluid volume element, molecular properties (polarizability, magnetizability, etc.) are incompletely averaged with respect to the director; the average projection of these second-rank tensorial properties depend on Ω, the angle between **n** and the z axis of a laboratory xyz frame, the average orientation of the klm frame relative to **n** via $\underline{S}$, and the intrinsic molecular anisotropy of the polarizability, $\underline{\alpha}'$.

$$\alpha_{zz} = \alpha + 2\,\text{trace}(\underline{\alpha}'\underline{S})\,P_2(\cos \Omega)/3 \qquad (6)$$

Equation 6 reduces to the result for an ordinary isotropic liquid when $\underline{S} = 0$. In the nematic, $\underline{S} \neq 0$, and the projection of the polarizability takes on its extreme values, $\alpha_{\parallel}$ and $\alpha_{\perp}$ when $\Omega = 0°$ and $90°$, respectively. Consider a mesogen with cylindrical symmetry (*see* Figure 1) having a principal value of the polarizability along the long molecular axis, $\alpha_l \equiv \alpha_{\parallel}$, and a unique value transverse to the l axis, $\alpha_t \equiv \alpha_{kk} = \alpha_{mm}$. When the tensor product in equation 6 is expanded and its trace is taken, the results are given as follows:

$$\alpha_{\parallel} = \alpha + 2(\alpha_l - \alpha_t)S/3 \qquad \alpha_{\perp} = \alpha - (\alpha_l - \alpha_t)S/3 \qquad (7)$$

In equation 7, the principal value of the order tensor, $S_{\parallel}$, has been replaced with S. The results of equation 7, in conjunction with the relationship between the polarizability and the refractive index (equation 5, leaving aside complications associated with anisotropic internal field corrections), indicate that $n_{r\parallel} \neq n_{r\perp}$. In short, a nematic liquid may be readily distinguished from an ordinary liquid because it exhibits birefringence, $\Delta n_r \equiv n_{r\parallel} - n_{r\perp}$. Thus, a relationship is established between the liquid crystal and the molecular crystal: a nematic volume element is birefringent, albeit with refractive index differences attenuated by molecular disorientation (accounted for by the factor S in equation 7) relative to refractive index differences observed in a perfectly ordered single crystal ($S = 1$). Additionally, the changing value of the refractive index as the orientation of the director changes randomly throughout the bulk sample of a nematic very effectively scatters light in a manner reminiscent of a polycrystalline powder and, together with thermally excited fluctuations of the director (fluctuations of the value of $n_{r\parallel}$), accounts for the opaque, milky appearance of the nematic fluid. The opaqueness disappears at the nematic-to-isotropic transition, hence the term "clearing temperature".

In general, anisotropy can be expected in all macroscopic properties, Q, of the nematic. This macroscopic anisotropy, denoted $\Delta Q \equiv Q_\parallel - Q_\perp$ (the difference between the value of the bulk property measured parallel to **n**, $Q_\parallel$, and that normal to **n**, $Q_\perp$), is simply related to the order parameter S by equation 8:

$$\Delta Q = N(q_1 - q_t)S \qquad (8)$$

In this equation, $(q_1 - q_t) \equiv \Delta q$ is the molecular anisotropy [the difference between the principal longitudinal (q_1) and transverse (q_t) tensorial molecular properties] and N is the number of molecules per unit volume (7, 18). Equation 8 provides a very convenient measure of S if Δq is known (e.g., from single-crystal studies) and ΔQ is measured experimentally in a macroscopically aligned nematic.

Dichroism

The molecular anisotropy also manifests itself in a variety of spectroscopic techniques. Dichroism, the difference in absorption coefficients of linearly polarized light measured in orthogonal directions, is another phenomenon that gives information about the average molecular orientational order present in the nematic phase. The dichroic ratio, $D \equiv A_\parallel / A_\perp$ (*see* Chapter 7), is the ratio of the intensity of the absorption band of a characteristic transition measured with the polarized incident radiation parallel to **n**, $A_\parallel$, to that with the polarized incident radiation perpendicular to **n**, $A_\perp$:

$$D = \frac{\cos^2 \gamma \langle \cos^2 \beta \rangle + \tfrac{1}{2}\sin^2 \gamma \langle \sin^2 \beta \rangle}{\tfrac{1}{2}\cos^2 \gamma \langle \sin^2 \beta \rangle + \tfrac{1}{2}\sin^2 \gamma \langle 1 + \cos^2 \beta \rangle} \qquad (9)$$

In equation 9, γ is the angle between the symmetry axis of the molecule (**l** axis) and the direction of the transition moment **t** in the molecule-fixed klm frame. The averages over the molecular orientation (e.g., $\langle \cos^2 \beta \rangle$,

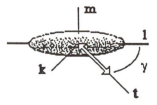

where β is the angle between **l** and **n**) may be rewritten in terms of the order parameter S. When the transition moment is parallel to **l** ($\gamma = 0°$), equation 9 reduces to a simple relationship between the dichroic ratio and the order parameter:

$$D = (1 + 2S)/(1 - S) \qquad (10)$$

Magnetic Resonance

Nuclear magnetic resonance (NMR) spectroscopy, in particular, deuterium NMR, is a valuable technique for determining the nature of molecular organization in liquid crystals. The utility of the ^{2}H NMR technique derives from the fact that the relevant NMR interactions are entirely intramolecular; that is, the dominant interaction is between the nuclear quadrupole moment of the deuteron and the local electric field gradient (efg) at the deuterium nucleus. The efg tensor is a traceless, axially symmetric, second-rank tensor with its principal component along the C–D bond. In a nematic fluid, rapid anisotropic reorientation incompletely averages the quadrupolar interaction tensor $\mathbf{q}$, the result of which is a nonzero projection similar to the result in equation 6:

$$q_{zz} = \frac{2}{3}\, \text{trace}\left(\underline{\mathbf{q}} \cdot \underline{\mathbf{S}}\right) P_2(\cos \Omega) \tag{11}$$

In a homogeneous nematic, the ^{2}H NMR spectrum consists of a resolved pair of resonances at frequencies ν_+ and ν_- centered about the Larmor frequency ν_L:

$$\nu_{\pm} = \nu_L \pm \frac{3}{4}\, q_{zz} P_2(\cos \Omega) \tag{12}$$

In an aligned nematic with $\Delta\chi_m > 0$, Ω, the angle between the magnetic field and $\mathbf{n}$, is zero, and for a molecule with assumed cylindrical symmetry (i.e., we ignore any molecular biaxiality, $S_{kk} - S_{mm}$), the frequency separation between the two transitions in equation 12, $\Delta\nu = (\nu_+ - \nu_-)$, the quadrupolar splitting, is given simply in terms of the average orientation of the molecular long axis $\mathbf{l}$ by equation 13:

$$\Delta\nu = \frac{3e^2qQ}{2h}\, P_2(\cos \gamma)\, S \tag{13}$$

Equation 13 is the explicit expression for the principal value of the quadrupolar interaction tensor in terms of the electrostatic charge, e; the electric field gradient at the deuterium nucleus, eq; the deuteron quadrupole moment, Q; and Planck's constant, h. The angle that the C–D bond (principal value of the interaction tensor, $\underline{\mathbf{q}}$) makes with the molecular symmetry axis $\mathbf{l}$ is γ, and S gives the degree of order of the $\mathbf{l}$ axis in the nematic. When the mesogen has internal degrees of freedom (more than one conformation, i.e., a variable dihedral angle ϕ), the quadrupolar

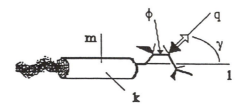

splitting is reduced to reflect the increased averaging of the efg by isomerization (i.e., rotations of about ϕ). Still with the oversimplified assumption of cylindrical symmetry, in the presence of internal motion, equation 13 would be modified by using $\langle P_2(\cos \gamma) \rangle$, where the angular brackets signify an intramolecular average over rapid isomerization. Hence, the magnitude of Δv is a direct measure of the efficacy of the motional averaging in the nematic and yields the order parameter when the molecular geometry (γ) is known; alternatively, Δv provides information about the internal flexibility of the mesogen $\langle P_2(\cos \gamma) \rangle$ when S is determined independently. For any real mesogen, in quantitative interpretations of NMR data, the simple symmetry implied in equation 13 cannot be assumed, and the total order tensor $\underline{\mathbf{S}}$ needs to be used. Moreover, when many conformations, $\{\phi\}$, exist, the order tensor for each conformer, $\underline{\mathbf{S}}\{\phi\}$, must be considered. Despite these complications, the NMR technique can be very valuable when molecular flexibility is present. For example, the rotational isomeric state approximation itself has been examined critically by analyzing carefully incompletely averaged NMR interactions (direct dipole–dipole couplings between proton pairs) exhibited by normal alkanes dissolved in a nematic solvent (*28*).

Field-Induced Director Reorientation

The macroscopic anisotropies in the electric susceptibility, $\Delta \chi_e$, where $\chi_e = 3(\varepsilon_r - 1)/(\varepsilon_r + 2)$, and the diamagnetic susceptibility, $\Delta \chi_m$, play an important role in alignment (and reorientation) of liquid crystals by external fields. This role also has its origins in the anisotropic average of molecular properties with respect to the director (equation 7). The potential energy when the field is parallel to $\mathbf{n}$ is different from when the field is perpendicular to $\mathbf{n}$. If an electric field, $\mathbf{E}$, interacts with the induced dipole moment per unit volume, $\mathbf{P} = \varepsilon_0 \chi_e \mathbf{E}$, the potential energy, U, is expressed as follows:

$$U = -PE \cos \theta = \varepsilon_0 \chi_e E^2 \tag{14}$$

In equation 14, θ ($= 0°$) is the angle between the induced moment and the field. Because equation 7 predicts that $\chi_{e\parallel}$ is different from $\chi_{e\perp}$ in a nematic volume element, the director will have a preferred low-energy orientation in the applied field. For positive dielectric anisotropy ($\Delta \chi_e > 0$), equation 14 indicates that the low-energy orientation of the director occurs when $\mathbf{n} \parallel \mathbf{E}$. Consequently, in a sufficiently strong field, all volume elements will assume the same director orientation, and a macroscopically aligned nematic results. In such an aligned sample, application of $\mathbf{E}$ normal to $\mathbf{n}$ will rapidly (on a millisecond time scale) drive a 90° rotation of the director in the fluid mesophase. This basic interaction between an external electric (or magnetic) field and the bulk anisotropy of the electric (or magnetic) susceptibility, $\Delta \chi_e$ (or $\Delta \chi_m$), in conjunction with the optical anisotropy of the nematic, Δn_r, may be exploited to reorient the director and change simultaneously the mesophase optical properties. This effect is the basis of field-generated electrooptic responses in LCD devices (*8*). Because of molecular structural similarities between mLCs and the monomers used in some mesogenic polymers (i.e., comparable molecular anisotropy values), the same phenomena may be observed in polymer liquid crystals.

Field-induced director reorientation with attendant optical changes has recently been used in a novel application with the potential for large-area LCDs: polymer-dispersed LCs (PDLCs). A PDLC is a microemulsion of mLC dispersed in a conventional transparent polymer film. In the "off" state, the refractive index of the mLC and that of the host polymer film are mismatched. Hence the dispersion of mLC droplets scatters light very effectively to give an optically opaque film (Figure 14, Left). On application of an external electric field (across a capacitorlike transparent metallic coating on both sides of the polymer film), the director assumes the same orientation in all of the microdroplets. If the refractive index along the director matches that of the polymer film host in the "on" state, the film suddenly switches from opaque to transparent (Figure 14, Right) to give a very economical, large-area "light valve".

Disclinations

In a bulk nematic, the director field is not uniform unless external influences (electric, magnetic or shear fields, surface alignments, etc.) are operative. At the junction of two differently oriented director fields, disclinations (the equivalent of a dislocation in a crystal) and domain walls ("grain boundaries") exist. The presence of these distortions in the director

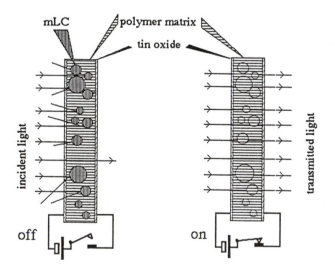

FIGURE 14. Polymer-dispersed liquid-crystal (PDLC) device consisting of a microdispersion of a low-molar-mass nematic fluid (mLC) in a conventional transparent polymer host matrix sandwiched between thin coats of transparent, conducting tin oxide. (Left) "Off" state with a refractive index mismatch between the dispersion and the host that scatters incident light. (Right) An external electric field aligns the nematic director matching the refractive indices of the dispersion and the host to yield an optically transparent medium.

field may be recognized readily and characterized with a polarizing microscope (*29*). The director field configurations for +1/2, +1, −1, and −1/2 strength line disclinations are illustrated in Figure 15. The strength is a function of the number of "dark brushes" meeting at a point while observed under crossed polars; the sign depends on the relative rotations of the brushes on rotating the polarizer. These patterns under crossed polars are useful in characterizing the phase type. For example, disclinations of ±1/2 are possible in nematic phases only and may be readily identified in photomicrographs such as that shown in Figure 6.

Strikingly clear visualizations of disclinations are found in transmission electron micrographs of replicas of solidified thermotropic polymer liquid crystal wherein the director field is mapped out by crystallite formation (*30*). Thus these textural features are exhibited by both polymer liquid crystals and mLCs. The density of disclinations may be increased by turbulent stirring of the nematic fluid. Disclinations with equivalent strengths and opposite signs may combine and annihilate one another to restore a uniform director field. The annihilation of disclinations, or "ripening process", in mLCs has recently been used as an example of the temporal evolution of complex systems in models of cosmological processes (*31*). Needless to say, watching the movement of disclinations and texture formation in quiescent LC polymers, fluids characterized by much larger viscosities and correspondingly smaller diffusion coefficients than those found in LC monomers, is not a very exciting spectator sport!

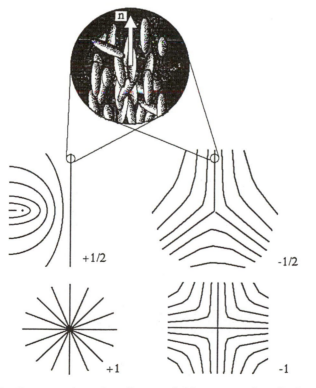

FIGURE 15. Patterns that the director field assumes for disclinations of strengths +1 / 2, +1, −1, and −1 / 2.

Director field distortions and disclinations are particularly important in commercial polymer mesophases, because the density of such defects is thought to play a significant role in both the rheological properties (ease of processing) of LCPs and the ultimate mechanical properties of polymeric solids derived from mesophases.

Elastic Properties

Deformation of the director field away from its equilibrium configuration increases the free energy density of the mesophase. The curvature strains (and associated restoring forces or the curvature stresses) are small and may be treated with continuum elasticity theory, because the scale over which the director changes orientation is very large compared to molecular dimensions. In fact, a variant of Hooke's law (stresses are proportional to the strains) may be used in conjunction with three distinct kinds of curvature strains and the associated elastic constants, k_{11}, k_{22}, and k_{33} corresponding, respectively, to splay, twist, and bend strains (Figure 16). In nematic mLCs, the k_{ii}s are approximately equal; splay is relatively difficult in smectics. These material constants are extremely small ($\sim 10^{-7}$ Nm^2) compared with the elastic constants of polymer networks and rubbers ($\sim 10^5$ Nm^2; *see* Chapter 1). Although these moduli appear negligile to polymer materials scientists, they provide the delicate elastic restoration

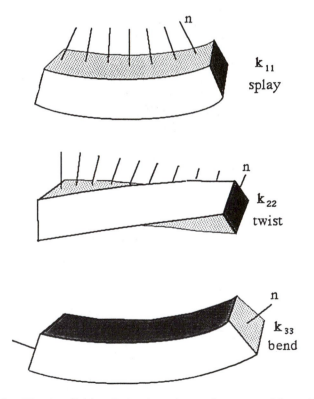

FIGURE 16. Director field patterns in volume elements subjected to splay, twist, and bend curvature strains.

of the initial director field in electrooptic devices; that is, the elasticity of mLCs drives the "off" state in a typical LCD. To "turn on" an LCD, a threshold voltage must be exceeded to overcome the elastic energy associated with the initial director field configuration. (The initial director configuration is established by various proprietary surface treatments that "anchor" the director field in LCD cells and PDLC microemulsions.) Under the influence of an external electric (or magnetic) field, the two competing forces, the elastic restoring force and the torque the applied field exerts on the anisotropic electric (diamagnetic) susceptibility, $\Delta\chi$, are related to the critical or threshold field, F_{crit}:

$$F_{crit} = (k_{ii}/\Delta\chi)^{1/2} \cdot \pi/d \qquad (15)$$

In equation 15, d is the distance scale over which the field-induced distortion of the director orientation takes place. In LCDs, electric fields are the usual method of switching; typical sample thicknesses (d) are $\sim 10^{-5}$ m, and critical field values (E_{crit}) are ~ 10 V.

Chiral Phases

The nematic phase is a translationally disordered fluid (with random, liquidlike positioning of the mesogen centers of mass) with long-range, uniaxial ordering of the molecular **l** axes. More subtle supramolecular arrangements are possible when the mesogen is chiral. A twisted nematic (cholesteric) phase forms with the local director normal to the **c** axis (Figure 17); by changing its orientation systematically in the mesophase, a helicoidal trajectory is traced over a large distance scale (over dimensions of 3000–8000 Å, which are $\sim 10^3$ times nominal intermolecular distances). This twisted nematic organization is symbolized by N* (the symbol N* is related to the chemical convention indicating a chiral center within a molecule by appending an asterisk to the chiral atomic site). Again, this twisted organization has dramatic macroscopic consequences. The periodic change in the electronic polarizability (or refractive index) as the director twists through the fluid establishes an optical grating when the pitch, P, is approximately equal to λ, the wavelength of visible light. In this case, particular wavelengths (colors) satisfying the Bragg equation will be diffracted to give these materials a beautiful iridescent sheen when examined in reflected light. This effect is the origin of the colors reflected from the surface of some insects; the major constituent of the beetle exocuticle, chitin, is an anisometric biopolymer aggregate, and it is deposited in a lyotropic cholesteric fluid form before the exocuticle congeals (*32*).

The reflected color from cholesterics is remarkable; the reflected light is circularly polarized with the same sense (right- or left-handed) as that of the N* helicoidal organization. (The opposite sense passes through the N* structure.) The strength of this supramolecular optical rotation of the reflected light is many times larger ($\sim 10^{4\circ}$) than optical rotations that chiral molecules exhibit in dilute solutions ($\sim 10^\circ$). Because the pitch, P, is established by a thermal average in the fluid N* phase, it will change with temperature. The temperature change changes the wavelength (color) of the selectively reflected light from the cholesteric phase, and therefore, an efficient temperature sensor is produced. Again, the forces underlying this phenomenon are extremely subtle. In the cholesteric fluid, the (motionally averaged) intermolecular forces between a pair of chiral molecules are slightly asymmetric, a situation prejudicing the quasiparallel alignment of molecular axes. In effect, chirality causes a very small (minutes of arc) average twist of the molecular **l** axis relative to its neighbor in the same sense (right- or left-handed, depending on the chiral center in the mesogen) that is manifested on a supramolecular scale as a unidirectional twist of the director **n**. The result is a structure of the type shown in Figure 17. (Frequently, the helicoidal structure is illustrated with a stack of planes, each with a rotated orientation of **n**. This representation has given the false impression that cholesteric phases possess a stratified supramolecular structure.) Cholesterics with opposite twist-handedness are obtained from L and D isomers of the mesogen; a racemic mixture of chiral mesogens yields a compensated (untwisted) N* phase. Cholesterics also may be compensated with external fields. For susceptibilities having positive anisotropy ($\Delta\chi > 0$), a sufficiently strong external field will untwist the helicoidal arrangement of the director and eventually align **n** parallel to the field. On the other hand, for $\Delta\chi < 0$, the spatial average of the susceptibility (over the helicoidal cholesteric arrangement) makes the

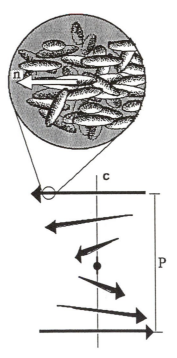

FIGURE 17.   The helicoidal director field of the twisted nematic (cholesteric) organization. The very small (minutes of arc) unidirectional twist from chiral mesogen to mesogen causes the spiral supramolecular organization of the director to turn through 180° after traversing a distance P (the cholesteric pitch).

low-energy arrangement one in which the cholesteric structure remains intact with the cholesteric **c** axis oriented parallel to the external field.

Molecular chirality also manifests itself in the technologically important class of chiral S_C mesogens. In the S_C phase, in addition to segregation of the molecules into smectic layers, the molecular **l** axis is on average tilted with respect to the layer normal (Figure 2c). A more realistic representation of the S_C organization is shown in Figure 18. In this arrangement, the mean direction of alignment is the local director **n**, and the magnitude of this unit vector **n** may be decomposed into two components: n_z, along the layer normal, and n_y, the component in the layer plane. In a chiral smectic C (S_C^*), the tilt exhibits a unidirectional twist as one moves from smectic layer to layer; n_y traces out a helicoidal path in the S_C^* phase (Figure 18). The molecular chirality breaks the local uniaxial symmetry about **n**, and hence, transverse molecular electric dipoles, μ_i, will be incompletely averaged by molecular rotations about the molecular long axis **l**. Consequently within a single S_C^* layer, a net residual electric polarization, $\mathbf{p}_x$ is oriented at right angles to the plane defined by n_z and n_y. This local ferroelectric structure is averaged by the twist of n_y to give a nonferroelectric bulk material. However, a ferroelectric monodomain S_C^* can be created, for example, by forming (cooling into) this phase in the presence of a strong magnetic field. The resulting texture has **n** (n_y) uniformly oriented throughout the bulk sample. The process of compen-

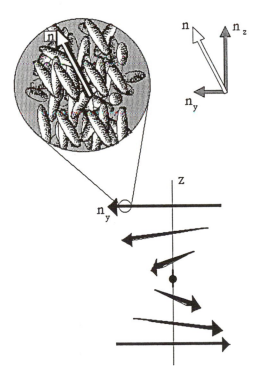

FIGURE 18. The chiral smectic C phase (S_C^*) has the molecular axes tilted, on average, with respect to the smectic layer normal, and the in-plane component of the local director n_y traces out a helicoidal path in the mesophase.

sating the natural twist of S_C^* yields a fluid with macroscopic polarization $\mathbf{p}_x$, a ferroelectric smectic liquid crystal. When this compensated structure is stabilized by surface treatments (that anchor the director field to give a monodomain sample), a surface-stabilized ferroelectric liquid crystal (SSFLC) is produced (33), which exhibits very fast electrooptic switching times ($\sim 10^{-6}$ s) and bistability: two states, i and ii, corresponding to the two orientations of $\mathbf{p}_x$, as shown in Figure 19. This fast switching is a consequence of the fact that the director $\mathbf{n}$ does not have to be reoriented; only the n_y component must be reoriented with an external electric field. When $\mathbf{E}$ is antiparallel to $\mathbf{p}_x$, the reorientation of $\mathbf{p}_x$ is readily accomplished by merely letting $\mathbf{l}$ travel on the surface of a cone to the new low-potential-energy orientation in the presence of the field (with $\mathbf{E}$ parallel to $\mathbf{p}_x$; Figure 19, bottom). This desirable rapid switching rate also can be realized in polymeric mesophases [e.g., when S_C^* mesogens are incorporated into polymers as side chains (34)], undoubtedly as a consequence of the very local nature of the motions required to reorient $\mathbf{p}_x$.

The intrinsic noncentrosymmetry of the S_C^* phase has potential uses. This symmetry is important for certain nonlinear optical (NLO; 35) appli-

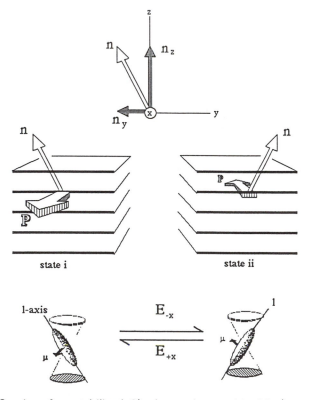

FIGURE 19. A surface-stabilized S_C^* phase adopts a bistable (state i or state ii) monodomain ferroelectric organization with a net polarization **P** (the three-dimensional arrow in the figure) derived from incompletely averaged (by molecular rotations about the **l** axis) transverse molecular dipoles. Application of an electric field antiparallel to **P** (E_{-x}) results in a reorientation of the director (and polarization) from state i to state ii by efficient rotation of the molecule over the surface of a cone (bottom illustration).

cations such as second harmonic generation (SHG; i.e., doubling the frequency of incident laser light). Consequently, this phase is being considered as a host for hyperpolarizable, organic (NLO-active) chromophore guest molecules. In many cases, the molecular structural attributes of NLO-active chromophores for SHG (linear, hyperpolarizable molecules) are close to those of calamitic mesogens themselves. Often, only small structural modifications (e.g., appending a sufficiently long alkyl chain) will convert such hyperpolarizable molecules into a mesogen. Hence, if glassy liquid-crystal textures with uniform alignment (homeotropic or planar) can be prepared (to minimize light scattering), mesogenic, NLO-active chromophores may be ideal materials for fabricating stable, organic (polymeric), optoelectronic devices requiring special local symmetries (*4, 35*).

Dynamics and Transport Properties

In isotropic liquids, molecular dynamics and transport phenomena are reasonably well understood. In the case of small prolate, ellipsoidal molecules, the rotational diffusion of molecules is very fast. The rotational correlation times, τ, corresponding to diffusion rates about the major axes of the ellipsoid are in the range 10^{-9}–10^{-11} s. Intramolecular transitions among conformers (i.e., isomers formed by rotating dihedral angles defined by three consecutive chemical bonds within flexible molecules) also are very fast (conformer life times of $\leq 10^{-10}$ s). The center of mass translational diffusion is isotropic and characterized by a self-diffusion coefficient, $D_{cm} \approx 10^{-10}$ m$^2 \cdot$ s^{-1}.

In the nematic phase, these same time scales are operative. Although conformer probabilities are slightly shifted from the equilibrium distribution in the isotropic liquid (more anisometric conformers are favored in the nematic), intramolecular isomerization rates are not influenced by the long-range orientational order. Incoherent, quasielastic neutron scattering gives the typically fast rotational diffusion about the principle axis **l**, $\tau_\parallel \approx 10^{-10}$–$10^{-11}$ s. Reorientational flips of the **l** axis itself are slower ($\tau_\perp \approx 10^{-7}$–$10^{-10}$ s) than the mean value for τ in the isotropic state. Thus, the long-range order retards large-scale mesogen reorientations that require cooperative movement of neighboring molecules. These differences between $\tau_\parallel$ and $\tau_\perp$ are increased in fluid smectic phases. Nevertheless, these correlation times indicate a fluid phase in which molecules exhibit rapid rotational diffusion. Self-diffusion is anisotropic in liquid-crystalline phases; diffusion along the nematic director, $D_{cm}^\parallel$, is more facile than diffusion in the transverse direction, and diffusion coefficients differ by a factor of two ($D_{cm}^\parallel \approx 2 D_{cm}^\perp$). Such anisotropy may reverse sign ($D_{cm}^\perp > D_{cm}^\parallel$) in smectic phases, especially for smectics with good layer definition wherein diffusion within a layer is easier than translation of the mesogen from one layer to another. In general, the magnitude of D_{cm} in mLC mesophases is about 10–100 times smaller than that observed for similar-sized molecules in the isotropic liquid state.

The macroscopic properties of polymer solutions and melts are fundamentally different from those of low-molar-mass materials, although the local dynamics (isomerizations and librations) operate on roughly the same time scales. Transport properties (D_{cm}) and viscosities are dramatically different in polymer fluids, as emphasized in Chapter 3. These characteristic similarities and differences between polymers and small molecules carry over when comparing mLCs and polymer LCs. D_{cm} for a polymer may be many orders of magnitude smaller than that observed in mLCs; hence, it takes longer for textures to develop, disclinations to annihilate, director fields to respond to externally applied fields, etc. However, aside from the more sluggish response, roughly speaking, the dynamical and transport phenomena in polymer mesophases parallel those observed in low-molar-mass materials.

Macromolecular Mesomorphism

For the purpose of discussing mesophase formation in polymers, it is convenient to group polymers into two categories and to use abbreviations that refer to these categories: (1) Polymerized liquid crystals, or PLCs, are derived from known, low-molar-mass monomer liquid crystals (mLCs). (2) Liquid crystalline polymers, or LCPs, are semiflexible, linear polymers that are structurally related to conventional engineering thermoplastics (i.e., polyesters, polyamides, polyimides, etc.).

Polymerized Liquid Crystals (PLCs)

When incorporating mLC secondary structures of the type shown in Figure 1 into polymers, two general types of topologies readily come to mind: (1) linear or main-chain polymers (mcPLCs) having covalently concatenated mesogenic cores (Figure 20a) and (2) side-chain polymers (scPLCs) having mesogenic cores attached covalently as side chains on a polymer backbone (Figure 20b). In both types of PLCs, the core is linked to the polymer via a flexible "spacer chain" (usually an alkyl, siloxane, or ethylene glycol chain). It is obvious, even from the limited number of secondary structures shown in Figure 20, that many variations on these topologies are possible (mcPLCs, scPLCs, and their combination). In a recent attempt to classify existing variations, Brostow (36) delineated more than 20 topologically distinct classes! All of these polymers merely exploit the intrinsic tendency of the core to form spontaneously a thermotropic mesophase. The only difference between PLCs and mLCs are the variable topological constraints that result from covalently embedding the cores into a macromolecule. In fact, if these polymers are considered in terms of their idealized shapes (Figure 1), the covalent linkages become irrelevant. In other words, at that extreme level of abstraction, behavior that is nearly the same as that exhibited by mLCs may be expected, and the polymer backbone and linkages may be regarded as insignificant diluents in an otherwise conventional mLC. This extreme picture is worthy of some consideration, because all of the static, equilibrium properties of thermotropic (lyotropic) mLCs may be realized in PLCs simply by polymerizing the appropriate mesogenic core and heating (dissolving) the polymer into the mesophase ($T_m < T < T_c$), particularly in scPLCs when the spacer linkage is very flexible and sufficiently long to "decouple" the behavior of the core from the polymer chain (6).

(a)

(b)

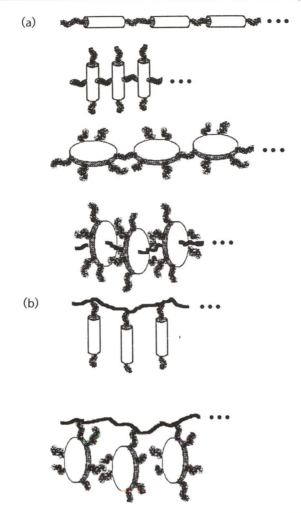

FIGURE 20. Some simple topologies of polymer liquid crystals (PLCs) derived from calamitic and discotic mesogenic cores. (a) Main-chain PLCs, copolymers with alternating cores and flexible spacers. (b) Side-chain PLCs with mesogenic cores attached by flexible spacers to a conventional polymer main chain.

Inserting a spacer chain in an mcPLC yields a regular alternating copolymer: ... –core–spacer–core–spacer– ... . This mcPLC topology exhibits additional features that require consideration of the resulting secondary structure: spacer chain parity (an even versus an odd number of atoms in the spacer chain) and spacer chain length may make one mesophase type more stable than others (e.g., nematic versus smectic). These features and others that will be introduced later obviously cannot be accounted for if discussion is restricted to the idealized mesogenic cores. For example, nematic (and N*) and smectic phases are readily formed by both main-chain and side-chain topologies. X-ray diffraction patterns from the nematic phases of PLCs are in some ways indistinguishable from those of mLCs. At the same time, however, having to accommodate the polymer backbone in the mesophase of scPLCs leads to a rich variety of structures (and quasiperiodic defects) discerned by X-ray diffraction, especially in smectic phases (37).

The magnitude of the scPLC core order parameter and, consequently, the size of the anisotropy in physical properties are comparable with those observed for mLCs. Differences between mLCs and the polymerized, linear analogues (mcPLCs) show up, however, in thermodynamic properties. Linear mcPLCs, for example, show transition temperatures that depend strongly on the degree of polymerization (dp) at low dp values (38). Asymptotic values of T_m and T_c are generally reached for dp values of ≥ 10. Within a homologous series of mLCs having terminal alkyl tails of successively longer lengths, the magnitudes of transition temperatures and thermodynamic quantities oscillate with the number of methylene units in the alkyl tails. This odd–even oscillation with spacer chain parity is exaggerated in linear oligomers and mcPLCs, because the connectivity of the cores in the polymer backbone reinforces core orientational correlations between successive cores for even spacer chains, whereas odd chain conformer geometries discourage such correlations (Figure 21a). Such effects of spacer chain parity are apparent in the order parameters exhibited by the mesogenic cores. Figure 21b shows the dramatic influence of dimerization and spacer chain parity on mesogenic core orientational order as delineated by ^{2}H NMR (39). Odd spacer chain parity in D7 yields values of S that are lower than those observed in mLCs M3 and M4, whereas even spacer chain parity (D8) enhances the core order parameter (on the scale used in Figure 21b, terminal chain parity has only marginal effects on ordering in mLCs). Thus, the connectivity of mesogenic cores in mcPLCs manifests itself by influencing the degree of order of the core, and in turn, the degree of order of the core influences the magnitude of the anisotropies of bulk properties (via equation 7).

Another example of the role of covalent connectivity of cores is suggested by the fact that the temperature range of mesophase stability for a particular mesogen can be increased by incorporating that mesogen into a polymer. Sometimes, a particular core that does not exhibit mesomorphism as a monomer will become mesomorphic when it is polymerized (40). Presumably in such situations, the spatial restrictions imposed on such nominally nonmesogenic cores when they are covalently linked together in the polymer enable them to maintain the required relative orientations for mesomorphism over some temperature range.

Polymerization of mLCs introduces covalent connectivity between mesogens and, therefore, orientational and translational restrictions into the mesophase of PLCs. Simultaneously, the director field generated by the spontaneously aligned mesogenic cores will, in turn, impose configurational constraints on the trajectory of the polymer chain. The constraints in this highly coupled system are introduced by considering the behavior of a flexible polymer chain dissolved in a liquid-crystalline solvent. Neutron scattering in conjunction with isotopic labeling can give insights into the overall shape of the solute chain in an anisotropic fluid (*see* Chapter 7). Recent studies of such solute–solvent systems suggest that the radius of gyration tensor, $\mathbf{R}^g$, of the solute chain conforms to the core organization in the mLC solvent (Figure 22). In nematic phases, $R_\parallel^g > R_\perp^g$, where $R_\parallel^g$ is the radius of gyrations along the nematic director. However, because

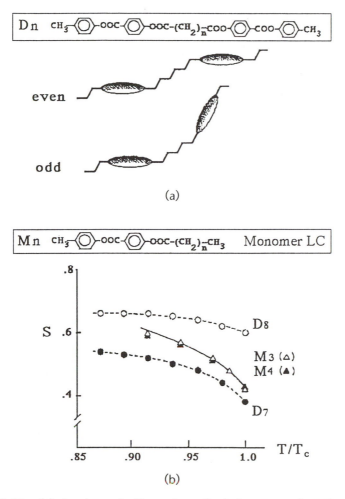

FIGURE 21. (a) A schematic illustration of relative core orientations in a covalently connected pair of mesogenic cores, a dimer LC(Dn), for even and odd spacer chain parities (all-*trans* conformation). (b) Dimer (Dn) and monomer (Mn) LC core order parameters (S) in the nematic phase.

of the severe entropic cost for chain deformation, the anisotropy ($R_\parallel^g - R_\perp^g$) is rather small. When the polymer chain is covalently linked to prolate mesogenic cores to form an scPLC, the influence of the director field alluded to for a free solute chain in a nematic may be overwhelmed by the covalent topological constraints in the scPLC. For the simple scPLC wherein the mesogenic core is connected to the backbone with a flexible spacer, both prolate and oblate chain trajectories have been observed. Oblate radii of gyrations ($R_\parallel^g < R_\perp^g$) are observed in scPLC nematics with a tendency for smectic fluctuations (i.e., nematics with a lower temperature smectic phase; *41*). In this situation, the chain trajectory persists in a plane normal to the director (Figure 23) and is denoted a nematic N_I phase by Warner (*42*). A prolate trajectory (designated N_{III}) with the polymer backbone parallel to the director is also conceivable for scPLCs. In smectic phases of scPLCs, the polymer backbone assumes an oblate trajectory

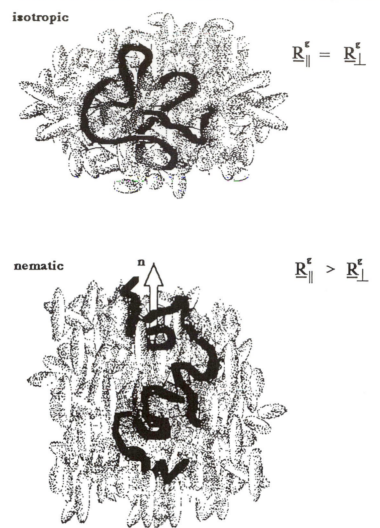

isotropic

$$\underline{R}^{g}_{\parallel} = \underline{R}^{g}_{\perp}$$

nematic   **n**

$$\underline{R}^{g}_{\parallel} > \underline{R}^{g}_{\perp}$$

FIGURE 22. Schematic illustration of the trajectory of a polymer solute chain in an isotropic (Top) and nematic (Bottom) solvent. Relative magnitudes of parallel and perpendicular (to the vertical direction and **n**, respectively) components of the radius of gyration tensor, $\underline{R}^{g}$, are indicated.

$(R^{g}_{\parallel} < R^{g}_{\perp})$. However, the anisotropy in the radius of gyrations is small $(R^{g}_{\parallel} \approx R^{g}_{\perp})$, a fact indicating that the backbone is not confined to the narrow interface between smectic layers. The backbone chain apparently traverses (ill-defined) layers to avoid the severe entropic penalty associated with confinement between smectic layers. Entropic consequences in PLCs, namely, the rubber elastic properties of liquid-crystalline polymer networks, will be discussed after a brief consideration of the solid state of PLCs.

One very striking attribute of polymeric mesogens is glass formation (*see* Chapter 4). Although only a few mLCs on vitrification retain in the solid state the molecular organization present in the mesophase (most

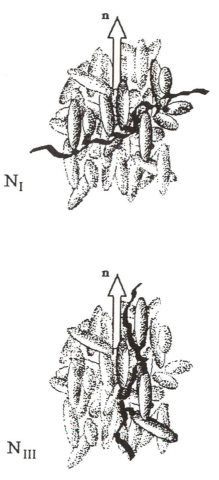

FIGURE 23. Two possible configurations of the polymer backbone relative to its mesogenic side chains, N_I and N_{III}, are illustrated.

mLCs crystallize), glass formation is a very common phenomenon in PLCs. The nematic or smectic glass thereby becomes a valuable solid host medium for orienting and exploiting optical properties of selected guest molecules. Moreover, the polymer becomes a vehicle for covalently including substantial numbers of guest molecules in the mesophase and the oriented solid state (*43*). [Only limited amounts of a nonmesogenic guest (≤ 10 mol %) can be accommodated by an mLC without depressing the mesophase; levels of ~40% of an anisometric guest can readily be tolerated in PLCs if the guest (comonomer) is covalently incorporated into a PLC.] The same rationale for glass formation in ordinary (isotropic) polymers applies to PLCs; that is, restricted reorientational mobility on cooling the (LC) melt traps nonequilibrium configurations and prevents crystallization. Figure 24 shows a generic DSC (differential scanning calorimetry) trace for PLCs. After the first heating, a supercooled nematic glass is formed on cooling; this glass may transform directly into the nematic fluid on subsequent heatings above the glass transition temperature, T_g. In addition to the T_g, for sufficiently long spacer chains, a T_m also

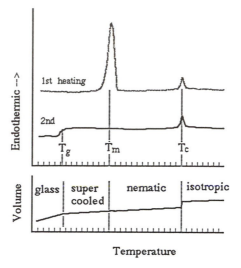

FIGURE 24. Schematic illustration of DSC and volume changes on heating a thermotropic polymer liquid crystal. After the first heating, the mesophase supercools and vitrifies; on subsequent heatings (lower curves), the DSC and volume changes show transformations from glass to mesophase at T_g. The mesophase goes isotropic at T_c.

may exist (not shown in Figure 24) that is associated with melting crystalline spacer chains.

Viscoelasticity is perhaps the most ubiquitous characteristic of high-molecular-weight polymers at temperatures above T_g. The implications of coupling rubber elasticity and mesomorphism may be considered by synthesizing covalent networks from conventional elastomers (siloxanes, isoprenes, etc.) and typical mLC mesogenic cores; such networks form thermotropic phases. At low levels of cross-link densities in the PLC network, transition temperatures (T_g, T_c, etc.) do not change appreciably from those of the non-cross-linked PLC (and its ancestral mLC; *44*). Below T_c, rather modest mechanical deformations (extension ratio, λ, of <1.5) may convert an initial random and disclination-ridden texture into a macroscopically uniform director field, that is, a single liquid crystal (*45*). In the isotropic melt of such an elastic network, application of a mechanical deformation will generate a stress field that will bias the orientational alignment of the mesogenic cores. When the temperature is lowered, such orientational biasing will predispose the melt for spontaneous organization of the cores into a mesophase. As described in more detail in a later section (*see Theories of Mesomorphism*), a mechanical deformation can effectively increase T_c and thereby induce a phase transition (I ↔ N) at a temperature T_c' that is higher than the zero-stress T_c (Figure 25, intermediate dotted curve). The order parameters at the clearing temperatures (S_c) and their temperature dependence are illustrated in Figure 25 for three situations: at zero external stress (thick curve), at an intermediate stress level (dotted curve), and at the critical stress level (thin line). At the critical stress level (deformation ratio, λ_{crit}), the orientational order in the isotropic melt should increase continuously as the temperature is lowered;

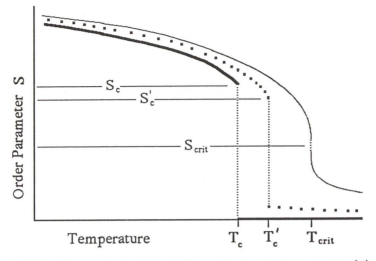

FIGURE 25. Elastomeric liquid-crystalline polymer order parameter (*S*) versus temperature in the absence of (bold solid line; Maier–Saupe-like theory) and in the presence of (dotted line and thin line) external stress. As the stress is increased, the first-order transition temperature (T_c) increases (dotted vertical lines); above some critical stress, *S* increases continuously as the temperature is lowered (thin curve).

that is, no first-order phase transition is observed as one lowers the temperature through T_{crit} if the sample is deformed above λ_{crit} (Figure 25, thin line).

Coupling mesophase transitions and director reorientations to mechanical deformation leads to some novel applications. Mechano-optical sensors and/or switches may be fabricated. Spatially localized deformations can induce uniform director fields (homogeneous nematic optical anisotropy) and thereby delineate waveguide pathways in nematic elastomers (*46*). Chiral monomer-based elastomeric PLCs that form N* and S_C^* polydomain textures can be converted to single-domain samples at low deformations ($\lambda \approx 1.3$); the supramolecular twist may even be compensated at high deformations ($\lambda > 3$) to yield, for example, a compensated N* or S_C^* phase (*44*). At low strains, these chiral elastomeric mesophases exhibit unique electromechanical phenomena that originate from the change in electric polarization caused by a mechanical deformation. The opposite interaction, that is, a change in elastomer dimensions, may be induced by applying an external electric field. The former piezoelectric effect also may be accompanied by pyroelectric and flexoelectric effects, induction of surface change by temperature changes, and director field distortions (*47, 48*). All of these (potential) phenomena have recently been recognized, but practical demonstrations have not yet been made.

Block Copolymers

Macroscopic anisotropy is observed in fluid, microphase-separated melts and solvent-swollen gels of block copolymers (*49*). By analogy with the hydrophobic–hydrophilic-driven aggregation of amphiphiles, di- and tri-block copolymers with the chemically different blocks exhibiting differential solubility in conventional solvents also exhibit mesomorphism. (Hydrophobic interactions also may be exploited in block copolymer mesophases; reference 50 provides an entry to recent literature on polymeric amphiphiles.) The insoluble block will aggregate at high polymer concentrations (Figure 26) and these aggregates (with the swollen block on the periphery of the aggregate) will pack in the excess solvent in various regular morphologies (cubic, hexagonal, lamellar, etc.). The long-range positional order in neat or swollen microphase-separated block copolymers confers anisotropic macroscopic properties (e.g., birefringence) to these fluid (gel-like) systems. However, the nature of the molecular (monomer) orientational order in such mesophases departs from that typically associated with mLCs. Generally the chain trajectory in each block will be random; the chain will try to maintain maximum-entropy configurations in the spatially distinct phases. At the same time, the fact that the chain trajectory begins with the chain contour normal to the interface separating the incompatible blocks requires "brushlike" configurations (51) with the trajectory extended within the interphase (Figure 26, inset). Similar kinds of biased chain trajectories would occur in semicrystalline polymers at the crystal–amorphous interface. However, the orientational anisotropy of the monomer for these intrinsically nonmesogenic systems is rather small ($S \approx 10^{-3}$ in deformed elastomers, for example) and leads to some confu-

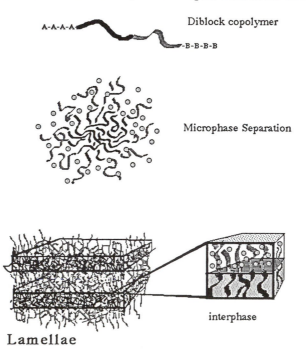

FIGURE 26. Linear (AB) block copolymer with preferential solvation of the B block. (Top) Schematic of diblock copolymer structure. (Middle) An "inverted" micelle. (Bottom) A swollen lamellar morphology. The insert shows perpendicularly oriented chain trajectories within the interphase region near the A–B interface in the fluid microphase-separated morphology.

sion when terminology such as "nematic order" (implying $S \approx 0.5$) is used to describe the biased chain trajectories. More recently, the term "interphase" has been introduced to acknowledge the departure of the polymer chains from a totally relaxed, equilibrium configuration near the interfaces in these systems. It is on a global scale (i.e., supra-aggregate scale) that mesomorphic phenomena are encountered in block copolymers, and consequently, mesomorphism is not dependent on anisometric monomer geometry as it is in mLCs.

Liquid-Crystalline Polymers (LCPs)

LCPs are commercially important lyotropic and thermotropic polymers that are structurally related to conventional engineering thermoplastics. These polymers are not based simply on idealized, monomeric structures having an obvious tendency towards mesomorphism as PLCs are; rather, they are derived from known classes of conventional monomers that exhibit desirable properties (thermal stability, solvent resistance, high strength, etc.). Liquid-crystalline polymers fortuitously have the requisite secondary structural rigidity and local chain anisometry to form spontaneously ordered fluid phases—if they can be solubilized or melted! Some characteristic structures are shown in Scheme I. The first three structures in Scheme I form lyotropic mesophases, and except for the helical polypeptide (first structure) very aggressive acid solvents are required to solubilize the polymers at the sufficiently high concentrations needed for liquid crystal formation [especially the poly(benzobisoxazole)s and poly(benzobisthiazole)s, second structure with X = oxygen and sulfur, respectively]. Poly(arylamide)s (third structure; e.g., DuPont's Kevlar and Akzo's Twaron) dominate the industrially important lyotropic LCPs.

The remaining structures in Scheme I form thermotropic LCPs. Generally these polymers are semiflexible with variable persistence lengths. Often the primary structure can be decomposed into mesogenic, corelike subunits (e.g., a couple of aromatic rings linked by an ester unit), which may be structurally related to known mLCs. Essentially all of the industrially important thermotropic LCPs are copolyesters with more than two comonomers, and in such copolyester thermotropics, heterogeneity in the primary structure is very likely. A few studies have addressed the impact of polyester structural heterogeneity (brought about by ester interchange in the melt) on mesophase properties (*52*). Heterogeneity (in the molecular weight distribution as well as the primary structure) could cause nanophase separation, that is, microscopic volume elements of isotropic melt dispersed in mesophase, and, correspondingly, the possibility of a single chain traversing both isotropic and liquid-crystal phases. This possibility suggests that nomenclature such as "degree of liquid crystallinity", by analogy with "degree of crystallinity" for semicrystalline polymers (*see* Chapter 4), might be appropriate. Because polyesters are multicomponent systems (a mixture of chain lengths and compositional heterogeneity), according to the phase rule, these LCPs invariably exhibit large two-phase regimes (isotropic + mesophase) when heated near the clearing temperature. The significance of phase equilibria in thermotropic LCPs has prompted heated discussions (*53*) in the literature, which have not been

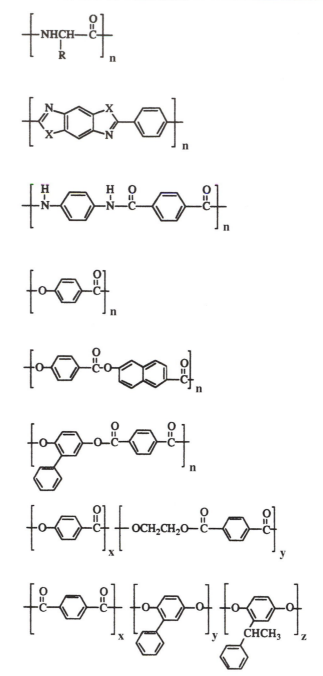

SCHEME I. Examples of chemical structures of main-chain liquid-crystal-forming polymers.

resolved satisfactorily. Even more important is the fact that, although sample homogeneity is critical to ultimate physical properties of thermotropic LCPs, little systematic work on the degree of liquid crystallinity using thermally stable LCPs has been reported.

Structural defects (bent or kinked monomers, e.g., 1,3-phenylenes and 2,6-naphthalenes, respectively) have been deliberately introduced into the backbones of polyesters in an effort to make thermotropic LCPs more thermally tractable. New monomers of intermediate tortuosity (e.g., 2,5-thiophene; *54*) may have an important role to play in this respect. Moreover, other monomer candidates could be identified if the theoretical foundation of mesomorphism in semiflexible polymers was more developed and predictive in scope.

Theories of Mesomorphism

Four principal theories describing the transformation of the isotropic fluid into a spontaneously organized nematic fluid, the I ↔ N transition, have been developed: (1) Onsager's density expansion of the free energy of anisometric particles, (2) Flory's estimate of the insertion probability for a rodlike (multisite) solute into a lattice, (3) Maier and Saupe's construction of a potential of mean torque experienced by mesogens (or solutes) in a nematic environment, and (4) de Genne's transposition of Landau theory to the I ↔ N transition. These theories will be examined briefly (in reverse chronological order) to emphasize their influence on more recent theoretical descriptions of polymer mesophases.

Landau Molecular Field Theory

The Landau molecular field theory as applied to LC phase transitions by de Gennes (*18*) assumes that the Gibbs free energy density, $g(P, T, S)$, is an analytic function of the order parameter, S. Expanding the nematic component of the free energy density, $g_{nem} \equiv g - g_{iso}$, in a power series (assuming that S is small) yields the following:

$$g_{nem} = AS^2/3 - 2BS^3/27 + CS^4/9 \qquad (16)$$

The coefficients A, B, and C are, in general, functions of P (pressure) and T (temperature). Equation 16 predicts a phase transition in the vicinity of the temperature T^* where A vanishes; A is assumed to have the following form:

$$A = A'(T - T^*). \qquad (17)$$

The discontinuous first-order phase transition occurs at a temperature T_c that is slightly higher than T^* (T^* is a second-order transition temperature). The first-order nature of the I ↔ N transition is due to the presence of the odd-order power of S. The associated clearing temperature, T_c, and the order parameter at the I ↔ N transition, S_c, may be obtained by minimizing g with respect to S (by setting $\partial g / \partial S = 0$):

$$T = T^* + B^2/27A'C \qquad S_c = B/3C \qquad (18)$$

As suggested by the thick solid curve in Figure 25, the order parameter changes discontinuously from $S = 0$ to a finite value S_c on lowering the temperature. If an external field, F, interacts with the anisotropic molecular susceptibility, $\Delta\chi$, an additional term, $-(N\Delta\chi)SF^2/2$ (N is number density), must be added to the expression for g (18, 20). This external alignment influence shows up as a shift in T_c to higher temperatures (T_c'; Figure 25, dotted curve).

The general Landau theory, although developed by de Gennes to describe critical phenomena in mLCs, has recently been applied to elastic networks comprising PLCs (42). The Landau formalism also enables one to interface with conventional orientation phenomena in nonmesogenic polymer networks. In particular, a mechanical deformation and its associated stress field, σ, influences g (and therefore T_c and S_c) with effects similar to those of external magnetic or electric fields. For a small (uniaxial) extension ratio, $\lambda = e - 1$, where e is the strain, the form of g in equation 16 is modified by the additional terms:

$$g = g_{\text{iso}} + g_{\text{nem}} - USe + \mu e^2/2 - \sigma e \qquad (19)$$

In equation 19, $-USe$ describes the coupling of strain to the nematic order, and μ is the modulus (U and μ would, for example, depend on cross-link density). These extra terms have remarkable consequences. Even in the absence of an external stress field ($\sigma = 0$), a spontaneous change in the macroscopic dimensions of a network cross-linked in the nematic state can occur as the free energy is minimized at the strain value, $e_{\text{min}} = US/\mu$. The resulting form for $g(e_{\text{min}})$ yields a higher transition temperature, $T_c' = T_c + U^2/2\mu A$ (A is defined in equation 17). The elastic PLC network will exhibit an I $\leftrightarrow$ N transition that is increased relative to that in the absence of cross-links, and T_c' will increase with increasing stress (Figure 25, dotted curve). Above some critical stress level, σ_{crit}, as the temperature is lowered, the network will pass continuously from isotropic to mesophase at a critical temperature, T_{crit} (Figure 25, thin curve). This regime cannot be realized in low-molar-mass mesogens with electric or magnetic stresses. However, the coupling between the mechanical stress field and the anisotropic excluded volume of the mesogenic cores is much stronger than the coupling between electric (magnetic) fields and mesogen anisotropic susceptibilities, $\Delta\chi$. Hence mechanical strains yield dramatic results in elastic PLC networks and allows access to regimes that are inaccessible in mLCs. Pretransition phenomena driven by the stress field (e.g., the increase in birefringence as the nematic phase is approached from $T > T_c$) have been reported (55). The field-induced birefringence, $\Delta n \propto (\Delta\chi)F^2/A$, diverges as $T \to T_c$ according to the definition of A in equation 17 [i.e., A is the principal source of temperature in the expression for g_{nem}: $\Delta n_r \propto A^{-1} \approx (T - T^*)^{-1}$].

These recent observations with PLC networks have reopened an old question in conventional elastic rubber networks (composed of nonmesogenic monomers such as isoprenes, butadienes, and siloxanes), which are treated classically in Chapter 1: Are orientational correlations (excluded-volume effects) among ordinary chain segments significant, and might there be coupling of segment orientation to the stress field in conventional elastic networks ("nematiclike interactions")? Recent theoretical work has suggested that deviations of experimental stress–strain data from classical descriptions of rubber elasticity could be accounted for by considering nematiclike interactions in such materials (56). These ideas also have been implicated in stress relaxation mechanisms for "isotropic" polymer melts (57) in an effort to answer, for example, the following question: Does the stress-induced orientational order in deformed melts influence the reptation trajectory?

Maier–Saupe Theory

The Maier–Saupe theory (58) posits a simple potential of mean torque that originates from an average over the interactions a given mesogen experiences because of its (oriented) neighbors:

$$V(\beta) = -wSP_2(\cos \beta) \tag{20}$$

The mean field, $V(\beta)$, satisfies the symmetry conditions of the apolar nematic fluid with its simple $P_2(\cos \beta)$ angular dependence (β is the polar angle between the mesogen molecular **l** axis and the director **n**); increases in importance with improved orientational ordering, S; and is parameterized by the coupling constant, w, a measure of the strength of the influence of the nematic mean field on the mesogen. As $V(\beta)$ disappears in the isotropic liquid ($S = 0$), it represents anisotropic interactions over and above those encountered in ordinary liquids. A self-consistent definition of the order parameter follows:

$$S = \frac{\int_0^1 P_2(\cos \beta) \exp[V(\beta)/k_B T] \, d(\cos \beta)}{\int_0^1 \exp[V(\beta)/k_B T] \, d(\cos \beta)} \tag{21}$$

Equation 21 may be solved numerically [S appears on both sides of equation 21 because $V(\beta)$ is a function of S] to give the temperature dependence of the order parameter (Figure 25, thick solid curve). Conventional statistical mechanical manipulations of the partition function show that a first-order phase transition is predicted at T_c. Other quantities are $S_c = 0.43$ and $T_c = 0.22 \ w/k_B$ at the I $\leftrightarrow$ N phase transition.

This theory has been used by Warner (42) to suggest more than one type of nematic phase in scPLCs depending on the relative magnitude (and signs) of the coupling constants w_{core} and $w_{backbone}$, representing the respective interactions of the side chain and the backbone with the mean field (*see* Figure 23). Except for scPLCs, the Maier–Saupe theory only makes an appearance as a supplement to the athermal excluded-volume interaction in the Flory lattice theory (*see* next section). However, although w was identified with anisotropic attractive interactions (dispersion forces) in the original theory, Gelbart (59) and Cotter (60) have shown that averaging isotropic attractive interactions over an anisotropic space (arising from excluded-volume considerations among rodlike mesogenic cores) leads to the same form for $V(\beta)$. That is, an apparent anisotropic attractive potential of mean torque may be generated from purely isotropic attractions when shape anisotropy is correctly factored into the averaged intermolecular interactions. Such local (effective) anisotropy derived from excluded-volume considerations may also underlie the physics of linear thermotropic LCPs having semiflexible mainchains.

Flory Lattice Model

The Flory lattice model (61) has received the most attention for polymeric LCs, although it lay dormant for more than a decade after it was introduced in 1956. It is ideally suited for lyotropic LCs consisting of solvent and rigid rods, although more recently, it has been considered (with modifications) in the context of semiflexible linear polymers and thermotropic monomer LCs. The crux of this model is the derivation of the partition function, Z, corresponding to the insertion of n_p rodlike solution particles (each comprising x segments) into a lattice with n_o sites; all sites are filled in the solution by inserting $n_s = n_o - xn_p$ solvent particles. Z is the product of two components: a combinational part (Z_c, equation 22) and an orientational part (Z_o, equation 23):

$$Z_c = \left(\frac{1}{n_p!}\right)^{n_p} \prod_{j=1} v_j \tag{22}$$

$$Z_o = \prod_y \left(n_p \sin \beta_y / n_{py}\right)^{n_{py}} \approx \left(\frac{\bar{y}}{x}\right)^{2n_p} \tag{23}$$

In Z_c, v_j is the insertion probability for the jth solute rod; it is a function of $y = x \sin \beta_y$, where β_y is the inclination of the rod from the local director (y also may be viewed as the number of "subrods" comprising a particle at inclination β_y; this deconstruction of the particle into subrods is a natural consequence of inserting the entire rod into a discrete lattice.) The insertion probability, v_j (and thus, Z_c), increases with increasing y and is a maximum for perfect order ($y = 1$). Thus the behavior of Z_c is opposite to that of Z_o, which is a maximum for isotropic configurations of the n_p particles on the lattice (n_{py} is the number of particles with inclination y). In this model, the length of the particle, x, divided by its width (one lattice site) is the aspect ratio of the solute rods ($x = L/d$). For a small axial ratio, x, and/or small concentrations of rods, n_p, Z_o dominates, and complete disorder ($y = x$) is the most stable state of the system of rods and solvent. For a large x and/or n_p, Z_c can compensate Z_o to yield a regime of free energy ($-\ln[Z]$) where partially ordered mixtures of rods and solvent are stabilized. The critical volume fraction, ϕ_I, signaling the appearance of the LC phase is simply a function of solute geometry (aspect ratio, x):

$$\phi_I \approx (8/x)(1 - 2/x) \tag{24}$$

The general predictions of equation 24 are in agreement with experimental results for various lyotropic LCPs (61). One consequence of this theory for a system polydisperse rods (a distribution of x values) is the possibility of fractionation (62): longer rods would distribute into the anisotropic LC

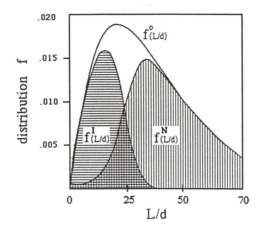

FIGURE 27. The computed partitioning of a most probable distribution, $f^o(L/d)$, of rod lengths into the isotropic phase, $f^I(L/d)$, and the coexisting nematic phase, $f^N(L/d)$. (Reproduced from reference 62. Copyright 1978 American Chemical Society.)

phase, whereas short ones would be relegated to the isotropic phase in the two-phase regime (Figure 27). The theory can also be extended (*63*) to make predictions about ternary mixtures (polymer rods, polymer random coils, and solvent) and is in agreement with experimental observations (*64*) that indicate the strong incompatibility of the two kinds of polymers. It has been adapted to treat semiflexible polymers; this problem is examined in more detail in the next segment.

Efforts have been made to treat mLCs with the Flory lattice theory by adding anisotropic attractive interactions of the type given by Maier and Saupe (equation 20) (*65*). The need to add intrinsically anisotropic attractive interactions appears to be an artifact of the way attractive interactions are handled in the lattice model itself: only interactions between nearest neighbors in the lattice site are independently summed. Consequently, anisotropic attraction can be observed only if intrinsic (site) anisotropy is added. However, if in addition to interactions between nearest neighbors, interactions between second, third, etc., neighbors are also incorporated into the lattice sums, then anisotropic attraction would naturally result from isotropic (site) interactions. The magnitude of the total attractive interaction between a pair of rods (in sequentially occupied sites) would be, in a longer range lattice summation scheme, a function of the relative angular orientation of a pair of rods (assuming that the interaction between two solute-occupied sites has a magnitude that differs from that between a solute site and a solvent-occupied site). This phenomenon is readily illustrated in the following two-dimensional square lattice fragments:

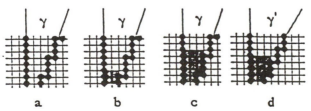

In **a**, nearest neighbor interactions only are considered and in the fragment shown, no nearest neighbor sites are occupied for the rod fragments oriented at an angle γ. In **b** and **c**, second and third nearest neighbor interactions, respectively, are illustrated with shaded regions. When γ is increased to γ' in **d**, the number of interactions decreases (from 13 in **c** to 11 in **d**); this decrease is indicative of an angular-dependent attractive interaction. This kind of angular-dependent attraction is derived from excluded-volume considerations in a manner reminiscent of the modeling (59, 60) of Cotter and Gelbart for mLCs, without resorting to intrinsic anisotropy at the sites occupied by the rodlike polymer.

Onsager's Virial Expansion

Onsager's virial expansion (66) was the first correct model of an athermal $I \leftrightarrow N$ phase transition. Its formulation was motivated by lyotropic mesomorphism in solutions of the rodlike tobacco mosaic virus particles. Other colloidal particles with anisotropic shape were identified as mesomorphic in the 1920s, including inorganic V_2O_5 particles (67); liquid-crystalline dispersions of poly(tetrafluoroethylene) "whiskers" (68) are a contemporary example of this phenomenon. In the Onsager model, the Helmholtz free energy (ΔF) of n_p polymer rods (axis ratio, $x = L/d$) in a volume, V, is given by equation 25:

$$\frac{\Delta F}{n_p k_B T} \approx \frac{\mu^0}{k_B T} + \ln\left(\frac{n_p}{V}\right) - 1 + \int f(\Omega) \ln[4\pi f(\Omega)] \, d\Omega + \rho B_2 + \cdots \quad (25)$$

In equation 25, the distribution function, $f(\Omega)$, gives the probability of finding a rod at orientation theta, k_B is the Boltzmann constant, T is temperature, μ^0 is the reference chemical potential of the solute, and ρ is the number density. The second virial coefficient, B_2, is given in terms of the cluster integral:

$$B_2 = -\frac{1}{2} \iint B(\Omega, \Omega') f(\Omega) f(\Omega') \, d\Omega \, d\Omega' \quad (26)$$

Onsager approximated $f(\Omega)$ as follows:

$$f(\beta) = \frac{\alpha}{4\pi \sinh \alpha} \cosh(\alpha \cos \beta) \quad (27)$$

In these equations, $d\Omega = 2\pi \sin \beta \, d\beta$, and α is a variational parameter determined by minimizing $\Delta F/n_p k_B T$. Onsager approximated the mutually excluded volume of a pair of cylinders as follows:

$$-B(\Omega, \Omega') \approx 2dL^2 |\sin \gamma| \quad (28)$$

In equation 28, L is the length, d is the diameter, and γ is the relative orientation. The coexistence of nematic (N) and isotropic (I) phases was

Table I. Critical Parameters from Different Trial Functions, f(Ω)

f(Ω)	xC_I	xC_N	S_{crit}
Onsager trial function	3.340	4.486	0.848
Gaussian function	3.450	5.120	0.910
Numerical iteration	3.290	4.191	0.792

computed by equating the concentration-dependent osmotic pressures (Π) and chemical potentials (μ):

$$\Pi_I(C_I) = \Pi_N(C_N) \qquad \mu_I(C_I) = \mu_N(C_N) \qquad (29)$$

In equation 29, C is concentration. The onset of liquid crystallinity (the product of C_I and the axial ratio, x), the concentration when the entire system is nematic (xC_N), and the critical order parameter (S_{crit}) are given in Table I for different trial functions, $f(\Omega)$.

The critical concentrations and order parameter are obviously sensitive to the nature of the rod orientational distribution. The Onsager description has been largely ignored by the polymer community; truncation at the second virial coefficient was thought to be too unrealistic for rod densities needed to obtain liquid crystals. Recently, however, schemes for decoupling translational and orientational degrees of freedom have extended the range of validity of the virial expansion to higher concentrations (and shorter axial ratios, x; 69). This decoupling also enables the introduction of rod flexibility into the model in a simple way. [Matheson and Flory (70) considered rod flexibility in the lattice model contemporaneously with Khokhlov and Semenov's (71) consideration of this phenomenon in the context of Onsager's model.] If the chain persistence length is R_x, a rigid rod exists when $L/R_x << 1$. The dependence of C_I and C_N on flexibility, L/R_x, has been simulated recently by DuPre and Yang (72); the necessity for larger polymer concentrations to get mesomorphism as the flexibility, L/R_x, increases is apparent (Figure 28). Excellent fits to experimental data for macromolecules exhibiting different inherent flexibilities are obtained on introducing the (additional) L/R_x parameter (72). The revival of interest in the seminal work of Onsager, with flexibility incorporated as introduced by Khokhlov and Semenov (71) and Odijk (73), is the subject of a comprehensive review by Vroege and Lekkerkerker (74).

Rheology

When anisotropy is present in the fluid state, the phenomena described in Chapter 3 become more complex. This complexity may be illustrated with a few aspects of viscoelastic behavior taken from recent literature on lyotropic polypeptide LCPs (rodlike α-helical macromolecules in helicogenic solvents). For example, when $C < C_I$, conventional viscosity versus shear rate behavior (shear thinning) is observed (*see* Chapter 3). Contrary to intuition, however, Figure 29 shows that the viscosity decreases with increasing polymer concentration after the polypeptide concentration, C,

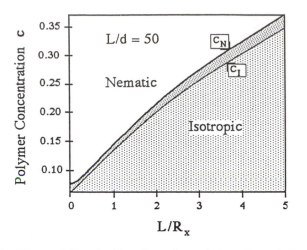

FIGURE 28. The variation in the phase boundaries, C_I and C_N, with rod flexibility, L/R_x. (Reproduced with permission from reference 72. Copyright 1991 American Institute of Physics.)

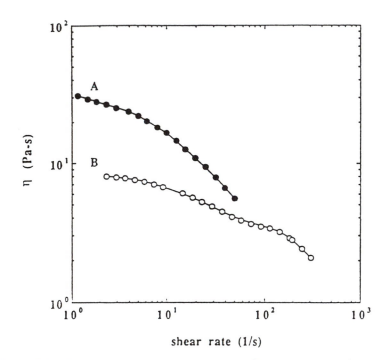

shear rate (1/s)

FIGURE 29. Dependence of shear rate of poly(benzyl glutamate) solutions on polymer concentration: (A) $C < C_I$ (isotropic solution) and (B) $C > C_N$ (mesomorphic solution.) (Reproduced from reference 77. Copyright 1991 American Chemical Society.)

exceeds that required for uniform LC formation, $C > C_N$. Superficially, when rigid rodlike polymers are added to a solution, the viscosity, η, depends very strongly on the rod concentration, ϕ. [For consistency with the original literature, the rod aspect ratio is used to transform the number concentration, C, to the volume fraction of rods, $\phi = (\pi/4)Cd^2L$.] Below the I ↔ N transition, η increases dramatically as ϕ^3; the viscosity exhibits a maximum in the two-phase regime ($\phi_I < \phi < \phi_N$), and then η decreases for $\phi > \phi_N$. Doi (75) has successfully described these qualitative observations by using a rheological constitutive equation based on a molecular kinetic equation that includes the role of mutual ordering of the rodlike solute. In this model, $\phi_I = 8\phi^*/9$, and $\phi_N = \phi^*$. The reduced steady-state viscosity, η/η^*, depends on the reduced concentration, ϕ/ϕ^*, and the degree of rod orientational order, S:

$$\frac{\eta}{\eta^*} = \left(\frac{\varphi}{\varphi^*}\right)^3 \left[\frac{(1-S)^4(1+S)^2(1+2S)(1+3S/2)}{(1+S/2)^2}\right] \quad (30)$$

S in turn depends on the density of rods in the anisotropic phase:

$$S = 0 \quad \text{for} \quad \varphi < \varphi^*$$

$$S = \frac{1}{4} + \frac{3}{4}(1 - 8\varphi^*/9\varphi)^{1/2} \quad \text{for} \quad \varphi > 8\varphi^*/9 \quad (31)$$

A plot of equation 30 shows the rise and fall of η/η^* in the isotropic and nematic phases, respectively (Figure 30a). The inset, Figure 30b, is an estimate of the viscosity behavior in the two-phase regime, $8\phi^*/9 < \phi < \phi^*$

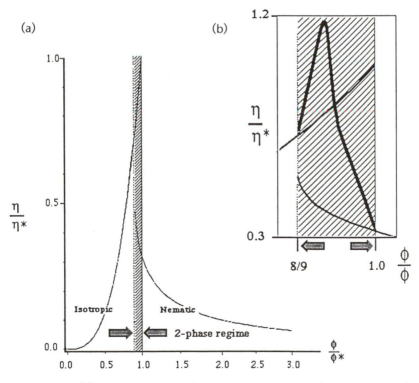

FIGURE 30. (a) Predicted reduced viscosity versus reduced concentration for rigid rods according to Doi (equation 20). (b) Reduced viscosity in the two-phase regime ($C_I > C < C_N$), which is anticipated to reach a maximum according to the classical behavior of two immiscible fluid phases. (Based on reference 76.)

(a mixture of isotropic and anisotropic phases is present), computed with the Taylor formula for a mixture of two immiscible fluids:

$$\frac{\eta}{\eta_0} = 1 + f\left[\frac{(5\eta_1/\eta_0 + 2)}{2(\eta_1/\eta_0 + 1)}\right] \tag{32}$$

The variable f is the fraction of anisotropic phase in the mixture, and η_0 and η_1 are the viscosities of the "host" and "guest" phases, respectively. When $f < 0.5$, the Doi values of the isotropic and anisotropic reduced viscosities are used for η_0 and η_1, respectively; these are interchanged when $f > 0.5$ (it is assumed that f changes linearly with the reduced volume fraction in the two-phase regime, from $f = 0$ at ϕ_I to $f = 1.0$ at ϕ_N). Thus the Taylor formula predicts (76) a maximum in η/η^* within the two-phase region, in agreement with experiments.

Larson and co-workers (77) have carefully studied the rheology of lyotropic polypeptide liquid crystals. The dependence of η on shear rate has been examined for isotropic and anisotropic (liquid-crystalline) solutions (Figure 29). Additionally the first and second normal stress differences, N_1 and N_2, have been shown to exhibit very unusual behavior; for example, N_2 is an oscillatory function of shear rate (Figure 31). The observations are in qualitative agreement with extensions of the Doi theory. (In a disclination-ridden mesophase, it is necessary to average over domain orientations, because the Doi theory pertains to monodomain samples, that is, a macroscopically uniform director field.) In recent theoretical work, Marrucci (78) explicitly considers the "tumbling regime" of the nematic fluid at low shear rates, that is, the rotational motion of the director and the resulting polydomain texture. (At high shear rates, a monodomain with a uniform director field exists.) The theory is in accord with experimentally observed (79) damped oscillations of the transient stress response to step strains.

Finally, band formation is a "serpentine" distortion of the director field caused by shearing (80). Although the mechanism of band formation is not fully understood [two orthogonal deformations of the director field appear to be necessary (81): simple shear and perhaps Frank elasticity orthogonal to the shear direction], band formation may be very important in processing LCPs. A periodic orientational distortion reminiscent of the banded texture is observed in monofilaments that are spun from lyotropic LCPs, and its associated influence on the ultimate mechanical properties of the fiber have been discussed (see Solid State Morphology and Properties).

Elasticity

In rigid rod mesophases where interactions are dominated by excluded-volume interactions, the elasticity contributions to the free energy are entropic in nature. The elastic constants, k_{ii}, are proportional to $\varphi^2(L/d)^2$ and order parameter-dependent factors; bend (k_{33}) is more important than splay (k_{11}), which is three times the twist elastic constant (k_{22}) (82). On the other hand, semiflexible polymers with a chain contour that could

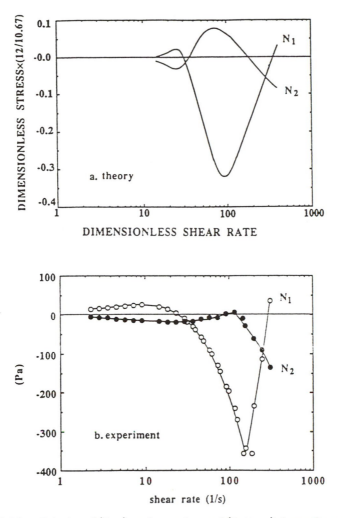

FIGURE 31. Calculated (Top) and experimental (Bottom) dependence of the first and second normal stress differences (N_1 and N_2) on shear rate for poly(benzyl glutamate) mesophase ($C < C_N$). The calculations are based on the extended Doi theory. (Reproduced from reference 77. Copyright 1991 American Chemical Society.)

follow local distortions of the director field suggest different relationships among the k_{ii}; k_{33} should be linear in φ and independent of chain length, L. Meyer (*83*) suggested that k_{ii} is linear in L/d and φ, and that $k_{11} > k_{33}$ for long chains; splay is thought to be dominated by changes in entropy associated with the chain ends. Lee and Meyer (*84*) delineated the distinctions between rigid and semiflexible LCPs by using the lyotropic polypeptide mesophase. They see a crossover from rigid rod behavior, $L/d < R_x$, to semiflexible behavior, $L/d > R_x$, with increasing polypeptide molecular weight. They observed that bending distortions of the chain contour play a significant part in determining the viscoelastic properties of LCPs and confirm the validity of some of the geometrical arguments used to model the elasticity of LCPs.

Solid-State Morphology and Properties

Generally speaking, the morphological features of the solid state of LCPs are derived from the macromolecular organizations exhibited by meso-morphic melts and solutions. A striking example is the hierarchical morphology found in fibers (Figure 32). Until the microscopic levels of this hierarchy are reached, the features of fiber morphology are similar in fibers obtained from both conventional flexible polymers and LCPs. In conventional flexible polymers, a two-phase morphology is present at the molecular level: a chain-folded crystal habit coexists with amorphous connecting chains (Figure 32, Right insert; *see* also Chapter 4). In LCPs, the chain persists for its entire length in an extended-chain crystal, which is essentially continuous along the length of the fiber (Figure 32, Left insert). Despite very similar morphologies, large differences in physical properties exist. These differences may be readily appreciated by contrasting the ultimate tensile moduli of polyethylene (PE) in fibers with the folded-chain habit (fcPE) with those having the extended-chain crystal

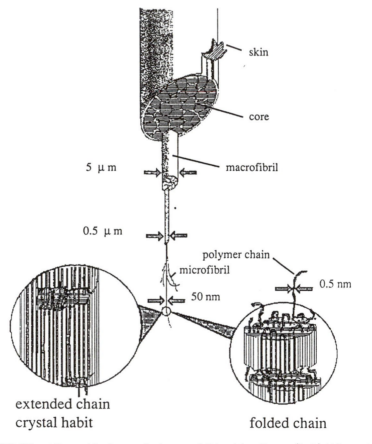

FIGURE 32. Hierarchical morphology exhibited by fibers. (Left) When rigid, rectilinear polymers are drawn from mesophases, an extended-chain crystal habit is adopted, with the chains running parallel to the fiber axis. (Right) In conventional flexible polymers, the semicrystalline folded-chain crystal habit exists.

(ecPE). The ecPE habit may be produced in nonmesomorphic polyethylene by gel spinning (*85*) or solid-state extrusion (*86*). In fcPE, the average modulus is ~ 80 GPa; it is obtained, by summing in series the contributions of the amorphous component (~6 GPa) and estimates of the crystal contribution (~300 GPa). In ecPE, a value of 220 GPa is observed for the tensile modulus; the theoretical maximum estimated for PE is 320 GPa (*87*). In fibers from lyotropic semiflexible poly(arylamide)s, moduli are ~185 GPa; for the more rigid poly(benzobisoxazole), moduli of 365 GPa are reported.

Fibers produced from lyotropic spinning dopes still appear to be limited in the ultimate physical properties because of higher order morphological defects (periodic, serpentine director orientation distortions alluded to earlier; *88*). In this context, much experimental and theoretical work remains to be done to delineate those parameters that control disclination textures and director patterns created by complex shear fields encountered in processing LCPs. As is typically the case, these difficulties appear to have been optimally minimized in natural systems: spiders spin nearly defect-free fibers from a mesomorphic form of silk (*89*). Consequently, efforts are underway to analyze the spinning process, (i.e., the spinner draw-down geometry and its associated shear field) in arachnids.

In melts and solutions of semiflexible LCPs, rheologically induced stress fields probably couple with polymer conformational changes; this additional complication needs to be investigated to resolve the origins of the hierarchical "skin core" morphologies observed in the solid state of LCPs, because these supramolecular considerations undoubtedly play a key role in determining the ultimate physical properties of high-performance LC polymers. In fact, lack of understanding of the processing of LCPs is one factor limiting the utility of this class of materials (*4, 90*); current monomer costs also are a significant consideration. Nevertheless, the polymer science community agrees that the role of LCPs in specialized applications that exploit anisotropy, in blends and (self-reinforcing) composites, and as processing aids, will be very important and, at the same time, technologically challenging.

References

1. Donald, A. M.; Windle, A. H. *Liquid Crystalline Polymers;* Cambridge University Press: Cambridge, England, 1991.
2. *Liquid Crystallinity in Polymers;* Ciferri, A., Ed.; Cambridge University Press: New York, 1991.
3. Collings, P. J. *Liquid Crystals;* Princeton University Press: Princeton, NJ, 1990.
4. *Liquid Crystalline Polymers* (National Materials Advisory Board Committee, Samulski, E. T., chair); National Academy Press: Washington, DC, 1990.
5. *Liquid Crystalline Polymers;* Weiss, R. A.; Ober, C. K., Eds.; ACS Symposium Series 435; American Chemical Society: Washington, DC, 1990.
6. *Side Chain Liquid Crystal Polymers;* McArdle, C. B., Ed.; Chapman and Hall: London, 1989.
7. Vertogen, G.; deJeu, W. H. *Thermotropic Liquid Crystals, Fundamentals;* Springer-Verlag: Berlin, Germany, 1988.

8. *Thermotropic Liquid Crystals;* Gray, G. W., Ed.; *Crit. Rep. Appl. Chem. 22;* John Wiley: New York, 1987.

9. *Recent Advances in Liquid Crystalline Polymers;* Chapoy, L. L., Ed.; Elsevier: London, 1985.

10. *Polymeric Liquid Crystals;* Blumstein, A., Ed.; Plenum: New York, 1985.

11. *Faraday Discuss. Chem. Soc.* **1985,** *79.*

12. *Liquid Crystal Polymers I;* Gordon, M., Ed.; Springer-Verlag: Berlin, Germany, 1984.

13. *Liquid Crystal Polymers II / III;* Gordon, M., Ed.; Springer-Verlag: Berlin, Germany, 1984.

14. *Polymer Liquid Crystals;* Cifferi, A.; Krigbaum, W. R.; Meyer, R. B., Eds.; Academic: New York, 1982.

15. Kelker, H.; Hatz, R. *Handbook of Liquid Crystals;* Verlag Chemie: Weinham, Germany, 1980.

16. *Liquid Crystalline Order in Polymers;* Blumstein, A., Ed.; Academic: New York, 1978.

17. Chandrasekhar, S. *Liquid Crystals;* Cambridge University Press: Cambridge, England, 1977.

18. DeGennes, P. G. *The Physics of Liquid Crystals;* Oxford University Press: Oxford, England, 1974.

19. *Introduction to Liquid Crystals;* Priestly, E. B.; Wojtowicz, P. J.; Sheng, P., Eds.; Plenum: New York, 1974.

20. Stephen, M. J.; Straley, J. P. *Rev. Mod. Phys.* **1974,** *46,* 617.

21. Sage, I. *Thermotropic Liquid Crystals;* Gray, G. W., Ed.; *Crit. Rep. Appl. Chem. 22;* John Wiley: New York, 1987; Chapter 3.

22. Cai, R.; Samulski, E. T. *Liq. Cryst.* **1991,** *9,* 617.

23. Demus, D.; Zasehke, H. *Flussige Kristalle in Tabellen*; VEB Deutscher Verlag für Grundstoffindustrie; Leipzig, 1984. Toyne, K. J. *Thermotropic Liquid Crystals;* Gray, G. W., ed.; *Crit. Rep. Appl. Chem. 22;* John Wiley: New York, 1987.

24. Chandler, D.; Weeks, J. D.; Anderson, H. C. *Science* **1983,** *220,* 787.

25. Vainshtein, B. Ka. *Diffraction of X-rays by Chain Molecules;* Elsevier: Amsterdam, 1966.

26. Oldenbourg, R.; Wen, X.; Meyer, R. B.; Caspar, D. L. D. *Phys. Rev. Lett.* **1988,** *61,* 1851.

27. Murthy, S. N.; Knox, J. R.; Samulski, E. T. *J. Chem. Phys.* **1978,** *65,* 4835.

28. Photinos, D. J.; Poliks, B. J.; Samulski, E. T.; Terzis, A. F.; Toriumi, H. *Mol. Phys.* **1991,** *72,* 333.

29. Viney, C.; Daniels, C. M. *Mol. Cryst. Liq. Cryst.* **1991,** *196,* 133, and references therein.

30. Wood, B. A.; Thomas, E. L. *Nature* **1986,** *324,* 655.

31. Chuang, I.; Durrer, R.; Turok, N.; Yurke, B. *Science* **1991,** *251,* 1336.

32. Neville, A. C.; Caveney, S. *Biol. Rev.* **1969,** *44,* 531.

33. Clark, N. A.; Lagerwall, S. T. *Appl. Phys. Lett.* **1980,** *36,* 899. Myrvold, B. O. *Mol. Cryst. Liq. Cryst.* **1991,** *202,* 123.

34. Naciri, J.; Pfeifer, S.; Shashidhar, R. *Liq. Cryst.* **1991,** *10,* 585.

35. Prasael, P. N.; Williams, D. J. *Introduction to Nonlinear Optical Effects in Molecules and Polymers;* John Wiley: New York, 1991. See also, Mohlman, G. R.; van der Vorst, C. P. J. M. *Side Chain Liquid Crystal Polymers;* McArdle, C. B., ed.; Chapman and Hall: London, 1989; Chapter 12.

36. Brostow, W. *Polymer* **1990,** *31,* 979.

37. Davidson, P.; Levelut, A. M. *Liq. Cryst.* **1992,** *11,* 469.

38. Ober, C. K.; Jin, J.; Lenz, R. W. *Liquid Crystal Polymers II / III;* Gordon, M., Ed.; Springer-Verlag: Berlin, Germany, 1984; p 104.

39. Toriumi, H.; Furuya, H.; Abe, A. *Polym. J.* **1985,** *17,* 895.

40. Blumstein, R. B.; Stickles, E. M. *Mol. Cryst. Liq. Cryst.* **1982,** *82,* 151.
41. Davidson, P.; Noirez, L.; Cotton, J. P.; Keller, P. *Liq. Cryst.* **1991,** *10,* 111, and references cited therein.
42. Warner, M. *Side Chain Liquid Crystal Polymers;* McArdle, C. B., Ed.; Chapman and Hall: London, 1989; Chapter 2.
43. Schmidt, H.-W. *Angew. Chem. Adv. Mater.* **1989,** *101,* 964.
44. Zentel, R. *Angew. Chem. Int. Ed. Engl. Adv. Mater.* **1989,** *28,* 1407.
45. Mitchell, G. R.; Davis, F. J.; Ashman, A. *Polymer* **1987,** *28,* 639.
46. Finkelmann, H. *Angew. Chem. Int. Ed. Engl.* **1988,** *27,* 987.
47. Wendorff, J. H. *Angew. Chem. Int. Ed. Engl.* **1991,** *30,* 405.
48. Meier, W.; Finkelmann, H. *Makromol. Chem., Rapid Commun.* **1990,** *11,* 599.
49. Skoulios, A. *Adv. Liq. Cryst.* **1975,** *1,* 169.
50. Yang, J.; Wegner, G. *Macromolecules* **1992,** *25,* 1786.
51. Milner, S. T. *Science* **1991,** *251,* 905.
52. Martin, D. G.; Stupp, S. I. *Macromolecules* **1988,** *21,* 1222. See also Jin, J.-I. *Liquid Crystalline Polymers;* Weiss, R. A.; Ober, C. K., Eds.; ACS Symposium Series 435; American Chemical Society: Washington, DC, 1990; Chapter 3; and Economy, J.; Johnson, R. D.; Lyerla, J. R.; Multilebach, A. *Liquid Crystalline Polymers;* Weiss, R. A.; Ober, C. K., Eds.; ACS Symposium Series 435; American Chemical Society: Washington, DC, 1990; Chapter 10.
53. General Discussion in *Faraday Discuss. Chem. Soc.* **1985,** *79,* 89.
54. Cai, R.; Preston, J.; Samulski, E. T. *Macromolecules* **1992,** *25,* 563.
55. Gleim, W.; Finkelmann, H. *Side Chain Liquid Crystal Polymers;* McArdle, C. B., Ed.; Chapman and Hall: London, 1989; Chapter 10.
56. Warner, M.; Wang, X. J. *Macromolecules* **1991,** *24,* 4932, and references therein.
57. Doi, M.; Pearson, D.; Kornfeld, J.; Fuller, G. *Macromolecules* **1989,** *22,* 1488.
58. Maier, W.; Saupe, A. *Z. Naturf.* **1958,** *A13,* 564; **1959,** *A14,* 1909; **1960,** *A15,* 282.
59. Gelbart, W. M. *J. Phys. Chem.* **1982,** *86,* 4298.
60. Cotter, M. A. *Philos. Trans. R. Soc. London* **1983,** *A309,* 127.
61. Flory, P. J. *Proc. R. Soc. London* **1956,** *A234,* 73. See also *Liquid Crystal Polymers I;* Gordon, M., Ed.; Springer-Verlag: Berlin, Germany, 1984; p 1.
62. Flory, P. J.; Frost, R. S. *Macromolecules* **1978,** *11,* 1126.
63. Flory, P. J. *Macromolecules* **1982,** *15,* 1286.
64. Aharoni, S. M. *Polymer* **1980,** *21,* 21. Bianchi, E.; Ciferri, A.; Tealdi, A. *Macromolecules* **1982,** *15,* 1268.
65. Flory, P. J.; Ronca, G. *Mol. Cryst. Liq. Cryst.* **1979,** *54,* 311.
66. Onsager, L. *Ann. N.Y. Acad. Sci.* **1949,** *51,* 627.
67. Zocher, H. *Anorg. Chem.* **1925,** *186,* 75.
68. Folda, T.; Hoffman, H.; Canzy, H.; Smith, P. *Nature* **1988,** *333,* 55.
69. Lee, S.-D. *J. Chem. Phys.* **1987,** *87,* 4972.
70. Matheson, R. R.; Flory, P. J. *Macromolecules* **1981,** *14,* 954. Flory, P. J. *Macromolecules* **1978,** *11,* 1141.
71. Khokhlov, A. R.; Semenov, A. N. *Physica* **1981,** *108A,* 546. *Macromolecules* **1984,** *17,* 2678. Khokhlov, A. R. *Phys. Lett.* **1978,** *68A,* 135.
72. DuPre, D. B.; Yang, S.-J. *J. Chem. Phys.* **1991,** *94,* 7466.
73. Odijk, T. *Macromolecules* **1986,** *19,* 2313.
74. Vroege, G. J.; Lekkerkerker, H. N. W. *Rpts. Prog. Phys.* **1992.**
75. Doi, M. *J. Polym. Sci., Polym. Phys. Ed.* **1981,** *19,* 229.
76. Samulski, E. T. *Physics Today* **1982,** *35,* 40. Private communication to Flory, P. J.; July 2, 1982.
77. Magda, J. J.; Baek, S.-G.; DeVries, K. L.; Larson, R. G. *Macromolecules* **1991,** *24,* 4460, and references cited therein.

78. Marrucci, G. *Macromolecules* **1991,** *24,* 4176. Marrucci, G.; Maffettone, P. L. *Macromolecules* **1989,** *22,* 4076.
79. Mewis, J.; Moldenaars, P. *Mol. Cryst. Liq. Cryst.* **1987,** *153,* 291.
80. Viney, C.; Windle, A. H. *Polymer* **1986,** *27,* 1325. Navard, P.; Zachariades, A. E. *J. Polym. Sci., Polym. Phys. Ed.* **1987,** *25,* 1089.
8ι. Maffettone, P. L.; Grizzuti, N.; Marrucci, G. *Liq. Cryst.* **1989,** *4,* 385.
82. Lee, S.-D.; Meyer, R. B. *J. Chem. Phys.* **1986,** *84,* 3443.
83. Meyer, R. B. *Polymer Liquid Crystals;* Cifferi, A.; Krigbaum, W. R.; Meyer, R. B., Eds.; Academic: New York, 1982; Chapter 6.
84. Lee, S.-D.; Meyer, R. B. *Phys. Rev. Lett.* **1988,** *61,* 2217.
85. Pennings, A. J.; Schouteten, C. J. H.; Kiel, A. M. *J. Polym. Sci.* **1972,** *C38,* 167.
86. Kanamoto, T.; Tsuruta, A.; Tanaka, K.; Takeda, M.; Porter, R. *Polym. J.* **1983,** *15,* 327.
87. Adams, W. W.; Eby, R. K. *MRS Bull.* **1987,** *12,* 22.
88. Northolt, M. G.; Sikkema, D. J. *Adv. Polym. Sci.* **1990,** *98,* 115.
89. Kerkam, K.; Viney, C.; Kaplan, D.; Lombardi, S. *Nature* **1991,** *349,* 596.
90. Economy, J. *Angew. Chem. Int. Ed. Engl.* **1990, 29,** 1256.

Some Characterization Techniques

Chapter 6

Spectroscopic Characterization of Polymers

Jack L. Koenig

Case Western Reserve University,
Department of Macromolecular Science,
10900 Euclid Avenue, Cleveland, OH 44106–7202

Theory of Polymer Characterization

The primary motivation for determining the structure of a polymer chain is to relate the structure to the performance properties of the polymer in end use. If the polymer chain is completely characterized and the structural basis of polymer properties is known, the polymerization can be controlled and optimized to produce the best possible properties from the chemical system.

Synthetic polymer chains can vary in a large number of different structural elements (List I). The list of structural variables is long, and some polymers exhibit a number of these structural variables simultaneously. Hence the number of possible structures for a single chain is very large, indeed.

The problem is complicated further by the fact that the nature of the distribution of the structural variables along the chain also influences the properties. The types of distributions of structural elements that can occur include random, partially disordered, and ordered distributions (such as alternating and block distributions). These structural distributions are determined by the nature of the polymerization process; that is, the disordered distributions arise from the type of stochastic processes involved in the polymerization. The alternating, ordered distribution can be obtained only by systematic chemical design, and the block structure can be obtained only by chemical control of the polymerization process.

So how does this distribution influence our characterization of the polymer? First, the chain structure of the polymer is highly variable,

2505–2/93/0263 $13.50/1

List I. Structural Variables for Synthetic Polymers

- Molecular weight
- Chemical defects
    Impurities in feed
    Monomer isomerization
    Side reactions
- Enchainment defects
    Positional defects
    Stereospecific defects
    Branches
    Cyclic isomers
- Chain conformations
    Stiff, ordered chains
    Flexible, amorphous chains
- Morphological effects
    Crystal phases
    Interfacial regions
    Entanglements

because the polymerization process is a statistical process determined by probability considerations. Thus, a polymer sample always represents a multicomponent, complex structural mixture. Second, detailed pictures of the structure cannot be obtained, because measurements will give only weighted averages of the structure.

Elements of Polymer Structure

The *composition* of a molecule defines the nature of the atoms and their type of bonding irrespective of their spatial arrangement. The *configuration* of chemical groups characterizes the chemical state of a molecule. Different configurations constitute different chemical individuals and cannot be converted into one another without rupture of chemical bonds. The *conformation* of chemical groups characterizes the geometrical state of a molecule. Different conformations of a molecule can be produced by rotation about single bonds without rupture of chemical bonds. Changes in conformation arise from physical considerations such as temperature, pressure, or stress and strain.

Polymer chains are made up of sequences of chemical repeating units that may be arranged regularly or irregularly on the backbone. The chemical microstructure is defined as the internal arrangement of the different sequences on the polymer chain.

Structural Model of Polymer Chain

For a microstructural analysis of the polymer chain, we are going to use a model made up of connected repeating units of similar or different structures. We will use letters to designate the different structural types of the repeating unit, that is, A and B. So the chains are structurally presented as sequences of similar or different units, that is, A, AA, AAA, B, BB, BBB, AB, AAB, BBB, etc. The other structural components can be indicated by the use of additional letters, C, D, E, and so forth. The application of this model to copolymers (A and B) and terpolymers (A, B, and C) is obvious. Polymers that differ only in positional conformation and configurational isomerism (stereoregularity), branching, and cross-linking will be considered as copolymer analogues, although they are not generated by copolymerization. The goal of polymer characterization is to generate the structural sequence distribution function necessary to calculate the highest possible weighted-average sequence structure that represents the chain. We want to relate these average sequence structures to the performance properties of the polymer under consideration, and we will seek the polymerization mechanism and parameters that generated the sequence structure so that we can optimize and control the polymer structure.

Measurement of Polymer Structure Composition

For simplicity, let us first consider the experimental measurements that are possible on the microstructure of a polymer chain made up of only two structures, A and B. Consider the following portion of a polymer chain:

$$\text{Chain A–B–B–A–B–B–B–A–B–A–Chain}$$

This portion of the chain has 10 repeating units. We will assume that this portion of the chain is representative of the complete chain. For simplicity, we will assume also that the molecular weight is sufficiently large so that the end groups may be neglected from consideration.

If we had infinite spectral resolution or the ability to detect the differences between the two structural elements A and B, what could we measure? The first and simplest property we would measure are the individual numbers of A and B structural elements in the polymer; that is, we would count the relative numbers of A and B elements taken one at a time, which we will term mono-ads, and write $N_1(A)$ as the number of mono-ads of A. For the model segment considered earlier, we have the following mono-ads for the two elements:

$$N_1(A) = 4 \text{ and } N_1(B) = 6$$

Spectroscopically, we measure the fraction of units in the chain. We can express this result in terms of the number fraction probabilities $P_1(A)$ or $P_1(B)$ for mono-ads by dividing by the total number of segments ($N = 10$):

$$P_1(A) = N_1(A)/N \tag{1a}$$

$$P_1(B) = N_1(B)/N \tag{1b}$$

For our system, $P_1(A) = 4/10 = 0.4$, and $P_1(B) = 6/10 = 0.6$. This result, $P_1(A)$ or $P_1(B)$, is readily recognized as the fractional composition in terms of A and B units of the polymer when we have assumed that the 10-segment portion of the polymer is representative of the total chain. For our model polymer, $P_1(A) + P_1(B) = 1$, because the polymer contains only the units A and B.

Measurement of Polymer Structure by Dyad Segments

Let us become slightly more sophisticated and analyze this 10-unit segment by counting sequences two at a time; that is, let us consider the number of dyads (2-ads) in the chain:

$$\text{Chain A-B-B-A-B-B-B-A-B-A-|-A-next sequence}$$

$$\text{Dyads AB-BB-BA-AB-BB-BB-BA-AB-BA-AA-}$$

The final AA dyad arises because the adjacent segment, which is identical to this segment, starts with an A. Four types of dyads are possible: AA, BB, AB, and BA. For this chain, $N_2(AA) = 1$, $N_2(BB) = 3$, $N_2(AB) = 3$, and $N_2(BA) = 3$.

Before we proceed, let us pose a question. Can we actually experimentally distinguish between the heterodyads AB and BA? They are differentiated in this case by the direction of the counting (right to left) that we made, but what if the chain segments were reversed? The AB dyads would become BA dyads and vice versa. Which direction of the chain is the proper one? Clearly, both directions are equally likely, and we will never know from examination of the final polymer chain which direction it grew during the polymerization. Therefore, AB and BA are equally likely and so $N_2(AB) = N_2(BA)$. The equality saves the day for spectroscopic measurements, because we have no spectroscopic method that can distinguish between AB and BA. From a spectroscopic point of view, AB and BA are indistinguishable. Consequently, any measurement of dyad units measures the sum $N_2(AB) + N_2(BA)$, which we will designate as $N_2(\underline{AB})$:

$$N_2(\underline{AB}) = N_2(AB) + N_2(BA)$$

With that problem settled, we can proceed to determine the number fraction of each dyad, $P_2(AA)$, $P_2(BB)$, and $P_2(AB)$, by using N_2 (total) = 10:

$$P_2(AA) = N_2(AA)/N_2 \tag{2a}$$

$$P_2(BB) = N_2(BB)/N_2 \tag{2b}$$

$$P_2(\underline{AB}) = P_2(AB) + P_2(BA) = N_2(\underline{AB})/N_2 \tag{2c}$$

For our 10-segment polymer chain,

$$P_2(AA) = 0.1, \ P_2(BB) = 0.3, \text{ and } P_2(\underline{AB}) = 0.6$$

The normalization condition is satisfied, which is $P_2(AA) + P_2(\underline{AB}) + P_2(BB) = 1$

Measurement of Polymer Structure by Triad Segments

We can proceed to dissect the polymer structure in terms of higher n-mers. Let us determine the structure of the polymer chain in terms of triads or 3-ads, that is, by counting the units as threes. For triads, a total of eight structures are possible:

Chain A–B–B–A–B–B–B–A–B–A–|–A–B–next segment

Triads–ABB–BBA–BAB–ABB–BBB–BBA–BAB–ABA–BAA–AAB–

Once again, the final two triads are recognized by the adjoining segment, which corresponds to the first two units in the beginning of this segment. For this chain, $N_3(AAA) = 0$, $N_3(AAB) = 1$, $N_3(ABA) = 1$, $N_3(BAA) = 1$, $N_3(ABB) = 2$, $N_3(BBA) = 2$, $N_3(BAB) = 2$, and $N_3(BBB) = 1$, with N_3 (total) = 10.

Two "reversibility relations" exist for the triads: $N_3(AAB) = N_3(BAA)$ and $N_3(BBA) = N_3(ABB)$. Again, a normalization condition represents the sum of all of the triads. We can now calculate the number fraction of each triad: $P_3(AAA) = 0$, $P_3(AAB) = 0.2$, $P_3(ABA) = 0.1$, $P_3(ABB) = 0.4$, $P_3(BAB) = 0.2$, and $P_3(BBB) = 0.1$.

Measurement of Polymer Structure by Higher n-*ad Segments*

We could go on to tetrads, pentads, etc., up to the 10-ad for the case at hand, and the process is the same. One could derive the necessary relations by observing that the total number of different combinations for the higher n-ads goes up as $2n$ where n is the n-ad length. Therefore, the number of combinations increases very rapidly, and the task gets tedious, although not difficult. Heptads have been observed in high-resolution NMR spectroscopy so it may be necessary to plow through the derivation process (or look it up in the literature).

The ability to count the fraction of the different n-ads is a function of the sensitivity of the spectroscopic method involved. In some cases, the spectroscopic method allows only the observation of isolated units; in others, dyads; and in some, triads, etc. The nature of the spectroscopic n-ad sensitivity and how it can be determined for the polymer system being studied will be discussed in later sections.

Relations between Various Orders of Sequences

This section addresses the relationships between the fractions of the n-ads of different order, that is, different length. All of the n-ad sequences are related to all the others in some fashion through the structural distribution function. The required relationships between the different levels of n-ads are easily derived.

Consider the simplest case first: the relationship between the 1-ads (isolated units or mono-ads) and the 2-ads (dyads). The dyads are formed from the 1-ads by addition of another unit, and there are only two possible ways of making the addition—before the unit or after. So consider starting with an A unit:

$$P_1(A) = P_2(\underline{A}A) + P_2(\underline{A}B) \text{ units added after the A unit}$$

$$P_1(A) = P_2(A\underline{A}) + P_2(B\underline{A}) \text{ units added before the A unit}$$

List II. Necessary Relations between Relative Concentrations of Comonomer Sequences of Different Lengths

- Dyad–mono-ad

$$P(AA) + (1/2)\underline{P(AB)} = P(A)$$
$$P(BB) + (1/2)\underline{P(AB)} = P(B)$$

- Triad–dyad

$$P(AAA) + (1/2)P(\underline{AAB}) = P(AA)$$
$$P(BAB) + (1/2)P(\underline{AAB}) + P(ABA) + (1/2)P(\underline{ABB}) = P(AB)$$
$$P(BBB) + (1/2)P(\underline{BAA}) = P(BB)$$

- Tetrad–triad

$$P(AAAA) + (1/2)\underline{P(AAAB)} = P(AAA)$$
$$P(BAAB) + (1/2)\underline{P(AAAB)} + (1/2)P(AABA) + (1/2)P(AABB) = \underline{P(AAB)}$$
$$(1/2)P(ABAB) + (1/2)P(BABB) = P(BAB)$$
$$(1/2)P(ABAB) + (1/2)P(AABA) = P(ABA)$$
$$P(ABBA) + (1/2)\underline{P(ABBB)} + (1/2)P(AABB) + (1/2)(BABB) = \underline{P(ABB)}$$
$$P(BBBB) + (1/2)\underline{P(ABBB)} = P(BBB)$$

So, experimentally we have the following relationships:

$$P_1(A) = P_2(AA) + 1/2\,P_2(\underline{AB}) \tag{3a}$$

$$P_1(B) = P_2(BB) + 1/2\,P_2(\underline{AB}) \tag{3b}$$

We have derived the reversibility relationship that we had previously alluded to: $P_2(AB) = P_2(BA)$.

The relationships between the various n-ads are shown in List II. These relationships are useful, because they represent a quantitative requirement to be met by the experimental measurements and a test of the structural assignments.

Calculation of Polymer Structural Parameters from Sequence Measurements

We now want to use the measured information on the n-ads to calculate structural information about the polymer, which will be useful in comparing one polymer system with another. Calculations of various average quantities will be useful in understanding the type of polymer we are studying in terms of the distribution of microstructure.

Structural Composition

First and foremost, we want to know the structural composition. The structural composition is the ratio of the various types of structure. If we can measure the number fraction of single units A or B, we can calculate the composition from the ratio of these measurements:

$$\text{Structural composition} = P_1(A)/P_1(B) \tag{4}$$

For our example, $P_1(A)/P_1(B) = 0.4/0.6 = 0.67$.

Sequence Order Parameter

If we can measure both mono-ads and dyads, we can determine a statistical parameter that will indicate whether the distribution of the structures is random or tends toward an alternating or block distribution. This order parameter, χ, is defined as follows:

$$\chi = 1/2\, P_2(\underline{AB})\,/P_1(A)\,P_1(B) \tag{5}$$

If $1/2\, P_2(\underline{AB}) = P_1(A)P_1(B)$ (or $\chi = 1$), we have independent probabilities, and the distribution is termed random. The values of χ are interpreted as follows:

$\chi = 1$, random distribution of A and B

$\chi > 1$, more alternating tendency

$\chi < 1$, more block tendency

$\chi = 2$, completely alternating A and B

$\chi = 0$, complete blocks of A and B

Number-Average Sequence Lengths

We are interested in the number of $N_A(n)$ fractions of the sequences of A units, which is defined as follows:

$$N_A(n) = P_{n+2}(BA_nB)\bigg/ \sum_{n=1}^{\infty} P_{n+2}(BA_nB) \tag{6}$$

It should also be noted that

$$P_2(AB) = \sum_{n=1}^{\infty} P_{n+2}(BA_nB) \tag{7a}$$

$$P_2(BA) = \sum_{n=1}^{\infty} P_{n+2}(BA_nA) \tag{7b}$$

Therefore,

$$N_A(n) = P_{n+2}(BA_nB)\,/P_2(BA) \tag{8}$$

The number-average sequence length, $\bar{l}_A$, is defined as follows:

$$\bar{l}_A = \left[\sum_{n=1}^{\infty} n N_A(n) \right] \bigg/ \left[\sum_{n=1}^{\infty} N_A(n) = \sum_{n=1}^{\infty} n N_A(n) \right] \tag{9}$$

Upon substitution, we obtain equation 10:

$$\bar{l} = \left[\sum_n nP_{n+2}(\mathrm{BA}_n\mathrm{B}) \right] \Big/ P_2(\underline{\mathrm{BA}}) \qquad (10)$$

Because $\sum_n nP_{n+2}(\mathrm{BA}_n\mathrm{B}) = P_1(\mathrm{A})$, we obtain the number-average sequence length in terms of simple measurable sequences:

$$\bar{l}_{\mathrm{A}} = P_1(\mathrm{A}) \Big/ 1/2\,P_2(\underline{\mathrm{BA}}) \qquad (11)$$

By using the required n-ad relationships, this result can be expressed in terms of the measured dyads only:

$$\bar{l}_{\mathrm{A}} = \left[P_2(\mathrm{AA}) + 1/2\,P_2(\underline{\mathrm{AB}}) \right] \Big/ 1/2\,P_2(\underline{\mathrm{BA}}) \qquad (12a)$$

$$\bar{l}_{\mathrm{B}} = \left[P_2(\mathrm{BB}) + 1/2\,P_2(\underline{\mathrm{AB}}) \right] \Big/ 1/2\,P_2(\underline{\mathrm{BA}}) \qquad (12b)$$

With triads, the number-average sequence length can be calculated by using the following:

$$\bar{l}_{\mathrm{A}} = \frac{\left[P_3(\mathrm{BAB}) + P_3(\mathrm{AAB}) + P_3(\mathrm{AAA}) \right]}{P_3(\mathrm{BAB}) + 1/2\,P_3(\underline{\mathrm{AAB}})} \qquad (13a)$$

$$\bar{l}_{\mathrm{B}} = \frac{\left[P_3(\mathrm{ABA}) + P_3(\mathrm{BBA}) + P_3(\mathrm{BBB}) \right]}{P_3(\mathrm{ABA}) + 1/2\,P_3(\underline{\mathrm{BBA}})} \qquad (13b)$$

If tetrads or higher n-ads are measured, corresponding relationships can be derived for calculation of the number-average sequence lengths.

Relating Polymer Structure to Polymerization Parameters

What factors in the polymerization reaction are responsible for the sequence structure of the polymer? If one can completely characterize the sequence distribution of the polymer chain, it should be possible to relate the structure to the basic polymerization process if proper control has been exercised (1). In the ideal circumstance, discovering the relationships between the chemistry and the structure is quite easy. However, in the real case, it is quite difficult but possible, particularly when using some of the modern computational techniques available (1).

Summary

This section has demonstrated, in the simplest possible terms, the approach that is necessary to go from spectroscopic measurements of the microstructure of the polymer chain to a determination of the fundamental parameters of the polymerization process. Although it would be interesting to know the position and the geometric relationships of every atom in a polymer chain, what would we do with the information? Spectroscopically, we can only measure the average structure of the polymer chains in the sample. Therefore, our only choice is to try to relate these structural parameters to the performance properties of the polymer and to the polymerization route that brought about the polymer structure.

Selection of Spectroscopic Method

Spectroscopic techniques for the study of polymers must yield high-resolution, narrow-line-width spectra that provide selectivity and structural information. Because polymer systems are always complex mixtures of structure and molecules, the spectroscopic probe must permit selective monitoring of more than one structural type at a time. It must possess sufficient sensitivity to detect and monitor very low levels of structure in the polymer, because small structural changes produce much larger effects on physical and mechanical properties. The spectroscopic probe must be very specific in its informational content, because we will need to determine not only the structure of the single repeat units but also how they are connected together and to what extent the units are ordered. The technique should be nondestructive and noninvasive, because the polymer samples will be evaluated by other methods besides the spectroscopic characterization. The probe should be capable of studying the polymer in its industrially useful form, be it fiber, film, composite, coating, or adhesive.

A number of factors determine the selection of the various spectroscopic methods for the characterization of polymers. The first criterion is the nature of the spectral information relative to the problem at hand. All of the spectroscopic techniques have strengths and weaknesses in certain aspects of their application so that no single technique is going to solve all of the problems. Vibrational spectroscopy (infrared and Raman) have advantages of being rapid and sensitive with sampling techniques that are easy to use. Yet vibrational spectroscopies are somewhat difficult to interpret unambiguously. Nuclear magnetic resonance (NMR) techniques (both high-resolution solid and solution techniques) are unrivaled in the detail of structural information that is obtained experimentally and in the relative ease of interpretation, but NMR spectroscopy is expensive, time-consuming, and requires analysts with considerable experimental skill to produce quantitative results. Mass spectroscopy is extremely sensitive, but the structural information is sometimes difficult to understand, particularly if the decomposition mechanisms are complex. Other analytical techniques are available that are suited for specific problems but do not have broad utility for the characterization of polymers.

Determination of Composition and Microstructure of Polymers

As indicated earlier, what differentiates polymer spectroscopy from ordinary spectroscopy is the need for information about the "connectivity" of structural elements in adjacent repeating units of the polymer chain. This connectivity information is required to measure the microstructural sequences and to determine the structural distribution along the polymer chain.

Dependence of Vibrational Spectra on Structural Sequences

In vibrational spectroscopy, the "vibrational coupling" of the modes results in frequency shifts that allow a determination of the microstructural sequences (2). The adjacent repeat units must have the same energy levels for coupling to occur. When coupling does occur, the frequency shifts are given by equation 14:

$$v_s^2 = v_0^2 + v_1^2 \ (1 + \cos \theta) \tag{14}$$

In equation 14, θ (the phase angle between units) $= s\pi(N + 1)$; $s = 1, 2, \ldots, N$; v_0 is the frequency of the uncoupled or isolated mode; and v_1 is the interaction parameter. Thus, if v_1 is large enough, frequency differences will be observed for BAB, $B(A)_2 B$, $B(A)_3 B$, $B(A)_4 B$, $B(A)_5 B$, etc., until the resolutional limitation of the instrument is reached. Although equation 14 suggests that there will be s bands observed for a sequence that is s units long, the experimental intensities are not observable for $s > 1$, except in rare instances. The general shape of the plot of the frequency dependence as a function of the number of units in sequence is shown in Figure 1.

For copolymers of ethylene with other monomers, IR bands arising from ethylene sequences are given in Table I. These vibrational modes have been used to characterize the sequence distribution of ethylene–propylene (E–P) copolymers (3). The IR spectrum of the methylene-rocking region is shown in Figure 2. The assignments of the peaks to the various methylene sequences are also shown. Quantitatively, head-to-tail P–P sequences can be identified by using the 810–815 cm^{-1} absorption band. The tail-to-tail P–P sequences are identified by the presence of the band

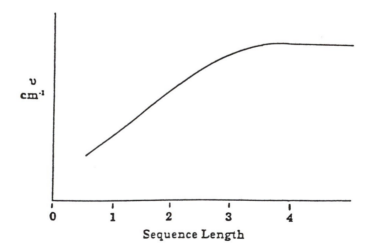

FIGURE 1.   General shape of the plot of the frequency dependence as a function of the number of units in the sequence.

Table I. Frequencies of Methylene Rocking Modes

Structure	Frequency of Rocking Mode (cm^{-1})
$XC-CH_2-CX$	815
$XC-(CH_2)_2-CX$	752
$XC-(CH_2)_3-CX$	733
$XC-(CH_2)_4-CX$	726
$XC-(CH_2)_5-CX$	722

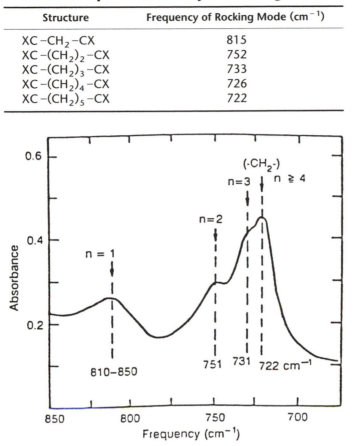

FIGURE 2. IR spectrum of the methylene-rocking region of an ethylene–propylene copolymer (54.3 wt % C_2). (Reproduced from reference 3. Copyright 1968 American Chemical Society.)

at 751 cm^{-1}, which is assigned to the methylene sequences with $n = 2$. The band at 731 cm^{-1} is assigned to the E–P sequences, and the line at 722 cm^{-1} is associated with the $(E-E)_n$ sequences. Quantitative measurements have been made for this system, and the results are shown in Figure 3 (*3*).

Dependence of NMR Spectra on Structural Sequences

For NMR spectroscopy, the spectral sensitivity for determining connectivity is based on the dependence of the chemical shifts on shielding parameters arising from the carbons adjacent to the measured carbons. The Grant and Paul equation for calculation of chemical shifts demonstrates simply this type of sensitivity (*4*). The empirical equation for predicting the chemical shifts is as follows:

$$\delta = C + \sum_i n_i A_i \qquad (15)$$

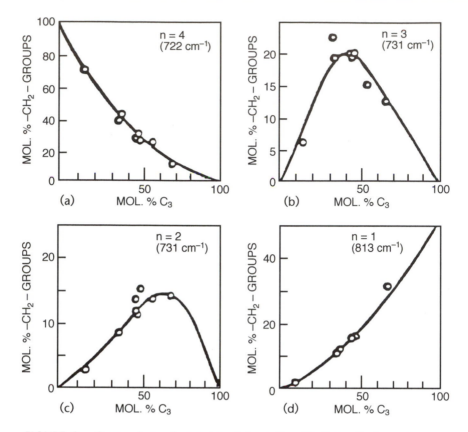

FIGURE 3. Comparison of experimental results with theoretical curves as a function of mole percent propylene: (a) four or more contiguous groups, (b) three contiguous groups, (c) two contiguous groups, and (d) isolated methylene groups. (Reproduced from reference 3. Copyright 1968 American Chemical Society.)

In Equation 15, C is a constant for hydrocarbons, n_i is the number of substituents in position i, A_1 is the α effect (9.1 ppm), A_2 is the β effect (9.4 ppm), and A_3 is the γ effect (-2.5 ppm).

A particularly appealing example of the sensitivity of high-resolution proton NMR to sequences is shown in the methoxyl resonance of acrylonitrile–methyl methacrylate copolymers shown in Figure 4 (5). In this case, the methoxyl resonance splits into three peaks. The center peak grows with decreasing mole percent of feed monomer of methyl methacrylate (B), while the peak at δ 3.74 decreases. Thus the peaks are assignable to ABA, ABB (and BBA), and BBB triads from lower to upper fields. The location of the BBB peak corresponds to the methoxyl peak in poly(methyl methacrylate).

A considerably more complicated example is shown in Figure 5, which is the proton-noise-decoupled ^{13}C NMR spectrum at 25.2 MHz of a 40/60 ethylene–propylene (E–P) copolymer in 1,2,4-trichlorobenzene at 120 °C (6). The peaks have been numbered for convenience, and the following methyl intensities can be defined uniquely in terms of the contribution of

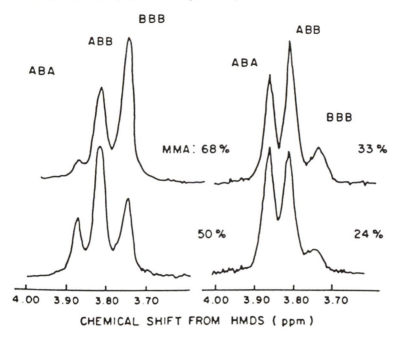

FIGURE 4. Methoxyl resonance in NMR spectra of acrylonitrile–methyl methacrylate copolymer initiated by benzoyl peroxide in tetrahydrofuran at 40 °C.

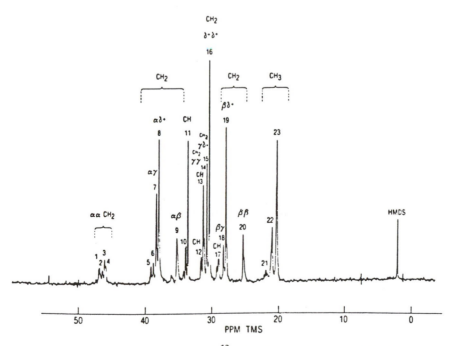

FIGURE 5. Proton-noise-decoupled ^{13}C NMR spectrum at 25.2 MHz of a 40/60 ethylene–propylene copolymer in 1,2,4-trichlorobenzene at 120 °C.

carbon sequences:

$$I_{1-4} = kN(\text{PEP}) \tag{16a}$$

$$I_9 = 2kN(\text{PEEP}) \tag{16b}$$

$$I_{20} = kN(\text{PEEEP}) \tag{16c}$$

$$I_{18} = 2kN(\text{PEEEEP}) \tag{16d}$$

In these equations, I_n is the intensity of the resonance of the methyl group at position n, and N is the number fraction of the sequence. The intensities of these resonances have been used to characterize the composition and number-average sequence length of the copolymer (6).

Dependence of Mass Spectra on Structural Sequences

Mass spectroscopy can be used to determine structural sequences, if the mechanism of degradation is appropriate. When the interlinks are weaker than the intrasequence links, the polymer should break up preferentially at these intersequence links, and homofragments should be observed. When the bond strengths of the copolymer chain are approximately equal, the appearance of homo- and heterosequence fragments should be in proportion to their frequency in the chain. When the intersequence bonds are stronger than the bonds in the homosequence, the fragments should consist of small homosequences together with larger heterosequences. In general, however, the decomposition processes are quite complicated, and the mass spectrum is not easy to interpret. Exceptions occur for copolymers of ethylene oxide and propylene oxide, whose fragment modes are particularly suited to analysis of the copolymer in terms of the number of ethylene oxide units, the number of propylene oxide units, and the order or randomness of the blocks (7).

Measurement of Orientation in Polymers

One of the most important factors in determining the mechanical properties of polymers is the orientation of the polymer chains in the sample. The classic and often used method for this purpose is X-ray diffraction. However, the limitation of this analytical technique is that only the crystalline phase can be characterized, and most polymers are multiphase systems. Therefore, it is necessary to have ways of measuring orientation by methods capable of examining multiphase systems. A number of spectroscopic techniques exist for such measurements and the principles of these methods will be discussed in this section.

Characterization of Molecular Orientation by Fourier Transform IR Spectroscopy

The absorption of infrared radiation by a molecule is given by equation 17:

$$A = \log_{10}(I_0/I)(\vec{M} \cdot \vec{E})^2 \tag{17}$$

In this equation, A is absorbance, I_0 and I are incident and transmitted intensities, M is the transition moment vector of the normal vibration, and E is the applied electric vector.

For gases and liquids, molecules are randomly distributed, and therefore, $\vec{M}$ becomes unity; the transition moment has no preferred direction. For solids and crystals, each molecule has a fixed direction. Therefore, the absorbance by a vibrational mode will depend on the direction of the transition moment. A maximum absorption will occur when $\vec{M}$ and $\vec{E}$ are parallel, and the vibration will have zero absorption if these vectors are perpendicular.

IR spectroscopy is a useful tool for the investigation of orientation. With IR spectroscopy, the orientation of a molecular group is characterized by the dichroic ratio D, which is the ratio of the intensities of the absorption band of a characteristic group measured parallel ($A_{\parallel}$) and perpendicular ($A_{\perp}$) to the polarization of radiation with respect to the direction of orientation:

$$D = A_{\parallel}/A_{\perp} \tag{18}$$

If α is the angle of the transition moment with the preferred direction, then $A_{\parallel} = \cos^2\alpha$ and $A_{\perp} \approx 1/2 \sin^2\alpha$. The factor $1/2$ is due to the two perpendicular components of the electric field. If molecules are perfectly oriented to the direction of preferred orientation, the dichroic ratio, D_0, is defined as follows:

$$D_0 = 2\cot^2\alpha \tag{19}$$

For a random system, D_0 is unity. One can determine the orientation angle of the molecular unit from D_0.

When oriented and unoriented fractions exist, the expression for the dichroic ratio D in terms of D_0 and f is given by equation 20:

$$D = \frac{1 + 1/3(D_0 - 1)(1 + 2f)}{1 + 1/3(D_0 - 1)(1 - f)} \tag{20}$$

In equation 20, f is the fraction of oriented molecules in the system (the fraction of unoriented molecules is $1 - f$). This equation can be simplified to equation 21:

$$f = (D - 1)(D_0 + 2)/(D + 2)(D_0 - 1) \tag{21}$$

The equation has two unknowns so an independent method must be used to determine f.

For a uniaxially oriented sample, the orientation distribution function, f, is related to D as follows:

$$f = (D - 1)/(D + 2) \qquad (22)$$

Examples of such measurements for uniaxially drawn polypropylene are shown in the Table II (8).

The principal advantage of IR analysis is the capability of independently measuring the crystalline and amorphous phases. In this manner, the relative orientations of the phases can be determined.

The classical dichroic ratio is a two-dimensional measurement, but the actual sample is usually oriented in three dimensions. To characterize completely the orientation, three-dimensional IR measurements are required (9–11). The macroscopic coordinate directions of a uniaxially drawn film are defined by y in the stretching direction, x in the transverse direction, and z in the direction that is normal to the film (Figure 6). The orientation in three dimensions is completely characterized if the A_x component of the absorption is determined together with the A_y and A_z components. The classical dichroic ratio (R) in this reference system is expressed as follows:

$$R_{yx} = A_y/A_x \qquad (23)$$

If either R_{zy} or R_{zx} is known in addition to R_{yz}, the remaining ratio can be found as a function of the other two:

$$R_{zy} = R_{zx}/R_{yx} \qquad (24)$$

The important parameter to be determined is the structure factor A_0 (9):

$$A_0 = (A_x + A_y + A_z)/3 \qquad (25)$$

A_0 represents the absorbance of the band exclusive of contributions due to orientation of the polymer. This factor is proportional to the amount of chemical structure in the oriented solid giving rise to the IR absorption and is the same quantity measured in gases and liquids, which have no preferred orientation. The difficulty in measuring A_0 in solids is that it is necessary to know the absorbance, A_z, in the thickness direction, which

Table II. Dichroic Ratios (R), Orientation Function (f), and Average Angles of Orientation for Drawn Polypropylene Films

Draw Ratio[a]	R	f	θ (°)
1	1.015	−0.010	55.1
1.5	0.745	0.186	
4	0.536	0.366	40.6
7	0.056	0.918	13.5
8	0.51	0.926	12.9
9	0.036	0.947	10.8

Note: The frequency was 1220 cm^{-1}, and the band angle was 90°.
[a]Each value is the ratio of the length of the drawn film to the initial length of the film.

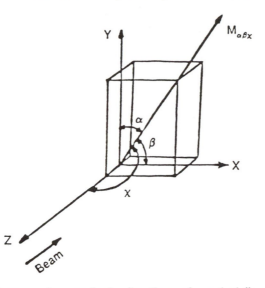

FIGURE 6. Macroscopic coordinate directions of a uniaxially drawn film as defined by y in the stretching direction, x in the transverse direction, and z in the direction normal to the film. (Reproduced with permission from reference 9. Copyright 1986 John Wiley and Sons, Inc.)

for polymer films is very small. A_z may be obtained by tilting the sample at some angle, α (Figure 7), and the component in the z direction is determined through the following relationship (*12*):

$$A_2 = A_\alpha \cos \alpha^2 A_y / \sin^2 \alpha_1 \tag{26}$$

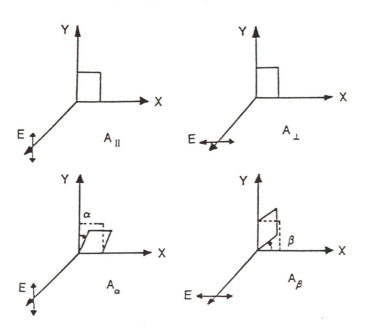

FIGURE 7. Determination of thickness-direction absorbance by tilting the sample at some angle, α, and determining the component in the z direction. (Reproduced with permission from reference 12. Copyright 1967 John Wiley and Sons, Inc.)

A_α and A_y are calculated from the spectra. The true angle, α_1, that the beam makes with the sample is found by correcting the measured angle for the refraction resulting from the refractive index (9).

For three-dimensional measurements, the orientation parameters $A_x/3A_0$ and $A_z/3A_0$ can be used to determine the fraction of molecules oriented in the three mutually perpendicular directions. In uniaxially oriented samples, which are often encountered by drawing in one direction, the structure factor is given by equation 27:

$$A_0 = 1/3(A_{\parallel} + 2 A_{\perp}) \tag{27}$$

Characterization of Molecular Orientation by Raman Spectroscopy

In general for Raman spectroscopy, the induced polarizability is not necessarily in the direction of the incident beam:

$$P_x = \alpha_{xx} E_x + \alpha_{xy} E_y + \alpha_{xz} E_z \tag{28a}$$

$$P_y = \alpha_{yx} E_x + \alpha_{yy} E_y + \alpha_{yz} E_z \tag{28b}$$

$$P_z = \alpha_{zx} E_x + \alpha_{zy} E_y + \alpha_{zz} E_z \tag{28c}$$

In these equations, P is polarizability; E is the magnitude of the incident beam; α_{xx}, α_{yy}, and α_{zz} are the components of the principal axes of the polarizability ellipsoid; and α_{xy}, α_{yz}, and α_{xz} are the other components. Consequently, the Raman-scattered light emanating from even a random sample is polarized to a greater or lesser extent.

In the usual Raman experiment, the observations are made perpendicular to the direction of the incident, plane-polarized beam. The "depolarization ratio" is defined as the ratio of the intensities of the two polarized components of the scattered light that are parallel and perpendicular to the direction of propagation of the (polarized) incident light. The polarization of the incident beam is perpendicular to the plane of propagation and observation. For this geometry, the depolarization ratio (d) is defined as the following ratio of intensities:

$$d = VH/VV \tag{29}$$

For the right-angle-scattering experiment, V is the intensity of the component perpendicular to the scattering plane, and H is the intensity of the component parallel to the scattering plane. An alternative notation expressed in terms of the laboratory coordinate system is A(BC)D, where A is the direction of travel of the incident beam; B and C are the polarization of the incident and scattered light, respectively; and D is the direction in which the Raman-scattered light is observed. Generally, the incoming beam is along the x axis, the scattered beam is along the z axis, and the y axis is perpendicular to the scattering plane.

Table III. Contribution to Spectrum for Three-Dimensional Isotropic Systems

Experiment	Geometry	Symmetry	Anisotropy
VV	X(YY)Z	δ^2	$4\gamma^2/45$
VH	X(YZ)Z	0	$\gamma^2/15$
HV	X(ZY)Z	0	$\gamma^2/15$
HH	X(ZX)Z	0	$\gamma^2/15$

For randomly oriented systems, the polarization properties are determined by the two tensor invariants of the polarization tensor, that is, the trace and the anisotropy. The depolarization ratio (δ) is always less than or equal to 3/4. For a specific scattering geometry, this polarization depends on the symmetry of the molecular vibration giving rise to the line.

To study the orientation of the sample, these experiments must be related to the molecular polarizabilities. For excitation, with its polarization along F, the scattered radiation with polarization F' will have intensity proportional to $[(\alpha_k)\text{ff}']^2$. For an isotropic system, the orientationally invariant terms δ and γ are defined as follows:

$$\delta = 1/3 \sum \alpha_{ii} \tag{30a}$$

$$\gamma^2 = 1/3 \sum_{i<j} \left[(\alpha_{ii} - \alpha_{jj})^2 - 6\alpha_{ij}^{2} \right] \tag{30b}$$

In these equations, the subscripts i and j refer to the axes of the molecular coordinate system. The contribution to the spectrum for three-dimensionally isotropic systems is shown in Table III (*13*). These results show that the intensities of the *VH*, *HV*, and *HH* experiments should be identical. These predictions are confirmed (*13*).

Theoretically, the depolarization ratio can have values ranging from 0 to 3/4, depending on the nature and symmetry of the vibrations. Nonsymmetric vibrations give depolarization ratios of 3/4. Symmetric vibrations have depolarization ratios ranging from 0 to 3/4, depending on the polarizability changes and the symmetry of the bonds in the molecule. Accurate values of the depolarization ratio are valuable for determining the assignments of Raman lines, and in conjunction with dichroic measurements in the infrared, they represent a powerful structural tool for polymers. Because the laser beam is inherently polarized and highly directional, polarization measurements can be made easily.

In solids, the problem of polarization is more complicated, but the results are more rewarding. In solids, the molecular species are oriented with respect to each other. Therefore, the molecular polarizability ellipsoids also are oriented along definite directions in the crystal. Because the electric vector of the incident laser beam is polarized, the directionality in the crystal can be used to excite and obtain Raman data from each element of the polarizability ellipsoid. With the laser polarization and collection along the z axis, a spectrum from the α_{zz} component of the tensor is obtained. By rotating the analyzer 90°, thereby collecting x-polarized light while still exciting along the z axis, α_{zx} is obtained.

Raman and IR spectroscopic techniques can determine the independent orientation of both the crystalline and the amorphous phases of semicrystalline polymers. Raman scattering has, however, some advantages over IR spectroscopy. First, it enables the determination of both the second and fourth moment terms, $\langle P_2(\cos\theta)\rangle$ and $\langle P_4(\cos\theta)\rangle$, respectively, of the expansion of the orientation distribution function. Second, for orientation determination, Raman spectroscopy does not require the use of thin polymer films, as is usually required for infrared absorption measurements.

The Raman-scattering intensities, I_s, are given by quadratic expressions of all components of the Raman tensor for the vibration studied, α_{ij} ($i, j = 1$–3), the coefficients being the direction cosines I_i and I_j' defining, respectively, the polarization directions of the incident and the scattered light with respect to a set of axes fixed in the sample:

$$I_s = I_0 \sum \left(\sum_{ij} I_i I_j' \alpha_{ij}\right)^2 \tag{31}$$

In equation 31, I_0 is a constant depending on instrumental factors and the intensity of incident light. Each ij can be expressed as a linear combination of the principal components 1, 2, and 3 of the derived polarizability or Raman tensor of the vibration investigated. The experimental values are of the form $I_0 \sum \alpha_{ij}\alpha_{pq}$ (14).

The following relation can be shown for uniaxial symmetry:

$$\sum \alpha_{ij}\alpha_{pq} = 4\pi^2 N_0 \sum M_{l00} A_{l00}^{ijpq} \tag{32}$$

where A_{l00}^{ijpq} is the sum of terms in α_1, α_2, and α_3 (15). N_0 is the number of segments, and M_{l00} is expressed in terms of Legendre polynomials $\langle P_l(\cos\theta)\rangle$:

$$M_{l00} = (1/4\pi^2)[(2l + 1)/2]^{1/2}\langle P_l(\cos\theta)\rangle \tag{33}$$

$\langle P_l(\cos\theta)\rangle$ is the coefficient of the lth term ($l = 1, 2, \dots$) in the expansion of the orientation distribution function for a set of symmetry axes. The angle between the chain axis of the structural unit and the stretching direction is θ. The following relations can be shown:

$$M_{000} = 1/4\pi^2(1/2)^{1/2} \tag{34a}$$

$$M_{200} = 1/4\pi^2(5/2)^{1/2}\langle P_2(\cos\theta)\rangle \tag{34b}$$

$$M_{400} = 1/4\pi^2(9/2)^{1/2}\langle P_4(\cos\theta)\rangle \tag{34c}$$

A sample with uniaxial symmetry has five unknowns: α_1, α_2, and α_3 and the orientation parameters $\langle P_2(\cos\theta)\rangle$ and $\langle P_4(\cos\theta)\rangle$. Therefore, five independent nonzero $I_0\sum\alpha_{ij}\alpha_{pq}$ sums are needed. These five independent $I_0\sum\alpha_{ij}\alpha_{pq}$ are $I_0\sum\alpha_{33}^2$, $I_0\sum\alpha_{32}^2$, $I_0\sum\alpha_{21}^2$, $I_0\sum\alpha_{22}^2$, and $I_0\sum\alpha_{22}\alpha_{33}$. These spectra are shown in Figure 8. The calculated results are shown in Table

IV. These results clearly show that the Raman tensor is not cylindrical ($\alpha_1/\alpha_3 \neq 1$), as expected from the planar zigzag structure of polyethylene. The sign changes demonstrate that the polarizability changes during the vibrations parallel to and perpendicular to the plane of the chains have opposite signs. The values of $\langle P_2(\cos\theta) \rangle$ increase with the draw ratio. The

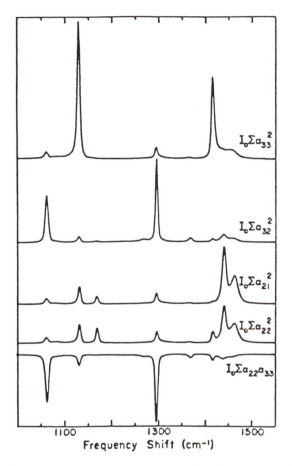

FIGURE 8. Five independent $I_0 \Sigma \alpha_{ij} \alpha_{pq}$ normalized spectra used for orientation determination of polyethylene (draw ratio, 11.7). (Reproduced from reference 14. Copyright 1991 American Chemical Society.)

Table IV. Ratios of the Principal Components of the Raman Tensors for the 1080-, 1130-, and 1170-cm^{-1} Bands of Polyethylene and Corresponding Orientation Coefficients

Frequency (cm^{-1})	λ^a	α_1/α_3	α_2/α_3	$\langle P_2(\cos\theta) \rangle$	$\langle P_4(\cos\theta) \rangle$
1080	7	-1.10	0.79	0.42	0.40
	11.7	-1.05	0.80	0.47	0.47
1130	7	-0.54	0.31	0.75	0.56
	11.7	-0.57	0.42	0.85	0.69
1170	7	-6.2	0.42	0.84	0.45
	11.7	-7.3	0.40	0.87	0.52

$^a\lambda$ is draw ratio.

values of $\langle P_4(\cos\theta)\rangle$ show that the distribution of crystal orientation is narrower at higher draw ratios and falls within the range of the most probable ones (14).

Characterization of Molecular Orientation by Solid-State NMR Spectroscopy

Another method for measurement of orientation is the measurement of the chemical shift anisotropy (CSA) line shapes (16). This approach has been used to study the uniaxial drawing of poly(tetrafluoroethylene) (PTFE). Figure 9 shows the ^{19}F NMR multiple-pulse spectrum of semicrystalline PTFE samples as a function of the angle, θ, between the direction of stress and the magnetic field (16). By varying the moments of these spectra about the isotropic chemical shift, an approximate orientational probability distribution of the chain axes about the direction of stress can be determined. These results are shown in Figure 10 for samples with different draw ratios (18). The increase in the intercept reflects the greater alignment of the chains with the draw direction.

Biaxial films of poly(ethylene terephthalate) also have been studied by the NMR CSA technique (19). A highly ordered component of the film is found in which the planes of the aromatic rings lie close to, but not in, the plane of the film. There is also a fast-relaxing component that is much less oriented and is assumed to indicate polymer chains that are relatively mobile, that is, the amorphous phase.

Characterization of Cross-linked Polymer Systems by High-Resolution, Solid-State NMR Spectroscopy

A number of experimental difficulties are encountered in the study of cross-linked or network polymers. First, one must work with solids, because cross-linked systems are insoluble. Second, high sensitivity is required, because the density of cross-links is usually low. Third, high accuracy is required, because the properties are substantially changed with small changes in the number of cross-links. Because the chemical nature of cross-links is not generally known, the techniques involved should have high spectral specificity and resolution. Additional, the structural differences between the cross-links and the polymer backbone are usually small.

Recent developments in solid-state NMR techniques have changed this situation, and the potential at this time is high for developing new insights into the chemical structure of cross-linked networks. These new structural insights should lead to a better understanding of structure–property relationships and the development of improved network systems.

Carbon-13 NMR spectroscopy is a powerful technique for the characterization of polymers in solution because of its simplicity of interpretation, sensitivity to subtle molecular changes, and relative ease of quantification. Much of this success is due to the sharpness of the resonances, with line widths typically on the order of a few hertz or less.

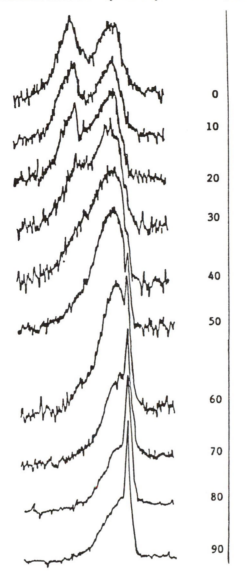

FIGURE 9. ^{19}F NMR spectra of deformed PTFE samples as a function of the orientation, β, of the stretch axis relative to the static magnetic field. (Reproduced with permission from reference 17. Copyright 1982 Butterworth Heinemann Ltd.)

Until recently, the application of ^{13}C NMR techniques to polymeric solids has been limited by the large line widths, which may be in excess of 20 kHz. The carbon–proton dipolar interaction and the chemical shift anisotropies broaden the lines in the solid-state NMR spectra. The major effect arises from the dipolar coupling of the carbon nuclei with neighboring protons. The large magnitude of dipolar ^{13}C–^{1}H coupling (up to 40 kHz) results in broad and structureless proton-coupled ^{13}C NMR absorptions. However, by using a combination of high-power decoupling, magic-angle spinning, and cross-polarization techniques, the line widths for

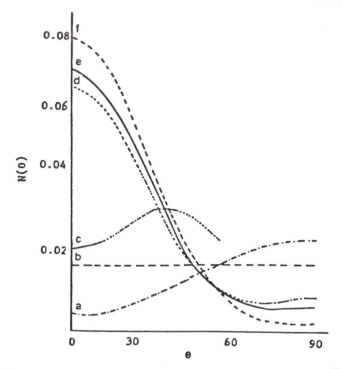

FIGURE 10. The distribution function, $N(\phi)$, derived from the ^{19}F NMR spectrum of PTFE at different draw ratios: (a) $\lambda = 0.82$, (b) $\lambda = 1.0$ (unoriented), (c) $\lambda = 1.40$, (d) $\lambda = 1.75$, (e) $\lambda = 2.10$, and (f) $\lambda = 2.40$. (Reproduced with permission from reference 18. Copyright 1983 John Wiley and Sons, Inc.)

solids may be reduced to 100 Hz, which allows the use of solid-state NMR spectroscopy for characterization of the chemical structure of solids.

Magic-Angle Spinning

For solutions or liquids, sharp lines in the NMR spectra are due to the rapid isotropic motions of the nuclei, which average to zero the anisotropic interactions, such as dipolar and quadrupolar, and effectively remove them from the spectrum. Generally, the motion in solids is insufficient to narrow the lines so that one imposes motion on the sample by rapidly spinning the solid sample at 54° 44′ (the so-called magic angle). Under these circumstances, each molecule experiences a continuous series of orientations with respect to the external magnetic field. The result is the observation of an isotropic average of the chemical shift for the solid similar to that observed for the same sample in solution.

Gated, High-Power Proton Decoupling

When dipolar interactions are present, as in the solid state in which molecular motion is insufficient to remove them, high-powered proton decoupling is used to average coherently these interactions to zero. By forcing the spins to change states at a faster rate compared with the frequency of the ^{13}C–^{1}H interaction, the local dipolar field at the ^{13}C nucleus is reduced to zero. The high-powered proton radio frequency (rf) field accomplishes this averaging and is particularly effective in reducing the line widths of ^{13}C resonances in the solid state. A decoupling field of 100 kHz corresponds to $1/100\,000 = 10^{-5}$ s. To decouple protons from carbon in solids, the magnitude of the decoupling field must be capable of exciting all proton transitions within a band width of 40–50 kHz, that is, a large band width compared with proton–proton dipolar coupling.

Cross-Polarization

Because of the relatively long ^{13}C relaxation times and the low abundance of ^{13}C nuclei, sensitivity problems are encountered in ^{13}C NMR spectroscopy. It is advantageous to build up the carbon magnetization via polarization transfer from the abundant proton spin system; cross-polarization (CP) is such a transfer process. Building up the carbon magnetization by the CP process has two major advantages: (1) the relaxation process is much more rapid, because it is determined by the much shorter proton T_1 (spin–lattice time), compared with the carbon T_1 (which allows short recycle times between pulse sequences), and (2) the intensity enhancement is up to four times greater than in the ordinary experiment because of the polarization transfer. These developments in NMR spectroscopy have made possible the study of a wide variety of solid materials, including insoluble networks. Although the resolution for the normal polymeric solid state is 100 times less than for solutions, it is still possible to detect the chemical shift resonances of most carbons in polymer networks, particularly for the motionally narrowed elastomeric systems. Each chemically distinct carbon generally gives a separate NMR resonance that reflects the electronic environment at that carbon.

Analysis of Elastomeric Systems by Solid-State NMR Spectroscopy

Elastomeric materials have specific dynamical properties, mainly the anisotropic reorientational motion above the glass transition and, from the point of view of NMR, relatively weak dipolar couplings, compared with those of the solid glassy polymers. Typical line widths under solution conditions in natural rubber are about 400 Hz. Thus, the line narrowing and the resulting high-resolution spectra can be obtained by applying spinning speeds and proton-decoupling power that are relatively lower than those used in the standard high-resolution, solid-state NMR method. The high-power decoupling narrows the spectral lines to 300 Hz; therefore, the contribution of dipolar interaction to the line width is only 100 Hz. This result reflects the presence of nearly isotropic motions that average the dipolar interactions and the chemical shift anisotropy. The application of magic-angle spinning (MAS) removes the chemical shift anisotropy and narrows the lines by 1 order of magnitude (30 Hz). The total effect of these techniques on the high-resolution, solid-state, 75-MHz NMR spectrum of polybutadiene is shown in Figure 11 (*20*). Figure 11 shows that the

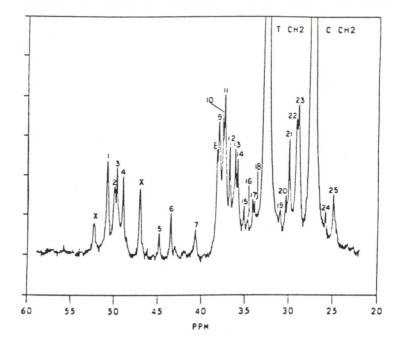

FIGURE 11. Aliphatic region of the C-13 GHPD / MAS spectrum of *cis*-BR highly vulcanized with sulfur (10 phr) at 150 °C for 30 min. (Reproduced from reference 20. Copyright 1989 American Chemical Society.)

unvulcanized polybutadiene has essentially only two resonance lines, with minor components due to structural impurities. Obviously, this multilined spectrum contains a wealth of information. Many new resonances have appeared that are apparently related to new structures in the vulcanizate. To use the chemical information provided by these spectra, chemical structures must be assigned to each new resonance.

Increased Mobility

Although sharper lines are obtained by using MAS and high-powered decoupling, the resonance lines of the vulcanized systems broaden as the cure proceeds, and the rubber becomes stiffer. When increased spectral resolution is desired for highly cured samples, it is possible to raise the temperature of the samples, by using a variable-temperature NMR probe. The idea is to induce some degree of increased molecular mobility in the samples, because increased mobility leads to narrower resonance lines. The same type of enhanced resolution can also be achieved by swelling the vulcanizate samples in an appropriate solvent. The NMR spectrum of the swollen sample is then subsequently obtained at room temperature.

NMR Method of Structure Determination for Polymers

The required step in determining structure from the NMR spectra is the assignment of the NMR resonances to specific structural features of the polymer. Most structural assignments are made by using one or more of the following methods:

- Comparison of the observed chemical shifts with those observed for analogous low-molecular-weight model compounds
- Calculation of chemical shifts by using derived additivity relationships
- Synthesis of polymers with known specific structural or compositional features and using them to establish resonance–structure relationships
- Synthesis of polymers with selectively enriched ^{13}C sites or deuterium substitution for protons and observing spectra
- Comparison of the intensities of structural sequences with predicted intensities calculated on the basis of assumed polymerization kinetics and statistical models
- One-dimensional spectral editing techniques, such as selective spin-decoupling experiments for the determination of the proton bonding of the carbons
- Two-dimensional NMR techniques

These techniques have been used to interpret the spectra of elastomeric systems.

NMR Imaging of Polymers

Conventional NMR spectroscopy is used to determine the chemical structure of a sample but cannot be used to locate the position of the stimulated nuclei in the sample. In NMR imaging, the stimulating signal is encoded so that an image can be reconstructed to show the spatial distribution of the nuclei in the sample. NMR imaging is performed similarly as standard NMR spectroscopy, except for the spatial encoding of the signal.

The key step in NMR imaging is the application of a linear magnetic gradient to the sample in the static magnetic field. In this way, spatial encoding of the stimulated signal is accomplished, and the frequencies of the nuclei reflect their position in the sample. To use an analogy, this process is similar to having a pianist play a chord on a piano. By frequency analysis of the resultant sound wave, it is possible to determine the positions of the piano keys the pianist struck.

The NMR-imaging technique relies on the detection of the nuclei in only a small and controllable region of the sample where the nuclear resonance frequency is matched to the rf signal. This selected region is systematically moved over the entire sample, and an image is obtained.

The NMR signals depend on the nuclear relaxation time constants that respond to the structural environments of the emitting nuclei. NMR imaging can measure the spatial distribution of the NMR parameters, including spin density and spin–lattice and spin–spin relaxation times. Thus NMR imaging can measure a variety of structural factors in situ. NMR imaging may be considered to be a type of chemical microscopy through which one can obtain images of heterogeneous samples.

Basis of NMR Imaging

In conventional NMR spectroscopy, the sample is placed in a homogeneous magnetic field, H_0, so that all chemically equivalent spins experience the same field and, hence, precess at the same Larmor angular frequency, ω_0:

$$\omega_0 = \gamma H_0 \tag{35}$$

In this equation, γ is the magnetogyric ratio. Under these circumstances, the observed proton NMR spectrum of water consists of a single, sharp line of intrinsic width of about 0.1 Hz.

In contrast with conventional NMR spectroscopy, NMR imaging involves placing a sample in a deliberately nonuniform magnetic field, which is achieved by modifying the homogeneous magnetic field with a system of gradient magnetic coils to generate linear gradients of the order of a few gauss per centimeter (10^{-2} T/m). The purpose of the nonuniform field is to label, or encode, different regions of the sample linearly with different NMR frequencies. Because the magnetic field is varied in a known manner at specific positions within the sample, the frequency of the NMR signal indicates the spatial position of the resonating nuclei (Figure 12). In one dimension (1D), the position of the sample is related to a frequency by the following relationship:

$$\Delta\omega_z = \omega_z - \omega_0 = \gamma G_z z \tag{36}$$

G_z is the magnetic field gradient in the z direction: $G_z = \partial B_z/\partial z$, where B_z is the magnetic field gradient strength in the z direction. A tailored rf pulse with a narrow frequency range is used to excite only those nuclei at corresponding positions in the z dimension. The amplitude of the NMR

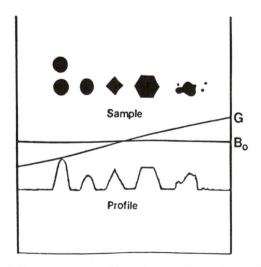

FIGURE 12. NMR-imaging experiment in one dimension with a linear magnetic gradient. The experiment generates a profile of the spin density of the sample.

signal received from the z axis line is a measure of the number of resonant nuclei on that line, and so the NMR spectrum represents a graph of spin density versus distance (with relaxation effects neglected). The field gradient is described by a tensor with nine components, but for a large H_0, only three components are required: $G\alpha = \partial B\alpha/\partial\alpha$, where $\alpha =$ x, y, and z.

In three dimensions, one operates in a three-dimensional gradient field. The frequency spectrum (still obtained by a Fourier transform of the free induction decay or FID) gives the number of resonating spins along a specific direction of the field gradient. In fact, each plane perpendicular to the direction of the field gradient has a different resonance frequency, and the signal intensity at that frequency will be proportional to the number of nuclei contained in that plane. In other words, the frequency spectrum is just a projection of the spin density (with relaxation effects neglected for the moment) along the field direction.

Relaxation Parameters in NMR Imaging

The spin densities and the molecular environments of the nuclei are reflected in the time variation of the amplitude of the measured rf signal and, hence, are reflected in the intensity of each voxel in the image. When the relaxation times, T_1s and T_2s, are different in the voxels of a heterogeneous sample, these differences can be exploited to develop contrast in the NMR images. The pulse sequence that is usually used to measure the T_2 relaxation phenomena in images is called multiple spin echo. At a given repetition time, TR, the NMR signal is measured at several different echo times, TE. These echos provide a measure of the T_2 relaxation. By repeating the process at different TR values, the T_1 relaxation also can be measured.

Because differences in relaxation times and spin densities determine image contrast, data on relaxation times are important in the selection of the optimal rf pulse sequence for imaging a selected sample. Relaxation times can be measured at any point on an image. The ability to quantify accurately relaxation rates is important in understanding and optimizing image contrast. Spin density and T_1 and T_2 images can be computed from measurements with pulse sequences with predetermined variations. These fundamental images represent the inherent data in the system and can be recombined to reconstitute computed images for a given pulse sequence.

Pulse Sequences for Generating Contrast in Imaging

Contrast in NMR imaging depends on both material-specific and operator-selected parameters. The material-specific parameters include the spin density and the relaxation times, T_1 and T_2. The operator-selected parameters include the pulse sequence (inversion recovery, spin echo, etc.) and the pulse delay and repetition times (timing parameters). For a given imaging system and pulse sequence, the delay and repetition times, in conjunction with the intrinsic material parameters, dictate the appearance of the final image. If the correct pulse sequence is used and the relaxation times of the two materials are known, it is possible to calculate the delay and/or repetition times that will produce the maximum difference in signal intensity for those materials.

Spin-Echo Images

The spin-echo (SE) technique is the most common pulse sequence applied in magnetic resonance imaging (MRI) today. Images are constructed by acquiring a multitude of projections (typically 256 per image), each with an identical setting of a readout gradient during which the sequence is sampled. Each projection is differentiated from the others by a phase difference that is produced by advancing the phase-encoding gradient.

As shown in Figure 13, the method consists of a series of rf pulses that are repeated many times to achieve a sufficient signal-to-noise ratio. Each projection is produced by a 90° pulse followed by a 180° pulse for induction of the spin echo. The 90° rf pulse tips the magnetization into the xy plane, where it begins dephasing. The 180° rf pulse is applied after a time, t, and forces the magnetization to refocus at a time $2t$ (also known as the echo time, TE) after the 90° rf pulse, at which time the data are collected. The frequency-encoding gradient, G_x, causes the spins to precess at different frequencies depending on their position in the static magnetic field. The phase-encoding gradient, G_y, is orthogonal to G_x. Varying the intensity of G_y causes the spins to dephase at different rates and provides the second dimension of a two-dimensional image. The slice-selection gradient, G_z, and the Gaussian shaped 90° rf pulse determine the position and the thickness of the region of interest. The data is Fourier transformed in two dimensions to produce the image of the selected slice. The time delay between the observation pulse and the observation is called the echo time (TE). The time between two consecutive pulse sequences is the repetition time (TR) and usually ranges from 250 to 2500 ms.

Spin-echo techniques have a unique position in NMR applications. The main problem with NMR imaging is the long data collection time, mainly due to the spin–lattice relaxation time, T_1. Each measurement necessitates a period of the order of T_1 (which is approximately 0.5 s for aqueous systems) for the system to return to equilibrium magnetization. By using spin-echo repetition, a large number of spin echos can be repeated within a T_1 or T_2 decay period.

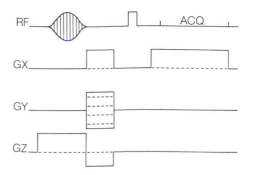

FIGURE 13.   Pulse sequence for spin-echo method.

The spin-echo experiment starts with a 90° rf pulse. A magnetic field gradient is switched on in the x direction. After some incremental time, a 180° pulse is applied. The phase of the different isochromates is then inverted. The gradient is switched on with an opposite sign and then is switched off after another $T_1/2$ period. At the end of the evolution period, the different components of the magnetization are dispersed only on the basis of their location. The dispersions due to either field inhomogeneity or chemical shift are refocused.

If *TR* is long compared with the longest T_1 in the voxels, the exponential e^{-TR/T_1} approaches zero, and the signal intensity becomes independent of T_1 and is modulated only by T_2 relaxation processes. If *TE* is short compared with T_2, then e^{-TE/T_2} approaches unity, and the signal is related only to T_1.

Multiecho Method

A variation of the spin-echo technique is the use of a series of echos. A nonselective 180° rf pulse is applied at a time *t*. A second echo is produced at time $2t$ after the second 180° rf pulse. The total echo time for the second echo is $2TE$. A second incremental gradient in the y direction after acquisition is used to rephase the spins before the next 180° rf pulse. The cycle is repeated until the desired number of echos is acquired. This procedure reduces the total time of the experiment by the total number of different echos acquired, because all echos are collected within one delay time (rather than one echo per delay).

The later echos produce progressively darker images, because the intensity of the echos gradually decreases from the T_2 decay. Thus, the intensity for all voxels decreases in each successive echo, with the signals from voxels with high T_2 values decaying more slowly than those with low T_2 values. Voxels with higher T_2 values, therefore, have higher absolute intensities in the later echos. The contrast at a given echo delay is more prominent for longer pulse repetition times, because the T_1 recovery at the time of the initial 90° rf pulse is more complete.

A T_2 image can be calculated from a single spin-echo scan without substantially increasing the scan time, provided that at least two echos are sampled. The signals from the first and second echos represent two points on the T_2 decay curve.

Applications of NMR Imaging to Polymers

Contrast in NMR images of polymers depends on differences in the environments of the nuclei. These environments are influenced by a wide variety of chemical and physical effects. If the relationships between the images and the differences in chemical and physical effects can be extracted from the NMR images, the potential applications of the NMR-imaging technique to materials are numerous. List III gives a preliminary list of such applications. In the following sections, some of these applications will be demonstrated, and others will be discussed in terms of the potential impact of NMR imaging.

List III. Applications of NMR Imaging to Polymers

- Adsorption and diffusion processes of fluids in materials
- Detection of internal defects and voids
- Characterization of heterogeneous mixtures of different materials
- Study of molecular interactions between materials
- Detection of internal concentration gradients in materials
- Determination of spatially distributed structural changes
- Composition profiling from surfaces and other sources
- Study of variations in molecular mobility throughout the sample
- Evaluation of homogeneity of mixing processes
- Determination of internal flow processes

Measurements of Internal Defects in Polymer Systems

Because NMR imaging allows one to obtain the image of a slice of a polymeric sample, internal defects can be measured if they are larger than the resolution of the technique (current resolution, >20 μm). An example is shown in Figure 14A for a polyurethane foam filled with water. The light portions of the image represent pores filled with water, and the dark areas arise either from solid foam or an absence of material (air). With an edge detection algorithm, one can observe the outline of the pores, as shown in Figure 14B. Obviously, it is possible to observe the distribution of the pore sizes. A histogram of the pore sizes versus the number of pores of that size can be constructed. Such a histogram can be correlated to the chemical-foaming process variables and could lead to an understanding of the mechanism. Such knowledge could promote better process control and an improved foamed product.

NMR imaging is a means of detecting internal material imperfections in fabricated articles. Its applications in the field of processing and fabrication of polymeric materials are diverse. They include the detection of subsurface defects, including interfacial flaws and microcracks, and the detection and characterization of areas modified by the introduction of mobile foreign substances, such as additives, degradation products, and contaminants.

Because of the sophisticated structure of polymeric engineering articles and the complexity of the process and fabrication procedures, noninvasive tests are needed to ensure the quality and integrity of the manufactured articles. Defective or damaged areas of polymeric materials can be made to appear in the NMR image by sorption of a mobile liquid such as water. Uniformity of the polymeric materials can be evaluated, because improperly manufactured engineering articles have different NMR images. A materials-acceptance criterion could therefore be written as a function of tolerances in the NMR images. Such an image-based materials-acceptance protocol would ensure proper manufacture and performance of polymeric engineering components.

In Figure 15, the two images shown were taken 0.5 cm apart through the same pultruded rod made with glass fibers and a nylon matrix (22). (Pultrusion is a new process for making composite rods by pulling the fibers through a tube with a polymer matrix.) The rod was soaked in water,

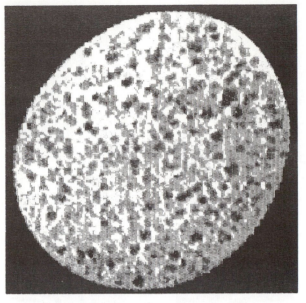

A

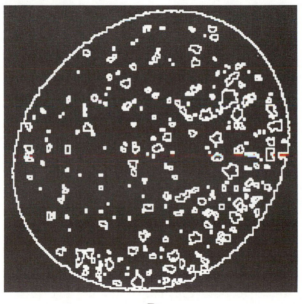

B

FIGURE 14. (A) NMR image of a polyurethane foam filled with water. (B) Image obtained after using an edge detection algorithm on the image in A. (Reproduced with permission from reference 21. Copyright 1989 John Wiley and Sons, Inc.)

and the light areas in the images represent void areas filled with water. The marker in the lower right-hand portion of the image (dark spot in white area) is 1 mm in diameter. Comparison of the sizes of the voids in the pultruded rod indicates that some of the voids approach the size of the marker. A comparison of the corresponding edge-enhanced images (Figure

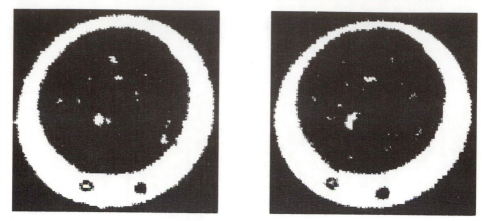

FIGURE 15. Two images taken through the same rod but 0.5 cm apart. (Reproduced with permission from reference 22. Copyright 1989 Gordon and Breach.)

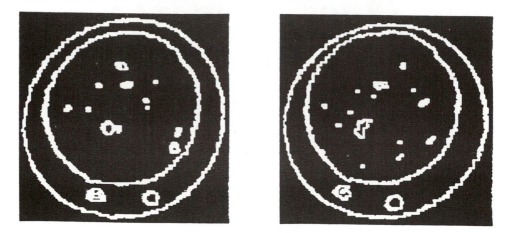

FIGURE 16. Profiles of a nylon RIM-pultruded composite rod showing the distribution in the location of the defects for images taken 0.5 cm apart.

16) shows that some of the voids in the images occur in the same location; therefore, the voids are connected or tubular in shape. Thus, a channellike void region is suggested over a length of 0.5 cm. Using a computer, one can represent the defects in the image (Figure 17). From the computer comparison of the two images taken 0.5 cm apart, it is possible to identify a tubular-shaped void running from one image to the other within the nylon rod. Such a void could be obtained if an air bubble was trapped in the matrix during the pultrusion process.

Diffusion of Liquids in Polymers

The diffusion of small molecules is important to polymer material properties, ranging from processing and production qualities to end use applications and shelf life. Polymer–penetrant systems exhibit a wide range of

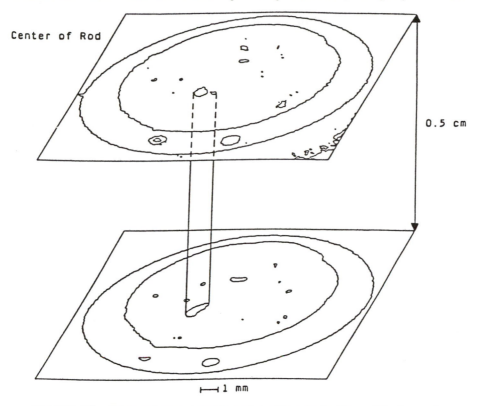

FIGURE 17. Edge-enhanced contour plot of images in Figure 16. A correlation of these images shows that some of the voids are tubular; that is, a channellike void region exists over the dimension of 1 cm. (Reproduced with permission from reference 22. Copyright 1989 Gordon and Breach.)

diffusion phenomena, from common Fickian to case II diffusion. These diffusion processes represent the extreme responses of polymers to a solvent and have substantially different characteristics.

Fickian diffusion is characterized by an exponential increase in concentration from the glassy polymer core to the fully swollen regions of the polymer. It is also characterized by weight gain that increases with the square root of time in a polymer sheet. The diffusion constant can be a function of concentration, C, in which case a series of relative concentrations versus distance, x, yields a master curve when C/C_0 is plotted versus $x/t_{1/2}$, where C_0 is the initial concentration and $t_{1/2}$ is the square root of time (23).

Case II behavior has the following four characteristics: (1) The concentration increases rapidly from zero in the glassy polymer core to an equilibrium value in the swollen regions of the sample, (2) the concentration throughout the swollen regions is constant, (3) the front advances through the glass at a constant velocity, and (4) a Fickian concentration profile precedes the sharp front as it advances through the glass.

The use of NMR imaging for the study of diffusion has some advantages. First, NMR imaging provides a visual representation of the spatial distribution of diluent by acquiring signals directly from the protons of the solvent without interrupting the diffusion process or destroying the sample. Second, regions of inhomogeneous diffusion can be detected and analyzed. With other techniques, these discontinuities are averaged in the data and add to the error. This advantage extends to sample geometries as well. NMR imaging is not restricted in geometry except for size. A three-dimensional image can monitor a diffusion process from all regions of a sample.

The geometry best suited for NMR imaging is a cylinder. A cylinder fills the sample chamber so that the best resolution per unit volume is obtained. It might be expected, however, that a sphere would be the best geometry, because a sphere has the most isotropic magnetic susceptibility. However, spheres present a problem in the slice-selection process, because signal contributions occur from diffusion that is not parallel to the slice plane when the slice thickness is small compared with the sphere diameter (24).

To study diffusion in polymers by NMR imaging, one must acquire an image that reflects accurately the quantity of solvent per unit volume—the spin-density image. However, to acquire a spin-density image, one must satisfy several conditions. First, the spins must return to equilibrium between subsequent repetitions of the pulse sequence to prevent saturation effects. To recover 99% of the original magnetization, the repetition time, TR, must be equal to five times T_1. Second, TE must be short compared with the spin–spin relaxation time, T_2, so that little signal loss occurs because of spin–spin interactions. To maintain 99% of the magnetization, TE must be less than 1% of T_2. By using computer simulation studies of case II diffusion (24), it has been determined that the image must be acquired in less time than it takes the case II front to traverse about 1/10 of an image pixel. This requirement severely limits the velocity range in which NMR imaging is effective. The maximum front velocity is 2.2 Å/s with the current hardware and the spin-echo method. Fortunately, other imaging experiments are available to reduce the imaging time. One such experiment, termed FLASH (fast, low-angle shot), reduces the imaging time by a factor of 10 by use of a small tip angle and a gradient echo (24).

Self-diffusion coefficients provide important information concerning the extent of the translation of solvent molecules in their environment. This information can be interpreted in terms of molecular organization and phase structure. The self-diffusion coefficients are sensitive to structural changes, as well as to binding and association phenomena. Measurements of the self-diffusion coefficients can be made by using the conventional Stejskal–Tanner pulsed-gradient, spin-echo (PGSE) pulse sequence, which is a spin-echo experiment (22). The first gradient pulse in this sequence causes the spins to dephase at a spatially dependent rate. After the 180° rf pulse, the second gradient causes the spins to refocus at the same spatially dependent rate. Under static conditions, all of the spins refocus at the correct time and generate a spin echo whose intensity is only attenuated by spin–spin effects. However, all molecules are subject to Brownian motions and move into regions of different effective fields. The result is that the spins do not refocus correctly, and the echo is attenuated according to the following:

$$M = M_0 \exp(-TE/T_2) \exp(-\gamma^2 G^2 \beta) \tag{37}$$

M is the magnetization of the echo, and M_0 is the initial magnetization. The first exponential term is the attenuation due to T_2 occurring during TE. The second exponential term accounts for the attenuation due to diffusion in the presence of a gradient pulse, G, which is on for a duration δ; β is the combination of the duration time of the gradient pulses and the interval between them, Δ:

$$\beta = \delta^2 \left(\Delta - \frac{\delta}{3} \right) \tag{38}$$

For spectra collected at the same TE, the first term in equation 37 cancels. A plot of the natural logarithm of the normalized signal versus $\gamma^2 G^2 \beta^2$ is linear, with the slope being the self-diffusion coefficient, D. Normal image-contrast measurements are made to visualize the effects of diffusion on NMR images (*21*). The imaging sequence is easily adapted to imaging by adding two more gradients in the x direction. The equations are the same as those for the spectroscopic version, and the read gradients have little effect if they are much smaller than the motion-probing gradients.

The diffusion coefficients (D) are quantitatively evaluated from a series of images recorded with different gradient field strengths. Analysis involves simulation of the effects of diffusion by using dynamic magnetization equations to calculate for each pixel the magnetization that ultimately yields an image whose intensities represent the spatially resolved diffusion coefficients. Finally, a true diffusion-constant image can be obtained if the calculated diffusion coefficients are encoded into an intensity scale (*25*). In this scale, high intensities correspond to fast diffusion. In this manner, the spatial diffusion of a liquid into a solid material can be characterized quantitatively.

Case II diffusion has been studied by NMR imaging for methanol diffusing into poly(methyl methacrylate) (PMMA; *25*). Figure 18 shows the image of PMMA in methanol after 48 h. The diffusion coefficient can be calculated by measuring the thickness of the sorbed layer as a function of time. With data-processing techniques, the measurements may be simplified by giving the images a three-level gray scale and then drawing a profile across the sample, as shown in Figure 19. The results are shown in Figure 20, where the thickness of the layer is plotted versus time. The linearity of this plot with time confirms that case II diffusion is occurring. The nonzero intercept at time zero indicates an initial Fickian diffusion process followed by case II diffusion. The constant level of methanol in the penetrant front also indicates case II diffusion.

NMR relaxation parameters are useful probes of molecular motions in polymers. Each correlation time represents the average value of the system with some distribution around that average. NMR imaging permits the determination of the spatial distribution of NMR relaxation times. This distribution provides information concerning the local motions of the

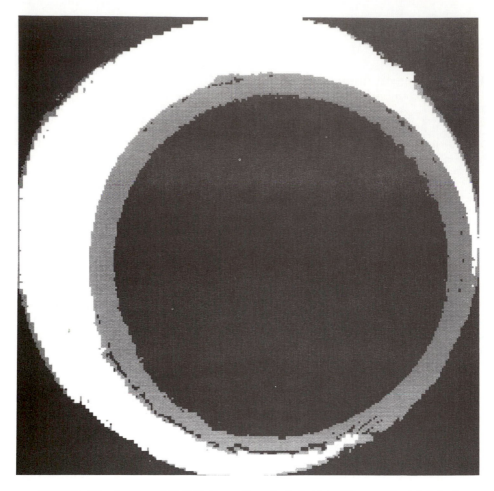

FIGURE 18. Image of PMMA in methanol after 48 h. The constant level of methanol in the penetrant front indicates case II diffusion. (Reproduced with permission from reference 26. Copyright 1989 John Wiley and Sons, Inc.)

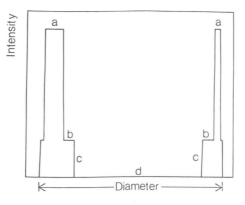

FIGURE 19. A three-level gray scale of image in Figure 18 obtained by drawing a profile across the sample. In this figure, a is the solvent, b is the imbibed region, c is the concentration of methanol in the imbibed region, and d is the core region of the PMMA rod. (Reproduced with permission from reference 26. Copyright 1989 John Wiley and Sons, Inc.)

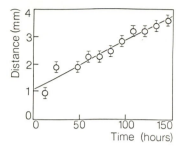

FIGURE 20. Thickness of methanol layer in PMMA versus time. (Reproduced with permission from reference 26. Copyright 1989 John Wiley and Sons, Inc.)

system. In this case, the polymer is partially swollen with solvent, and the spatial distributions of relaxation times reveal the interactions between the solvent and the polymer in the diffusion process.

Figure 21 shows a series of images of methanol in PMMA at different echo times from a multiecho experiment. The intensities are represented by colors, with red having the highest intensity and blue having the lowest. The T_2 attenuation of the intensity is more apparent for the longer echo times. Figure 22 is the calculated T_2 image from the images in Figure 21. In this case, red represents longer T_2s, and blue represents shorter T_2s. Figure 23 is a profile of the images in Figure 21; the profile shows an almost linear decrease of T_2s to the center core. Because the volume available to the methanol molecules is the same from the surface to the core, the relaxation time should be constant throughout the region if no other interactions exist. The change in T_2 from the surface to the core indicates that some physical effect other than the presence of free volume is affecting the mobility of the solvent. The polymer core effectively fixes the position of the chains at the interface between the swollen and the glassy regions of the polymer. The polymer chain dynamics change from the usual anisotropic motions with a variety of frequencies and amplitudes to anisotropic motions that depend on the position along the polymer chain. Frequencies and amplitudes decrease toward the fixed point of the chain, that is, the glassy core. This change in polymer motion influences the motion of the solvent in PMMA. The result is a change in the T_2s with distance from the glassy core.

The self-diffusion coefficient images are generated by using the magnitude images. The images are fit to equation 39 as a function of the gradient attenuation factor, d, where $d = -\gamma^2 G^2 \beta$:

$$I = I_0 \exp(d/B) \tag{39}$$

This fitting generates an image based on a dummy variable B: $B = 1/D$. The inverse of the image is taken to generate the self-diffusion coefficient imaged based on D. Figure 24 is a series of images of methanol in PMMA taken at different gradient strengths. Each successive image is the result of a more intense motion-probing gradient. The first image has a more intense signal on the outer regions of the rod, with the intensity decreasing toward the core. Figure 25 is the calculated self-diffusion coefficient image

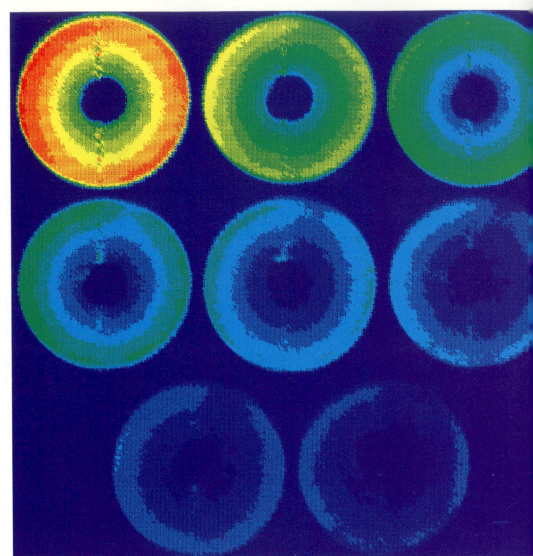

FIGURE 21. Images of methanol in PMMA at different echo times from a multiecho experiment. *TE* values (from top left to lower right) are 5.25, 10.50, 15.75, 21.00, 26.25, 31.50, 36.75, and 42.00 ms. (Reproduced from reference 25. Copyright 1990 American Chemical Society.)

generated from the images in Figure 24. The region within 100 μm of the glassy core exhibits a self-diffusion coefficient of 3.2 (± 0.9) $\times$ 10^{-7} cm^2/s. The outer region of the swollen polymer exhibits a self-diffusion coefficient of 9.2 ($\pm$ 0.9) $\times$ 10^{-7} cm^2/s. Acetone swells PMMA to a greater extent than methanol, and the self-diffusion coefficients of the system are about 2 orders of magnitude greater than those of the methanol–PMMA system. This greater swelling is apparently due to the increased volume available to the acetone molecules. The self-diffusion coefficients decrease by 36% from equilibrium in the outer regions to the region near the glassy core.

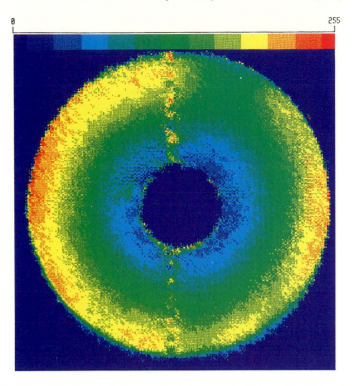

FIGURE 22. Calculated T_2 image from images in Figure 21. The color scale at the top of the image represents T_2 values ranging from 7 ms (red) to 24 ms (dark blue). (Reproduced from reference 25. Copyright 1990 American Chemical Society.)

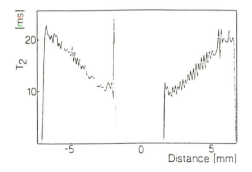

FIGURE 23. Profile of images in Figure 21, showing a nearly linear decrease of T_2 values to the center core. (Reproduced from reference 25. Copyright 1990 American Chemical Society.)

The decreasing motions of the polymer chains as the core is approached reduce the solvent mobility, as reflected in the self-diffusion coefficients.

NMR imaging of the diffusion process should be useful for multicomponent systems based on differences in relaxation times or chemical shifts. We have carried out a recent study of multicomponent imaging for acetone and water in PMMA with most interesting results (*26*).

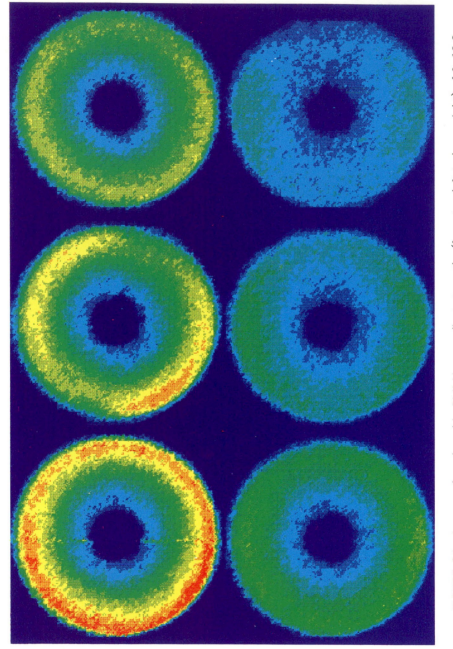

FIGURE 24. Images of methanol in PMMA at gradient strengths (from top left to lower right) of 0, 13.9, 27.6, 41.4, 55.1, and 68.8 Gy/cm. (Reproduced from reference 25. Copyright 1990 American Chemical Society.)

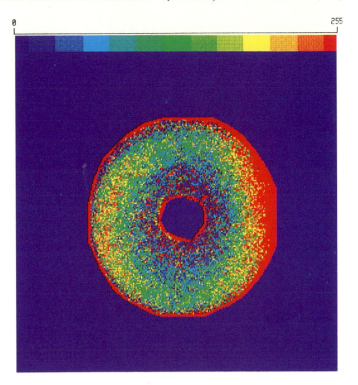

FIGURE 25.   Calculated self-diffusion image generated from images in Figure 24. (Reproduced from reference 25. Copyright 1990 American Chemical Society.)

The major limitation of NMR imaging for diffusion studies is the long measurement time, which restricts the measurements to relatively slow diffusing systems. When diffusion is too fast, the spins in the diffusion gradients are misregistered, and artifacts are produced. A number of schemes have been proposed for rapid NMR imaging of molecular self-diffusion coefficients, and one such method has been used that allows the measurement of a 256 × 256 image in 15 s (*25*).

Swelling Behavior of Polymer Systems

Solvent absorption and swelling behavior have been used to determine cross-link density in elastomeric systems. The basis of the method is that with a higher cross-link density, less solvent is imbibed in the system and the degree of swelling is lower. NMR imaging allows one to pursue this idea further by examining the homogeneity of the swelling process. The intensities of the mobile protons of the swelling agent probe the homogeneity and spatial distribution of the cross-links of the network system (*27*). Figure 26 shows the benzene proton image produced from a spin-echo pulse sequence of a highly cross-linked sulfur-vulcanized rubber sample that has been swollen in benzene for 2 days. The black spot in the image is an air bubble artifact. Figure 27 shows a three-level magnified contour plot of a portion of the swollen rubber image. This contour plot

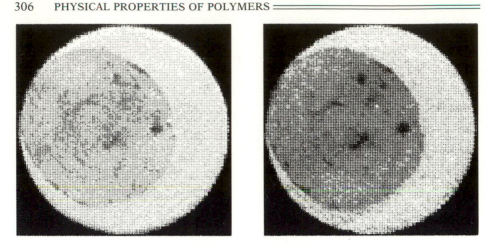

(a)   (b)

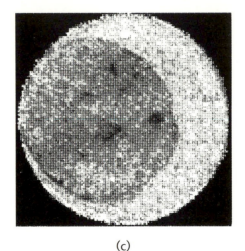

(c)

FIGURE 26. Benzene proton image from a spin-echo pulse sequence of a highly cross-linked sulfur-vulcanized rubber sample that has been swollen in benzene for 2 days. (Reproduced with permission from reference 27. Copyright 1989 John Wiley and Sons, Inc.)

indicates that there is a benzene background: regions of an intermediate level of benzene indicate a moderate level of cross-linking, and regions of little benzene indicate a high level of cross-linking. This rubber sample, obviously, has considerable inhomogeneity in cross-linking. Such inhomogeneities could arise from improper mixing, thermal gradients, or variations in vulcanization chemistry.

Samples of cured butyl rubber swollen in cyclohexane have been obtained that demonstrate poor mixing of the formulation (Figure 28). The image in Figure 28 shows regions of high cross-link density and entrapped air bubbles (*28*).

Because of the high mobility and narrow proton line width of swollen elastomer samples, images may be obtained directly from the rubber

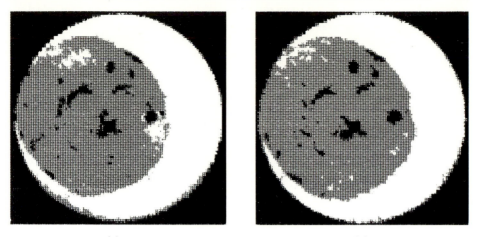

(a)   (b)

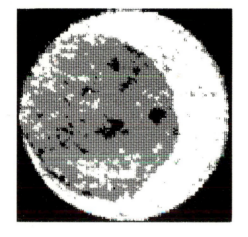

(c)

FIGURE 27. Three-level magnified contour plot of a portion of the swollen rubber image in Figure 26. (Reproduced with permission from reference 27. Copyright 1989 John Wiley and Sons, Inc.)

portion of the sample. Single and multiecho (T_2-resolved) images have been obtained of samples of *cis*-1,4-polybutadiene highly vulcanized with tetramethylthiuram disulfide (TMTD) (*28*). In preparation for imaging, all samples were extracted to remove nonnetwork material and then swollen with deuterated cyclohexane. Consequently, the NMR images of the swollen materials have been obtained through spatial variations in the proton signal intensities of the cross-linked rubber. The images were obtained by using standard Carr–Purcell spin-echo (selective 90°/nonselective 180°) pulse sequences. These T_2-weighted images are shown in Figure 29 for a polybutadiene sample containing TMTD (2 parts per hundred parts of rubber, or phr) and cured for the times indicated. The signal intensity of each pixel characterizes the degree of segmental motion of its contents.

FIGURE 28. Proton image of vulcanized butyl rubber sample curved for 25 min and swollen in cyclohexane. The right image is a two-color plot of the left image cut at 100 (on a 0–256 scale). The sample was swollen in a 20-mm-diameter tube, and the resolution was 78 μm/pixel. (Reproduced from reference 28. Copyright 1991 American Chemical Society.)

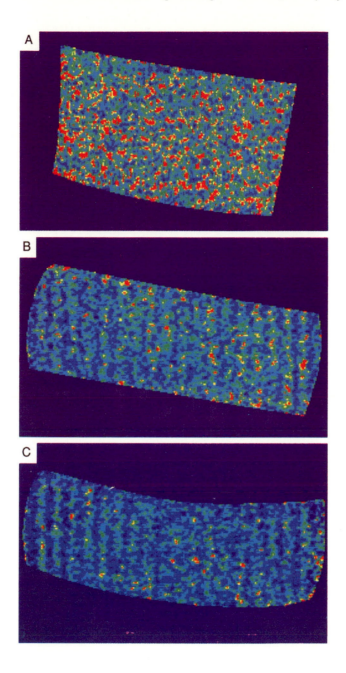

FIGURE 29. T_2-weighted images for a polybutadiene sample containing TMTD at 2 phr and cured for the times indicated: (A) 2 min, (B) 5 min, (C) 10 min. (Reproduced from reference 28. Copyright 1991 American Chemical Society.) *Continued on next page.*

For example, the decline in the S/N ratio of the images with increasing cure demonstrates that as the average length of chain segments between effective cross-link sites decreases so does the extent of their relative motion. Histograms can be obtained that yield the number of pixels with a

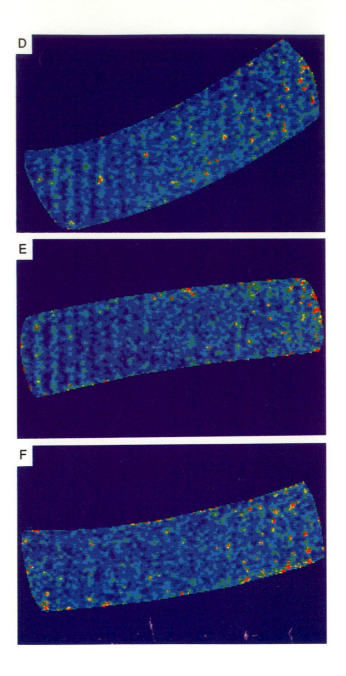

FIGURE 29.   Continued. Sample cure times: (D) 30 min, (E) 1h, and (F) 2h.

particular value of T_2, as shown in Figure 30. A significant feature of these histograms is the skewed nature of the distributions at all cure times favoring the long T_2 range. The histograms sharpen with cure time to reflect the loss of uncured chains. These histograms can be interpreted in terms of the distribution of cross-links along the chain (29).

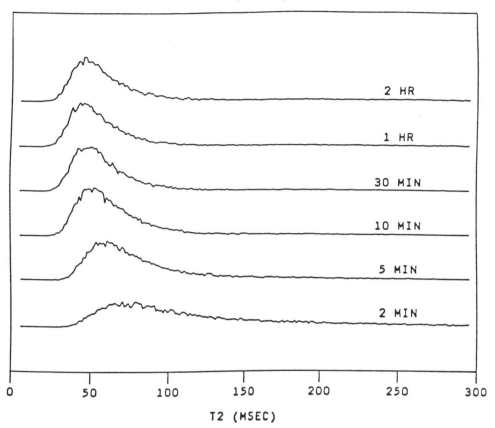

FIGURE 30. Histogram of T_2 values for TMTD-cured polybutadiene for the six sample images in Figure 28. (Reproduced from reference 28. Copyright 1991 American Chemical Society.)

References

1. Koenig, J. L. *Spectroscopy of Polymers*; American Chemical Society: Washington, DC, 1992.
2. Koenig, J. L. *Chemical Microstructure of Polymer Chains;* John Wiley: New York, 1982 (Reprinted by Kreiger, 1990).
3. Drushel, H. V.; Ellerbe, J. L.; Cox, R. C.; Love, L. H. *Anal. Chem.* **1968,** *40,* 370.
4. Grant, D. M.; Paul, E. G. *J. Am. Chem. Soc.* **1964,** *86,* 2984.
5. Kinsinger, J. P.; Fischer, T.; Wilson, C. W., III *J. Polym. Sci.* **1964,** *B4,* 379.
6. Randall, J. *Polymer Sequence Determination: Carbon-13 NMR Method;* Academic: New York, 1977.
7. Lee, A. K.; Sedgwick, R. D. *J. Polym. Sci. Polym. Chem. Ed.* **1978,** *16,* 685.
8. Mirabella, F. M., Jr. *J. Polym. Sci. Polym. Phys.* **1987,** *25,* 591.
9. Fina, L.; Koenig, J. L. *J. Polym. Sci. Polym. Phys.* **1986,** *24,* 2509.
10. Fina, L.; Koenig, J. L. *J. Polym. Sci. Polym. Phys.* **1986,** *24,* 2522.
11. Fina, L.; Koenig, J. L. *J. Polym. Sci. Polym. Phys.* **1986,** *24,* 2541.
12. Koenig, J. L.; Cornell, S. W.; Witenhafer, D. E. *J. Polym. Sci. Part A2* **1967,** *5,* 305.
13. Rabolt, J. In *Analytical Raman Spectroscopy;* Grasselli, J. G.; Bulkin, B. J., Eds.; Wiley Interscience: New York, 1991.

14. Pigeon, M.; Prud'homme, R.; Pezolet, M. *Macromolecules* **1991,** *24,* 5687.
15. Bower, D. I. *J. Polym. Sci. Polym. Phys. Ed.* **1972,** *10,* 2135; Bower, D. I. *J. Phys. B: At. Mol. Phys.* **1976,** *9,* 3275.
16. Inglefield, P. T.; Amici, R. M.; O'Gara, J. F.; Hung, C.; Jones, A. A. *Macro-molecules* **1983,** *16,* 1551.
17. Brandolini, A.; Apple, T.; Dybowski, C. *Polymer* **1982,** *23,* 40.
18. Brandolini, A. J.; Dybowski, C. *J. Polym. Sci. Polym. Lett.* **1983,** *21,* 427.
19. Roy, A. K.; Jones, A. A.; Inglefield, P. T. *Macromolecules* **1986,** *19,* 1356.
20. Clough, R.; Koenig, J. L. *Rubber Chem. Technol.* **1989,** *62,* 908.
21. Perry, B. C.; Koenig, J. L. *J. Polym. Sci. Part A Polym. Chem.* **1989,** *27*(10), 3433.
22. Hoh, K.-P.; Perry, B.; Rotter, G.; Ishida, H.; Koenig, J. L. *J. Adhes.* **1989,** *27,* 245.
23. Blackband, S.; Mansfield, P. *Solid State Phys.* **1986,** *19,* L49.
24. Weisenberger, L. A.; Koenig, J. L. *Appl. Spectrosc.* **1989,** *43,* 98.
25. Weisenberger, L. A.; Koenig, J. L. *Macromolecules* **1990,** *23,* 2445.
26. Grinsted, R.; Koenig, J. L. *Macromolecules* **1992,** *24,* 1235.
27. Clough, R. S.; Koenig, J. L. *J. Polym. Sci. Lett.* **1989,** *27,* 451.
28. Krejsa, M. R.; Koenig, J. L. *Rubber Chem. Technol.* **1991,** *64,* 635.
29. Smith, S.; Koenig, J. L. *Macromolecules* **1991,** *24,* 3496.

Chapter 7

Scattering Techniques with Particular Reference to Small-Angle Neutron Scattering

George D. Wignall

Solid State Division, Oak Ridge National Laboratory,
Oak Ridge, TN 37831-6393

Scattering techniques have been used since the beginnings of polymer science to provide information on the spatial arrangements of macromolecules (*1*). The first (X-ray) measurements were made in the 1920s and were concerned primarily with the determination of crystal structures via the Bragg law:

$$n\lambda = 2D \sin \Theta \qquad (1)$$

D is the distance between crystallographic planes, λ is the wavelength of the radiation used, n is the (integer) order of reflection, and 2Θ is the angle of scatter. The intensity is conventionally measured as a function of the momentum transfer, Q, which is related to 2Θ as follows:

$$Q = 4\pi\lambda^{-1} \sin \Theta \qquad (2)$$

However, several different symbols have been used to denote the momentum transfer in the literature (e.g., Q, K, h, k, s, q, μ, etc.). Combining equations 1 and 2 gives equation 3

$$D = 2\pi/Q \qquad (3)$$

2505–2/93/0313 $15.90/1

which indicates the distance scale probed by a measurement at a given value of Q. Assuming $\lambda = 1.54$ Å for the Cu K$_\alpha$ X-ray line and $D \approx 5$ Å for a typical intermolecular spacing, equation 1 gives $2\Theta \approx 18°$ for $n = 1$, $2\Theta \approx 36°$ for $n = 2$, etc.

Experiments in the range $0.6 < Q < 15$ Å^{-1} provide most of the information relevant for the determination of unit cell dimensions and are conventionally referred to as wide-angle X-ray scattering (WAXS), which probes (equation 3) a distance scale in the range $\sim 0.4 < D < 10$ Å. For semicrystalline polymers, WAXS methods also can provide information about crystallinity indices, lattice distortions, orientation distributions, etc., as described in standard textbooks (2). In the amorphous state, the intermolecular correlations are more diffuse, and the information available from X-ray or electron scattering is less precise. A widely used approach is Fourier transformation of the data to give a radial distribution function (RDF), which is a weighted sum of interatomic pair correlation functions $g_{ij}(r)$. These functions express the probability of finding atomic species i and j separated by a distance r and have been used extensively in structural and theoretical analyses of the liquid state. For crystalline regions of polymers, the $g_{ij}(r)$ functions reduce to a series of delta functions defining the interatomic distances in the unit cell. For amorphous materials, both intra- and intermolecular distances contribute to the RDFs, which are generally featureless for $r > 10$ Å (see Contrast). Such studies either with X-rays (3) or other radiations (4) have generally been used to argue for the absence of long-range intermolecular order between neighboring chains in amorphous polymers.

An alternative approach is to compare the experimental data with model calculations and to work in reciprocal rather than real space (i.e., to calculate the scattered intensity as a function of Q). This approach depends on being able to separate the intra- and intermolecular contributions to the data, and to achieve this separation, samples are oriented below the glass transition temperature, T_g. Although this method can give more detailed configurational information than does the RDF approach, the level of information in the scattering patterns is low, and different models can sometimes fit the data (5). A cross-check on the validity of a given model may be made by using a different incident radiation (e.g., neutrons) that highlights different features of the structure by varying the scattering power of different components of the system (6). A further cross-check can be achieved by using different techniques [e.g., IR (7) or NMR (8)] that are sensitive to the details of the local structure.

Although Bragg's law (equation 1) does not apply to amorphous materials, the Fourier or inverse relationship between the structure in real-space (r) and the scattering in Q-space means that equation 3 may be applied to a good approximation for all types of scattering. Thus, data at lower Q values probe longer length scales in the system, and X-ray scattering has been used for decades to give information on lamellar spacings (long periods) in crystalline polymers, chain dimensions in dilute solution, etc. These measurements are conventionally referred to as small-angle X-ray scattering (SAXS), although it is the Q range that determines the size of objects studied, and radiations with other wavelengths (e.g., light, neutrons) can obviously provide similar information in different angular ranges. In addition, measurements in the range $0.1 < Q < 0.6$ Å^{-1} are sometimes referred to as intermediate-angle scattering as distinguished from small- or wide-angle measurements, although no sharp boundary exists between such ranges, and the limits are to some extent arbitrary. For example, the minimum Q is instrument dependent, and as pointed out earlier, it is the Q range (rather than the angular range) that determines the length scale probed. In this chapter, these demarcation limits are adopted, and the terms small-angle scattering and wide-angle scattering are used for $Q < 0.1$ Å^{-1} and $Q > 0.6$ Å^{-1}, respectively.

Over the past 2 decades, small-angle neutron scattering (SANS), based on the availability of high fluxes of cold neutrons (wavelengths of 4–20 Å), has proven to be one of the most important tools for the evaluation of polymer chain conformation. Because of a fortuitous combination of several factors, high bulk-penetrating power, the ability to manipulate local scattering amplitudes through deuteration (isotopic labeling or staining) or an appropriate choice of solvent (contrast variation), and resolution ideally suited to the dimensions of interest in polymer studies, SANS has developed into an extremely powerful tool for the study of polymers. A great deal of new information provided by the technique has already stimulated significant advances in our understanding of polymers, leading in turn to proposals for further experiments.

Because of space limitations, it is not possible in a chapter of this length to survey all contributions to our understanding of polymer structure by all types of scattering (elastic, inelastic, quasielastic, etc.) for different types of radiation (neutrons, X-rays, electrons, light, etc.) in different Q ranges (small angle, intermediate angle, wide angle, etc.). Because many of these areas have been adequately described previously (*see* introductory and survey treatments listed in reference 9), this chapter will focus on the information provided by SANS and will demonstrate how SANS has complemented and expanded upon the data available from other scattering techniques. SANS is an example of predominantly elastic scattering, and thus, the theory will be developed for this case, and analogies to and differences from photon (e.g., X-rays and light) scattering will be pointed out at the appropriate point. The aim of this chapter is to aid potential users who have a general scientific background but no specialist knowledge of scattering to use the technique to provide new information in areas of their own particular interests. This article will therefore concentrate on explaining the essential physics of scattering with the minimum of unnecessary detail and mathematical rigor.

Energy and Momentum Transfer

Scattering in the context of this chapter means the deflection of a beam of radiation (neutrons, X-rays, etc.) from its original direction by interaction with the nuclei or electrons of polymer or solvent molecules in a sample. In an experiment with neutron radiation, a proportion of the incident beam is scattered, and the remaining fraction is transmitted through the sample. The intensity of the scattered neutrons is measured as a function of the scattering angle and/or energy. The kinetic energy of a typical neutron (particle velocity, ~750 m/s) with a wavelength (λ) of 5.3 Å is 4.7×10^{-15} ergs or ~3 meV (*10*). Such energies are of the same order as the vibrational and diffusional energies of molecular systems, and exchanges of energy between the particle (neutron) and molecule give rise to inelastic scattering, which depends on the dynamics of the system studied.

Whereas the angular dependence of the scattering of both X-rays and neutrons is easily measured, the energies of molecular vibrations ($\sim$3 meV) are much lower than those of incident photons ($\sim$10 keV), and thus, energy transfers resulting from X-ray scattering are difficult to detect. By contrast, the energy transfers resulting from neutron scattering are easily resolved and permit the elucidation of dynamic processes (10). Neutrons are thus a unique probe for studying the condensed state in that they simultaneously have both the appropriate wavelength and the energy required to investigate the structure and dynamics of polymers and other molecular systems.

Figure 1 shows the vector diagram for an incident neutron of initial wavelength λ_0 and initial velocity v_0. The neutron is scattered through an angle 2Θ to give a final wavelength of λ and a final velocity of v, and the energy gained (ΔE) by the target (and lost by the neutron) is given by equation 4

$$\Delta E = \frac{m}{2}\left(v^2 - v_0^2\right) = \frac{\hbar}{2m}\left(k^2 - k_0^2\right) = \hbar\omega \tag{4}$$

where $\mathbf{k}_0$ and $\mathbf{k}$ are the initial and final wave vectors ($k = 2\pi/\lambda$), respectively; m is the neutron mass; and $\hbar$ is $h/2\pi$, where h is Planck's constant. The momentum transfer is given by equations 5 and 6:

$$\hbar\mathbf{Q} = \hbar(\mathbf{k} - \mathbf{k}_0) \tag{5}$$

$$\hbar|\mathbf{Q}| = \hbar\left(k^2 - k_0^2 - 2kk_0\cos 2\Theta\right)^{1/2} \tag{6}$$

If energy is transferred in the scattering process ($\Delta E \neq 0$), the process is termed inelastic. If no energy change takes place ($\Delta E = \hbar\omega = 0$; $\lambda = \lambda_0$), the scattering is termed elastic, and $|\mathbf{Q}|$ is defined by equation 7:

$$|\mathbf{Q}| = 4\pi\lambda^{-1}\sin\Theta \tag{7}$$

If ΔE is small compared with the incident neutron energy ($\Delta E << E_0$), the scattering is termed quasielastic. Most of the neutron-scattering measurements on polymers have involved neutrons scattered at small values of the momentum transfer ($Q \to 0$). This type of measurement is usually referred to as small-angle (rather than small-Q) neutron scattering, al-

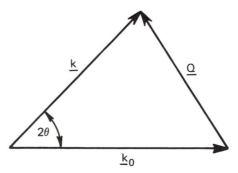

FIGURE 1. Relationship between the momentum transfer (**Q**), the scattered wave vector (**k**), and the incident wave vector (**k**$_0$) in a neutron-scattering event.

though the terms are equivalent for long wavelengths ($\lambda > 4$ Å). Equation 5 shows that, for long wavelengths, $Q \to 0$ implies $k \to k_0$, and the scattering is predominantly elastic, because any neutron scattered with a large energy transfer, ΔE, could not satisfy both energy and momentum conservation at small values of Q (*10*). Small-angle neutron scattering (SANS) experiments give information about the time-averaged structure and conformation of polymer molecules and form the bulk of the work described in this chapter. Less work has been done on quasielastic and inelastic processes, although such experiments give valuable information about polymer dynamics (*10*).

Scattering Length and Cross Section

Scattering theory is usually developed by considering a single atom that is fixed at the origin and hence cannot accept energy from the neutron (*11–12*). The interaction between the neutron and the nucleus is very short ranged ($\sim 10^{-7}$ Å) compared with the wavelength of the neutron (~ 5 Å). Because of this difference, it is shown in standard texts (*11–12*) that the scattering can contain only components with zero angular momentum. This has the important consequence that the scattering is isotropic for slow neutrons, and there is no angle-dependent form factor, as in the case of X-rays. Thus, if a plane wave of neutrons incident in the z direction is described by a wave function of unit density

$$\psi_0 = \exp(ik_0 z) \tag{8}$$

the scattered wave from a fixed nucleus (Figure 2) will be spherically symmetrical and of the following form:

$$\psi_1 = -\frac{b}{r}\exp(ikr) \tag{9}$$

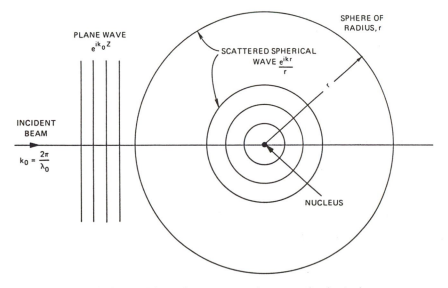

FIGURE 2. Incident plane wave and scattered spherical wave.

The quantity b has the dimension of length and is called the scattering length. Although the value of b is in principle dependent on the incident neutron energy, the variation is negligible for the energies normally encountered in neutron-scattering studies, and b may be regarded as a real (known) constant for a given nucleus (isotope). The scattered neutrons may be envisaged as originating from a sphere with radius r centered on a nucleus (Figure 2), and by using the scattering length, a scattering cross section, σ, may be defined (11) for the nucleus:

$$\sigma = \frac{\text{number of neutrons scattered}}{\text{incident neutron flux}} \tag{10}$$

Hence, the single-atom cross section is given (10–11) by equation 11:

$$\sigma = 4\pi b^2 \tag{11}$$

Equation 11 indicates that σ has the dimensions of area.

To a first approximation, the cross section may be regarded as the effective area that the target nucleus presents to the incident beam of neutrons for the elastic-scattering process. The cross section as defined by equation 11 is usually called the bound-atom cross section, because the nucleus is considered to be fixed at the origin. When the atom is free to recoil, however (e.g., in the gaseous state), the corresponding cross section is called the free-atom cross section (11). The bound-atom cross section is generally relevant to polymer studies, which are virtually always conducted on samples of macroscopic dimensions in the solid or liquid state.

Coherent and Incoherent Cross Sections

The magnitude of b varies from nucleus to nucleus and is typically of the order of 10^{-12} cm. This gives rise to the usual unit for a cross section, which is called a barn (10^{-24} cm^2). Unlike the X-ray scattering factor, f, which increases with the atomic number of the atom, no general trend is observed throughout the periodic table in the values of b, which vary from isotope to isotope and from nucleus to nucleus of the same isotope if it has nonzero spin. Because the neutron has spin 1/2, it can interact with a nucleus of spin I to form one of two compound nuclei with spins ($I \pm 1/2$), each of which has a different scattering length, b^+ or b^-, which is associated with the spin-up or spin-down states, respectively. For a given spin state J, the number of orientations is ($2J + 1$), and thus, the total possible orientations for the compound with spin states of ($I + 1/2$) and ($I - 1/2$) are $2(I + 1)$ and $2I$, respectively. The total number of spin states is $2(2I + 1)$, and because the probabilities of the spin states are equal, the statistical weights are ($I + 1$)/($2I + 1$) and $I/(2I + 1)$, respectively, for the spin-up and spin-down states. The average coherent scattering length, $\langle b \rangle$, is defined as follows:

$$\langle b \rangle = \left(\frac{I + 1}{2I + 1} \right) b^+ + \left(\frac{I}{2I + 1} \right) b^- \tag{12}$$

The brackets, $\langle\ \rangle$, represent a thermal average over the spin state population. The coherent scattering cross section (σ_{coh}) for each isotope is defined as follows:

$$\sigma_{coh} = 4\pi\langle b\rangle^2 \tag{13}$$

The total scattering cross section (σ_{tot}) is defined as follows:

$$\sigma_{tot} = 4\pi\langle b^2\rangle \tag{14}$$

The difference between σ_{coh} and σ_{tot} is the incoherent scattering cross section (σ_{inc}):

$$\sigma_{inc} = \sigma_{tot} - \sigma_{coh} = 4\pi\left[\langle b^2\rangle - \langle b\rangle^2\right] \tag{15}$$

If the isotope has no spin, then $\langle b^2\rangle = \langle b\rangle^2$, because $\langle b\rangle = b$ and there is no incoherent scattering. Only the coherent scattering cross section contains information about interference effects arising from spatial correlations of the nuclei in the system, that is, the structure of the sample. The incoherent scattering cross section contains no information about interference effects and forms an isotropic (flat) background that must be subtracted off in SANS structural investigations. The incoherent component of the scattering does, however, contain information about the motion of single atoms (particularly hydrogen), which may be investigated via energy analysis of the scattered beam (*10*). Although most of the atoms encountered in neutron scattering from polymers are mainly coherent scatterers (e.g., carbon and oxygen), there is one important exception. For hydrogen (^{1}H) the spin-up and spin-down scattering lengths have opposite signs ($b^+ = 1.080 \times 10^{-12}$ cm; $b^- = -4.737 \times 10^{-12}$ cm), and because $I = 1/2$, the following values are obtained:

$$\sigma_{coh} = 1.76 \times 10^{-24}\ cm^2 \tag{16}$$

$$\sigma_{tot} = 81.5 \times 10^{-24}\ cm^2 \tag{17}$$

$$\sigma_{inc} = 79.7 \times 10^{-24}\ cm^2 \tag{18}$$

For photons, there is no strict analogue of incoherent scattering of neutrons due to nonzero spin in the scattering nucleus. Compton scattering, which occurs for X-rays, is similar in that it contains no information about interference effects (i.e., the structure of the sample) and forms a background to the coherent signal. However, to a good first approximation, this background approaches zero in the limit $Q \to 0$ and is usually neglected in SAXS studies.

Table I gives the cross sections and scattering lengths of atoms commonly encountered in synthetic and natural polymers. These cross sections refer to bound protons and neglect inelastic effects arising from interchange of energy with the neutron. For coherent scattering, which is a collective effect arising from the interference of scattered waves over a large correlation volume, this approximation is reasonable, especially at low Q values when recoil effects are small. However, for incoherent scattering, which depends on the uncorrelated motion of individual atoms, inelastic effects become increasingly important for long-wavelength neu-

Table I. Bound-Atom Scattering Lengths and Cross Sections for Typical Elements in Synthetic and Natural Polymers

Atom	Nucleus	b_{coh} $(10^{-12}$ cm)	σ_{coh} $(10^{-24}$ cm^2)	σ_{inc} $(10^{-24}$ cm^2)	$f_{X\text{-ray}}$ $(10^{-12}$ cm)
Hydrogen	^{1}H	−0.374	1.76	79.7	0.28
Deuterium	^{2}H (D)	0.667	5.59	2.01	0.28
Carbon	^{12}C	0.665	5.56	0	1.69
Nitrogen	^{14}N	0.930	11.10	0	1.97
Oxygen	^{16}O	0.580	4.23	0	2.25
Fluorine	^{19}F	0.556	4.03	0	2.53
Silicon	^{28}Si	0.415	2.16	0	3.94
Chlorine	Cl^a	0.958	11.53	5.9	4.74

^{a} Values are for the naturally occurring element and are an average over the mixture of isotopes; $f_{X\text{-ray}}$ is given for $\Theta = 0$.

trons with the result that the incoherent cross section of hydrogen, and hence the sample transmission, is a function of both the incident neutron energy and sample temperature (*13*). Thus, the bound-atom cross section (Table I) cannot simply be used to calculate the incoherent background, because the incoherent cross section of hydrogen ($\sigma_{inc} = 79.7 \times 10^{-24}$ cm^2), although widely quoted in the literature, almost never applies to real polymer systems. For example, the effective incoherent cross-section, σ_{inc}, changes by ~30% for poly(methyl methacrylate) as λ changes from 4.7 to 10 Å (*13*). In addition, because of inelastic effects due to torsion, rotation, and vibration, the effective incoherent cross section is a function of the particular chemical group (methyl, hydroxyl, etc.) in which the proton is situated (*14*). Table I indicates that the difference in the coherent scattering lengths of deuterium and hydrogen is large and that the value for hydrogen is actually negative. This arises from a change of phase of the scattered wave and results in a marked difference in the scattering power (contrast) of polymer molecules synthesized with deuterium or hydrogen atoms along the chain.

The basic scattering experiment (Figure 3) consists of an incident neutron beam with energy E_0, which is scattered by an assembly of nuclei into solid angle $d\Omega$ with energy change dE recorded by a neutron detector. The double-differential scattering cross section for a unit volume of sample, $d^2\Sigma/d\Omega\, dE$, is defined as the number of neutrons scattered per second into a solid angle $d\Omega$ with energy change dE divided by the incident neutron flux (neutrons per second per unit area). In this chapter, σ denotes the cross section of a single nucleus, and Σ denotes the cross section for an assembly of nuclei (except when it denotes a standard summation sign, as in summing over i, j in equation 19). For such an assembly, the double-differential scattering cross section is given by standard scattering theory (*11, 15*):

$$\frac{d^2\Sigma}{d\Omega\, dE} = \frac{k}{2\pi k_0} \int_{-\infty}^{+\infty} dt\, \exp(-i\omega t)\left\langle \sum_{ij} b_i{}^* b_j F_{ij}(Q,t)\right\rangle$$

$$= \frac{k}{k_0} S(Q,\omega) \tag{19}$$

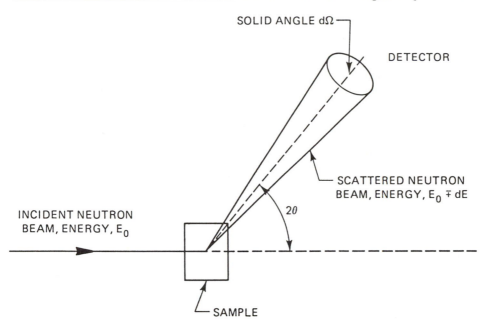

SOLID ANGLE dΩ

DETECTOR

SCATTERED NEUTRON
BEAM, ENERGY, $E_0 \mp dE$

INCIDENT NEUTRON
BEAM, ENERGY, E_0

2θ

SAMPLE

FIGURE 3.   The basic scattering experiment.

In equation 19, $F_{ij}(Q, t)$ is defined as follows:

$$F_{ij}(Q, t) = \exp[-i\mathbf{Q} \cdot \mathbf{R}_i(0)] \exp[i\mathbf{Q} \cdot \mathbf{R}_j(t)] \qquad (20)$$

In equation 19, the symbol $*$ denotes a complex conjugate and $\langle \ \rangle$ denotes a thermal average over all configurations of scatterers at position vectors $R(t)$ at time t. $S(Q, t)$ is called the scattering law or scattering function. Equation 19 may be separated into coherent and incoherent components of the cross section, as shown in equations 21 and 22, respectively:

$$\frac{d^2\Sigma_{\text{coh}}}{d\Omega\, dE} = \frac{k}{2\pi k_0} \int_{-\infty}^{+\infty} dt\, \exp(-i\omega t)\langle \sum_{ij} \langle b_i \rangle * \langle b_j \rangle F_{ij}(Q, t)\rangle \qquad (21)$$

$$\frac{d^2\Sigma_{\text{inc}}}{d\Omega\, dE} = \frac{k}{2\pi k_0} \int_{-\infty}^{+\infty} dt\, \exp(-i\omega t)\langle \sum_{i} |b_i - \langle b \rangle_i|^2 F_{ii}(Q, t)\rangle \qquad (22)$$

Equations 21 and 22 indicate that the coherent scattering cross section contains information about correlations between different nuclei (i, j) and hence gives information about the relative spatial arrangement of atoms in the system (e.g., the structure). The incoherent scattering cross section, on the other hand, contains information about correlations between the same nucleus (i, i) and hence gives information about the time dependence of the position of an individual atom (e.g., vibration and diffusion). The angular or Q-dependence of the scattering is related, via a Fourier transform, to the spatial variation of the structure. Similarly, the energy dependence of the scattering is related to the time dependence of the structure.

The majority of neutron-scattering experiments with polymers fall into the category of SANS from a fraction of deuterated chains in a matrix of normal hydrogenous polymer, and such experiments are examples of predominantly coherent elastic scattering peaked at $Q = 0$. In this region, incoherent scattering forms a flat background correction to the signal from the D-labeled molecules, which gives information about time-averaged structure (e.g., chain configuration or orientation in the bulk and polymer compatibility and segregation.) Similarly, for X-ray scattering, the energy changes are much less than the incident photon energy, so SAXS and WAXS are effectively elastic processes, which give complementary information (e.g., lamellar spacings, chain configuration in solution, and crystal structures.)

SANS Instrumentation

The initial applications of this technique to polymer science were made in the early 1970s, when the first SANS instruments (*16, 17*) suitable for the study of polymers were built on the research reactors of the Kernforschungsanlage (KFA) in Julich (Germany) and the Institut Laue-Langevin (ILL) in Grenoble (France). These instruments pioneered the use of long-wavelength neutrons and large distances between the entrance slit, sample, and detector. These facilities are long (≥ 40 m), and this length is a direct consequence of the low luminosity (L) or brilliance (particles per second per sterad per unit area) of current neutron sources, which is many orders of magnitude below the luminosity or brilliance produced in conventional or synchrotron X-ray sources (*18*). To compensate for this difference, large sample areas (~ 1 cm^2) must be used, and therefore, the overall size of the instrument must be large to maintain resolution, which can be varied in the 10–2000-Å range.

At the time of writing, more than 30 SANS instruments are in operation or under construction worldwide. This high number is due in large part to the successful application of SANS to the study of polymer structure and the new information resulting from this technique. In addition to instruments on reactor sources, SANS instrumentation on pulsed (spallation) sources has recently been developed (*19, 20*). This section will outline briefly the operation of a SANS instrument, although space limitations preclude a description of all types of instruments. In practice, details of instrumental design, operation, calibration, etc., are the responsibility of instrument scientists, and a knowledge of all these areas is not needed to use the technique. The use of SANS has spread far beyond recognized experts in the field, and much of the work described in this chapter has been undertaken by nonspecialists, who have applied the technique in areas of their own particular interests. This involvement has been made possible by the development of national and international facilities that routinely provide technical assistance and access to scattering facilities to a wide spectrum of outside users.

Most SANS instruments currently in operation are reactor-based, and a block diagram is shown in Figure 4, along with typical ranges of Q and 2Θ scanned. Fission neutrons are produced in the core, which is surrounded by a moderator (e.g., D_2O or H_2O) and a reflector (e.g., Be or graphite), which reduce the neutron energy. A typical moderator–reflector temperature is 310 K, which produces a Maxwellian spectrum of wave-

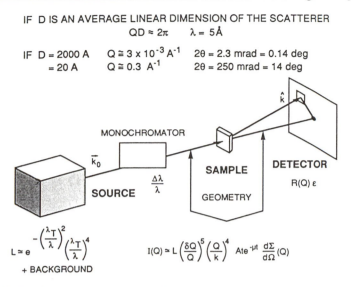

IF D IS AN AVERAGE LINEAR DIMENSION OF THE SCATTERER

$$QD \approx 2\pi \qquad \lambda = 5\text{Å}$$

IF D = 2000 A $Q \cong 3 \times 10^{-3}\,\text{A}^{-1}$ $2\theta = 2.3\ \text{mrad} = 0.14\ \text{deg}$

$\quad$ = 20 A $Q \cong 0.3\ \text{A}^{-1}$ $2\theta = 250\ \text{mrad} = 14\ \text{deg}$

$\hat{k}$

MONOCHROMATOR

$\bar{k}_0$

SAMPLE   DETECTOR

SOURCE $\dfrac{\Delta\lambda}{\lambda}$ GEOMETRY   R(Q) ε

$$L \approx e^{-\left(\frac{\lambda_T}{\lambda}\right)^2}\left(\frac{\lambda_T}{\lambda}\right)^4$$

+ BACKGROUND

$$I(Q) \approx L \left(\frac{\delta Q}{Q}\right)^5 \left(\frac{Q}{k}\right)^4 A t e^{-\mu t}\,\frac{d\Sigma}{d\Omega}(Q)$$

FIGURE 4. Block diagram of a SANS spectrometer for a reactor source and typical ranges of Q and 2Θ scanned.

lengths, which is a function of a characteristic wavelength λ_T (Figure 4) and peaks at $\lambda \approx 1$ Å (thermal neutrons). Because of the λ^{-4} factor that enters into the calculation (*18*) of the scattering power for a given resolution ($\Delta Q/Q$), it is highly advantageous to use long wavelengths and to increase the flux in this region. The increase in flux may be accomplished by further moderating the neutrons to a lower temperature by means of a cold source containing a small volume of liquid hydrogen at a temperature of ~20 K. This cold source is placed near the end of the beam tube and can give flux gains of up to an order of magnitude at $\lambda \approx 4$–15 Å. Alternative refrigerants include liquid deuterium and D_2O–ice.

The SANS cameras on the KFA (*16*) and ILL (*17*) reactors were the first to use the combination of a cold source and a large overall instrument size, and both cameras made extensive use of neutron guide tubes (*21*), which operate by total internal reflection and are used to transport the neutron beam from the cold source to the instrument. Reactor sources also produce appreciable background (e.g., fast neutrons and γ-rays), which can also be recorded by area detectors. By introducing some curvature into the guides, it is possible to separate out this component, which is not reflected as efficiently as thermal ($\lambda \approx 1$–3 Å) or cold ($\lambda \approx 4$–20 Å) neutrons. Such technology has now been installed on many subsequent facilities.

Monochromatization of the incident neutrons is usually achieved via a rotating velocity selector (*16, 17*) or via Bragg reflection with one or more pairs of crystals (*22*). The former method has the advantage of higher flux, due to the higher wavelength range transmitted ($\Delta\lambda/\lambda \approx 12$–20%), compared with ~4–6% for crystal monochromators; the latter system, however, has no moving parts and is therefore less susceptible to mechanical failure. The distance between the source and sample is typically 2–20 m, and this parameter may be varied by using beam guides to maximize the sample flux for a given resolution ($\Delta Q/Q$). The sample–detector distance (SDD) is usually in the 2–20-m range.

The size of the beam at the sample is usually defined by slits (irises) made of neutron-absorbing materials (e.g., cadmium or boron). The ratio of scattering to absorption for such elements is virtually zero, and several materials can be used for sample containment (e.g., quartz) that have very little absorption or scattering for neutrons. The result is that neutron beams can be very well collimated with very little parasitic scattering background from the slits or sample container. For SAXS, on the other hand, materials that have high absorption (to define a SAXS beam) also have high scattering power, because both parameters are strongly dependent on the atomic number; therefore, parasitic scattering is usually higher for SAXS. Similarly, samples are much harder to contain in a SAXS camera, because most materials have substantial absorption, which attenuates the beam. Thus, the high penetrating power of neutrons makes it relatively easy to contain samples in furnaces, cryostats, etc., with a minimum of instrumental backgrounds.

To reduce data acquisition times, most SANS instruments are equipped with position-sensitive detectors that collect data over a wide range of angles simultaneously. The majority of such detectors are multiwire proportional counters (17, 23), with an element size of ~1 cm^2, which is chosen to be of the same order as the sample size to equalize the various contributions to the instrumental resolution (18). In general, the detector response function, $R(Q)$, is a Gaussian distribution with a width of ~1 cm (FWHM, full width at half maximum), and the detector efficiency (ε) is a function of the position in the array (typically ~64 × 64 cm^2 or ~60-cm diameter). The spatial variation of the efficiency, $\varepsilon(x, y)$, is usually measured via an incoherent scatterer (light water or vanadium), which has an angle-independent intensity in the Q range measured. Any variation in the measured signal can then be attributed to the detector efficiency and used in the data analysis software to correct for this effect along with instrumental backgrounds. For singly scattered neutrons, the scattered intensity, $I(Q)$, is proportional to the sample thickness (t) and transmission ($T = e^{-\mu t}$, where μ is the linear attenuation coefficient) and is maximized for $\mu t = 1$. Thus, the optimum thickness is ~1–2 mm for H_2O and hydrogenous polymers (H-blanks) and ~1 cm for D_2O and deuterated polymers (D-blanks). The measured intensity is strongly dependent on the resolution, as indicated in Figure 4, and Schelten (18) has pointed out that a reduction of a factor of 2 in ΔQ will reduce the scattered intensity by over 3 orders of magnitude.

High-Concentration Labeling

In SANS experiments designed to determine the chain configuration in the bulk polymer, no energy discrimination is used, and the detector integrates over all energies. The scattered intensity, $I(Q)$, is measured as a function of angle, and for isotropic (nonoriented) samples, Q is a scalar quantity. The differential cross section is obtained by integrating equation 19 over all energies. Furthermore, for the typical elements contained in polymers, the scattering lengths may be treated as real; therefore, the complex conjugate is dropped, and assuming that SANS is a predominantly elastic process ($k \approx k_0$), then the differential scattering cross section per unit solid angle per unit sample volume is given by equation 23 (*10*):

$$\frac{d\Sigma}{d\Omega}(Q) = \left\langle \sum_{ij} b_i b_j F_{ij}(Q,0) \right\rangle + \text{INC} \tag{23}$$

Nuclear cross sections have the dimensions of area, and because sample cross sections are normalized to unit volume, $d\Sigma(Q)/d\Omega$ has the dimensions of inverse length and is typically given in units of reciprocal centimeter (cm^{-1}). The incoherent background (INC) arises from nuclei with nonzero spin (e.g., hydrogen; *see Coherent and Incoherent Cross Sections*). Because of multiple scattering effects, this background is a function of sample dimensions, transmission, etc., and thus cannot be expressed as a true cross section (*see* Figure 14). However, INC is usually smaller than the coherent signal and may be subtracted with good accuracy by empirical methods (*24*).

For a bulk polymer sample with N molecules per unit volume of pure (unlabeled) component, the coherent scattering length (a_H) of a monomer unit is defined as follows:

$$a_H = \sum_k b_k \tag{24}$$

The summation runs over all atoms (k) in an unlabeled monomer unit, and a similar equation may be written for the coherent scattering length of a labeled monomer unit, a_D. If the two polymers are blended so that X_H is the mole fraction of the unlabeled polymer component and X_D is the mole fraction of the labeled polymer component, the cross section is given by equation 25 (*10, 25, 26*):

$$\frac{d\Sigma}{d\Omega}(Q) = X_H X_D (a_H - a_D)^2 NZ^2 P(Q)$$

$$+ (X_H a_H + X_D a_D)^2 NZ^2 [P(Q) + NQ(Q)] + \text{INC} \tag{25}$$

$P(Q)$ is the intrachain signal (form factor), which originates from monomer pairs belonging to the same chain [$P(0) = 1$]; $Q(Q)$ is the interchain signal resulting from monomer pairs on different chains; and Z is the polymerization index. The second term on the right of equation 25 is proportional to the scattering from density fluctuations and may be neglected for some systems (*27–29*), for example, amorphous polymers, where this component results from thermal vibrations, etc., and is virtually zero ($P + NQ \approx 0$). For systems containing large density fluctuations (e.g., crystalline polymers) or with appreciable amounts of heterogeneities (catalyst residues, impurities, stabilizers, etc.), this term can produce a measurable intensity, which must be subtracted along with the incoherent background. The remaining term is the coherent cross section after subtraction of the density fluctuations (term 2) and the incoherent background (term 3). It is directly analogous to the Rayleigh ratio, used in light scattering (*30*), and contains information about the single-chain (intramolecular) scattering function, $P(Q)$. Equation 25 is also based on the assumptions that deuteration of the hydrogenous molecule has a negligible effect on monomer–monomer interactions (*see Isotope Effects*) and that both chains have the same polymerization index (Z) and the same number of molecules per unit volume (N). The coherent scattering is governed by the single-chain form factor $P(Q)$, and the mole fraction of each component modulates the scattered intensity, with the maximum coherent scattering of the blend occurring at a 50–50 mixture of the two components. Thus $P(Q)$ may be obtained from the measured coherent intensity at labeling levels of up to 50%. Although equation 25 is essentially the same equation given in 1918 by Von Laue (*31*) for random binary alloys, the result was not appreciated in the earliest SANS studies of bulk polymers and concentrated solutions. These early studies relied on analogies with light and X-ray scattering, for which the limit of zero concentration was required to eliminate interchain interference. For $X_D \ll 1$, $X_H \approx 1$, and the cross section is proportional to the mole fraction or concentration (*10*, *32*), as assumed in the Guinier (*33*) and Zimm (*10*, *33*) approximations.

Contrast

The quantity $(a_D - a_H)^2$ is related to the difference in scattering power between labeled and unlabeled chains and is called the contrast factor. In general, radiation incident on a medium whose scattering power is independent of position is scattered only into the forward direction ($\Theta = 0$). For every volume element (S) that scatters radiation through an angle $\Theta > 0$, another volume element (S') scatters exactly out of phase (by 180°; Figure 5, where P is a point on the incident wavefront with the same phase as S' and $PS - S'S = \lambda/2$). Therefore, all scattering cancels unless the scattering power at S is different from that at S'; that is, the scattering power fluctuates from point to point in the sample. X-rays and light photons interact with electrons in the sample and hence are scattered by fluctuations in electron density. Neutrons, on the other hand, have no interaction with electrons (apart from unpaired spins, in which case the interaction arises from the magnetic moment of elements such as rare

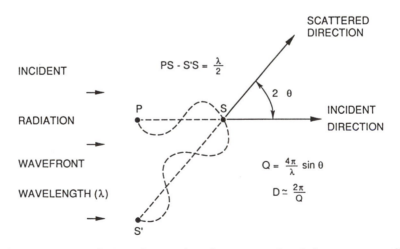

FIGURE 5. Cancellation of scattering. For every point S that scatters radiation through an angle $2\Theta > 0$, there is another point S' that scatters radiation exactly 180° out of phase. Therefore, all scattering cancels unless the scattering power at S is different from that at S' (i.e., fluctuates from point to point in the sample).

earths and transition metals). Because organic polymers do not contain such elements with unpaired spins, the only interaction with neutrons is via nuclear scattering, which arises from differences in scattering-length density (SLD). Because each nucleus has a different scattering amplitude (Table I), the SLD is defined as the sum of coherent scattering lengths over all atoms lying in a given volume element ΔV divided by ΔV (*10, 15*). For example, in partially labeled polymer blends, the SLD is given by the coherent scattering length (equation 24) divided by the monomer volume. The coherent cross section of a system of uniform SLD is zero, although fluctuations may be introduced by means of isotopic substitution to give rise to a finite cross section that is proportional to $(a_H - a_D)^2$. To produce observable SAXS contrast, which can be used to give direct information about $P(Q)$, the electron density of the chain must be changed. Such experiments have been performed by Hayashi et al. (*34*), who labeled polystyrene chains statistically with iodine atoms, and by this means, concentrated solutions and bulk polymers were investigated by X-ray scattering. This method of labeling relies on changing the chain chemistry, and in general, it produces a greater perturbation on the configuration than does deuterium labeling. The method seems to give reasonable results after extrapolation to infinite dilution of the iodine-labeled molecules, although it is unlikely that labeling levels of up to 50% could be used, and the vast majority of studies of polymer configurations have used deuterium labeling in conjunction with SANS.

The contrast-variation methods that have found wide application in structural biology can sometimes be used to remove a component of the scattering by matching its scattering power with that of the medium in which it is dispersed; such matching removes the fluctuation giving rise to the scattering. This principle is illustrated in Figure 6, prepared by D. Engelman (Yale University). Both tubes shown in Figure 6 contain two Pyrex beads embedded in glass wool, which has a refractive index lower than that of Pyrex glass. When light shines on tube B, both the beads and the glass wool scatter light, but only the glass wool can be seen, because it dominates the scattering. To observe the Pyrex beads, tube A was filled with a solvent that has the same refractive index as the glass wool. Thus, the electron density, and hence the scattering power of the glass wool, was matched with that of the solvent, thus eliminating this component of the scattering and making the glass wool transparent to light. This principle can be used in SANS experiments via isotopic solvent mixtures (e.g., H_2O-D_2O) to adjust the SLD of the medium, as for example in studies of core–shell polymer latexes (*see Latex Structure and Film Formation*).

Figures 7 and 8 illustrate some of the information available from SANS and SAXS via intensity slices of data about the beam stop ($Q = 0$). The solid circles show the SANS cross section for high-density polyethylene. The data are superimposed on a flat incoherent background of ~0.5 cm^{-1}, which is the third term in equation 25. The open circles show the extra coherent cross section produced by adding 2% deuterated molecules, which is the first term in equation 25 and is proportional to the contrast difference, $(a_H - a_D)^2$, between deuterated (PED) and hydrogenous (PEH) polyethylene molecules. Departures from the flat incoherent background of the PEH sample (solid circles) are caused by density fluctuations in the sample, and it is just possible to see the peaks at $Q \approx 0.025$ Å^{-1} due to the periodic stacking of crystalline lamellae alternating with amorphous re-

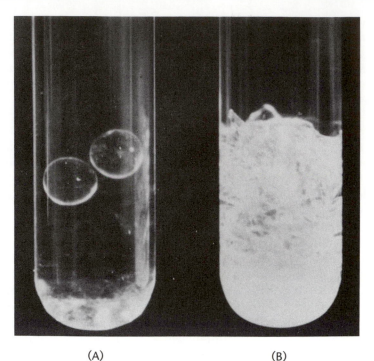

(A)   (B)

FIGURE 6. Principle of contrast variation demonstrated by two tubes containing Pyrex beads in glass wool and solvent: tube A, refractive index of solvent matches that of glass wool, and tube B, refractive index of solvent is different from that of glass wool or Pyrex beads and scattering from the glass wool dominates.

gions. The extra SANS coherent signal in PEH is very weak, however, because of the cancellation of the scattering lengths of carbon and hydrogen (Table I), which makes the prefactor, $(X_H a_H + X_D a_D)$, of the second term in equation 25 almost zero.

The SAXS signals for the protonated (PEH) and the partially deuterated (2% PED) samples are the same, because there is no (electron density) contrast between deuterated and protonated molecules and virtually no incoherent background; therefore, there is no equivalent of the first and third terms in equation 25 in the X-ray cross section. The density fluctuation scattering is much stronger, because there is no cancellation between the scattering from the H- and D-labeled molecules, and the cross section is proportional to the square of the electron density difference between the amorphous and crystalline regions (lamellae). Thus, both the interlamellar peak at $Q \approx 0.025 \text{ Å}^{-1}$ and the upturn as $Q \to 0$ show much more strongly (Figure 8) than does the SANS coherent signal, and such data have long been used to investigate the morphology of crystalline polymers. For unoriented samples, the scattering is symmetrical and may be radially (azimuthally) averaged for a given Q to produce the intensity as a function of $|Q|$. For oriented samples, the data must be analyzed in sectors or slices as a function of Q. As mentioned earlier, several symbols have been used for the scattering vector or momentum transfer (Q, K, h, k, s, q, μ, etc.). When literature data using a different nomenclature is discussed in this chapter, the alternative symbol will be displayed along

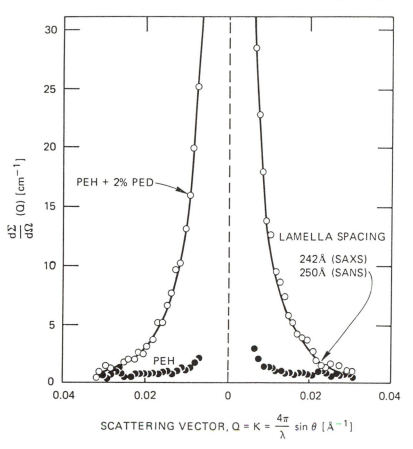

FIGURE 7. $d\Sigma/d\Omega(Q)$ versus Q from PEH (blank) and 2% PED in PEH matrix for slice of the two-dimensional scattering pattern.

with Q for clarity (e.g., in Figure 7). It should be borne in mind, however, that each symbol denotes $4\pi\lambda^{-1}\sin\Theta$ and that all are equivalent.

As described earlier, WAXS data probe the structure on a smaller length scale, and it is instructive to show how the radial distribution function (RDF) approach can complement the information available from SAXS for amorphous and crystalline samples. If ρ is the number density of atoms in a macroscopically isotropic system, the atom pair RDF, $g(r)$, is defined so that $\rho g(r)\mathbf{dr}$ is the number of atoms in a volume element $\mathbf{dr}$ at a distance r from an origin atom. Figure 9 shows the RDFs (*35*) describing the correlations between carbon atoms, $g_{cc}(r)$, for a sample of high-density polyethylene in the crystalline (powder) and amorphous (melt) states, obtained by Fourier transformation of the intensity data in the range $0.3 < Q < 15$ Å^{-1}. The peaks at 1.54 and 2.54 Å represent the intramolecular distances corresponding to the two skeletal C–C distances in the polyethylene chain, which are independent of conformation and, hence, are identical for the crystalline and amorphous states. In addition, the RDF for the crystalline (powder) state shows regular oscillations reflecting the periodicity of the polyethylene lattice; the oscillations extend to radial distances of ~50 Å, as expected for materials with lamellae of ~100-Å thickness. For the amorphous (melt) state however, the inter-

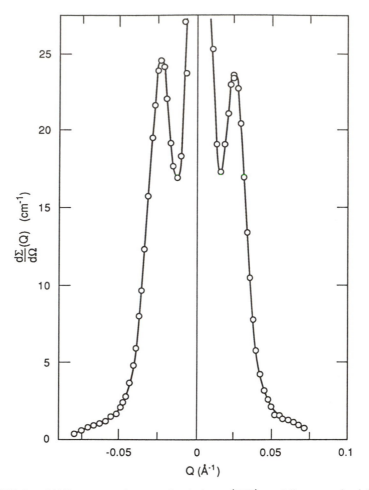

FIGURE 8. SAXS pattern from polyethylene (PEH) rapidly quenched from the melt.

atomic correlations are indistinguishable from the noise for $r > 12$ Å; similar results have been found for many amorphous polymers (*3, 4*). Such data have been used to argue that local interatomic ordering is short-ranged and liquidlike and extends only to second nearest neighbor contacts between chains. This finding is consistent with the underlying assumptions of the random (Gaussian) coil model, which has also received strong support from SANS experiments. The parameter used to describe the overall size of a polymer chain is the radius of gyration, R_g, the root mean square distance from the center of gravity of a polymer chain:

$$R_g^2 = \Sigma f_k r_k^2 / \Sigma f_k \qquad (26)$$

The summation runs over all scattering elements (k), which are the electrons in the case of SAXS. For SANS, the summation runs over all nuclei and is weighted by the scattering length of each atom. Thus in principle, the R_g values may be different when measured by different techniques. However, in practice, each monomer has the same scattering power for a given incident radiation, and so for large polymerization indices, the

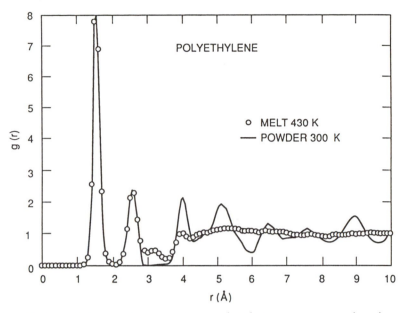

FIGURE 9. Radial distribution function (RDF) for amorphous (melt) and crystalline (powder) polyethylene. (Reproduced with permission from reference 35. Copyright 1989. American Institute of Physics.)

differences between SANS and SAXS radii are negligible. R_g may be derived from equation 25 by expanding $P(Q)$ in a power series for low Q values ($Q < R_g^{-1}$) and plotting $d\Sigma^{-1}(Q)/d\Omega$ versus Q^2 (*10, 33*). Alternatively, these parameters may be obtained by plotting $\ln[d\Sigma(Q)/d\Omega]$ versus Q^2 at low Q values (*33*). These types of plots are conventionally referred to as Zimm and Guinier plots, respectively; the Zimm plot is generally used for investigating polymer configurations, because it is linear over a wider Q range. The first measurements in bulk amorphous polymers and concentrated solutions were generally performed in the limit of low relative labeling and extrapolated to zero concentration. In this range, the cross section is given by equation 27:

$$\frac{d\Sigma^{-1}}{d\Omega}(Q) = \frac{m_D^2}{c(a_H - a_D)^2 M_{wD} N_A}\left[1 + \frac{Q^2 R_g^2}{3} + \cdots\right] \quad (27)$$

In this equation, c is the concentration (g/cm^3), N_A is Avogadro's number, m_D is the molecular weight of a deuterated repeat unit, and M_{wD} is the weight-average molecular weight (M_w) of a deuterated chain (*10, 32*). The realization that the same information could be obtained with greater accuracy at much higher levels of labeling was made (*25, 36, 37*) and verified independently by several groups (*25, 26, 37–39*). Figure 10 shows the variation of R_g with the concentration of labeled molecules in amorphous polystyrene (*38*); the measured values are independent of the level of labeling. Similarly, Figure 11 shows that the cross section extrapolated to $Q = 0$, $d\Sigma(0)/d\Omega$, is proportional to the product $X_D X_H$ for amorphous polystyrene (*38*), as expected from equation 25. This equation was derived with the assumption of equal polymerization indices for the labeled and

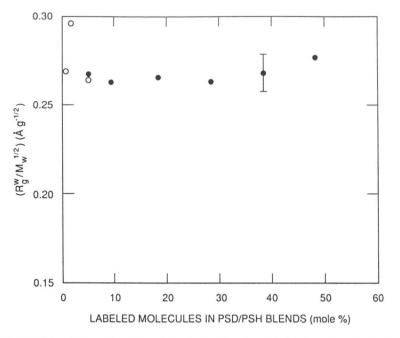

FIGURE 10. Determination of R_g from blends with high concentrations of deuterium-labeled molecules. Data are from references 38 (●) and 42 (○).

unlabeled chains. The effect of unequal indices has been considered by Boué et al. (*40*), who showed how the measured R_g and $d\Sigma(0)/d\Omega$ were perturbed by this mismatch. In this case, the mole fractions, X_H and X_D, in equation 25 are replaced by volume fractions, ϕ_H and ϕ_D, respectively. An alternative treatment of mismatch effects has been given by Tangari et al. (*39*). Similarly, the theory was developed for monodisperse polymers, and corrections to SANS data due to polydispersity have been reviewed by Boothroyd (*41*). The R_g measured via Zimm–Guinier analysis is a z-average over the configuration and is often converted to a weight-average value, R_{gw}, for comparison with model predictions (*10, 41, 42*).

Applications of Scattering Techniques

Amorphous Polymers and the Glassy State

An exhaustive review of all SANS applications is beyond the scope of this chapter, and the examples given in this section to illustrate the scope of the technique were chosen because of their impact on polymer science in general and on the areas reviewed in other chapters of this book in particular. For example, the random or Gaussian coil model has been used extensively to provide a theoretical basis for the understanding of rubber elasticity, polymer melts, and viscoelasticity (*see* Chapters 1–3). However, before the development of the SANS technique, there was no way of directly measuring the molecular configuration in bulk polymers; other models based on collapsed coils (*43*) and quasiparallel arrangement of molecules (*44, 45*) were also advanced. The possibility of using the con-

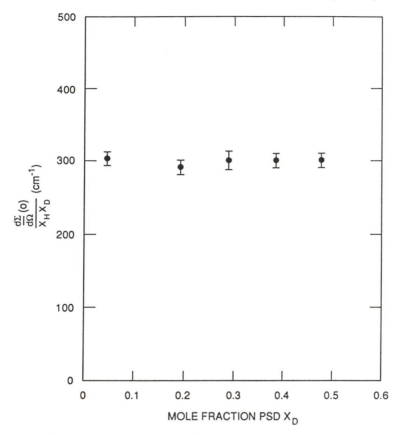

FIGURE 11. Variation of $d\Sigma/d\Omega(0)/X_H X_D$ with X_D for blends of deuterated and normal (hydrogenous) polystyrene.

trast between deuterated and normal (hydrogenous) molecules to determine $P(Q)$ was suggested in the late 1960s (*46, 47*) and demonstrated in practice in the early 1970s (*42, 48–50*). According to the random coil model, R_{gw} should be proportional to $M_w^{0.5}$, with the same constant of proportionality in the bulk as in an ideal theta solvent. SANS measurements, summarized in reference 10, demonstrate that this prediction holds remarkably well for amorphous polymers (Table II).

Although these results gave strong support to the random coil model, they were not in themselves conclusive, because it was subsequently shown for crystalline polymers that R_g values are very similar for molecules in the molten (amorphous) and solid (crystalline) states (*see The Crystalline State*). Thus, the finding that molecules exhibit the unperturbed (theta-solvent) dimensions in the molten or glassy amorphous state does not in itself rule out significant parallelism for an appreciable fraction of the molecules. The radius of gyration (R_g) is an average (equation 26) over the configuration, and hence, quite different molecular trajectories can have the same values of R_g. To test how far the local molecular conformation, as opposed to the overall R_g, is described by the various models, measurements have been extended to higher values of Q. As described earlier, the scattered intensity at a given Q is sensitive to fluctuations in the scattering length density (SLD) on a distance scale, D, of $\sim 2\pi/Q$;

thus, as Q increases, the scattering is increasingly determined by the local chain conformation. This may be calculated for the random coil model by using rotational isomeric statistics (51), and hence, the scattered intensity may be estimated numerically and compared with experiment. The comparison is accomplished by measuring the scattering in the intermediate-angle Q range (0.1 < Q < 0.6 Å^{-1}), which is sensitive to the local conformation of the chain over distances of ~10–50 Å. Figure 12 shows intermediate-angle neutron scattering (IANS) for atactic (glassy) polystyrene (42)

Table II. Molecular Dimension in Bulk Amorphous Polymers

Polymer	State	$R_{gw}/M_w^{0.5}$ [Å / (g / mol)$^{0.5}$]	
		Bulk	Theta Solvent
Polystyrene, atactic	Glass	0.265–0.280	0.27–0.28
	Melt	0.275	0.27–0.28
Polyethylene	Melt	0.45–0.46	0.45
Poly(vinyl chloride)	Glass	0.40	0.37
Polyisobutylene	Glass	0.31	0.30
Poly(ethylene terephthalate)	Glass	0.39	0.42
Poly(methyl methacrylate)			
Atactic	Glass	0.25–0.27	0.25
Syndiotactic	Glass	0.29	0.24
Isotactic	Glass	0.30	0.28
Polybutadiene	Melt	0.35	0.34–0.42

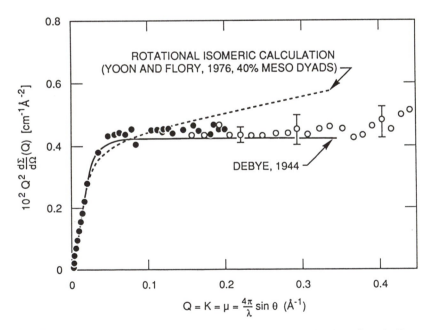

FIGURE 12. Kratky plot for atactic polystyrene. The broken line indicates results of rotational isomeric calculation for 40% meso dyads (52). The continuous line indicates the Debye model for a coil with a Gaussian distribution of chain elements (53).

compared with the rotational isomeric state (RIS) calculation of Yoon and Flory (*52*) and the Debye model for a coil with a Gaussian distribution of chain elements (*53*):

$$P(Q) = 2\left[R_g^2 Q^2 + \exp\left(-R_g^2 Q^2\right) - 1\right]/R_g^4 Q^4 \qquad (28)$$

The data are plotted as $Q^2 d\Sigma(Q)/d\Omega$ versus Q (conventionally referred to as a Kratky plot), because this representation enhances the scattering at higher Q values and facilitates comparison with different models. Figure 12 shows that $d\Sigma(Q)/d\Omega$ varies with Q^{-2} in this region, leading to a plateau in the Kratky plot, which is closely fitted by the Gaussian coil function (equation 28). Surprisingly, the Debye model, which is based on general assumptions independent of the local chain structure, provides a better fit to the data than does the RIS model, which reflects the local architecture (covalent bond lengths, angles, etc.) of the particular chain. The data shown in Figure 12 are for fully deuterated (D8) polystyrene (PSD) molecules, although subsequent measurements on polystyrene labeled only in the chain backbone (D3) indicate that the Kratky plot does exhibit a positive slope as predicted by the RIS model (*54*, *55*). Thus the agreement of the data with the Debye model (Figure 12), resulting from the cancellation of this (positive) slope with the (negative) trend due to the finite lateral dimensions (~4 Å) of a fully labeled chain, is probably fortuitous.

Results with poly(methyl methacrylate) (PMMA) indicate that the scattering in the intermediate Q range is markedly dependent on chain tacticity, which influences the local packing in the size range probed by IANS. Comparisons with theory (*30, 56, 57*) indicate that the shape of the curves as a function of tacticity is accounted for by the RIS model (*57*). Figure 13 shows IANS data for polyisobutylene. In Figure 13, $F_n(Q)$ is the absolute Kratky function (*58*), which is the same as $Q^2 d\Sigma/d\Omega$ except for a multiplicative constant (*see The Crystalline State*). The data for the bulk

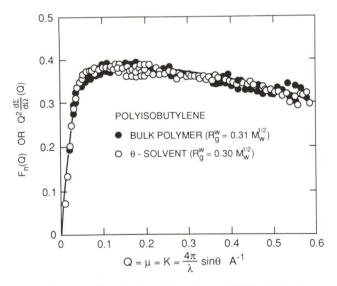

FIGURE 13. Absolute Kratky functions from IANS of hydrogenous polyisobutylene (PIB) in PIB-D$_8$ (bulk amorphous polymer) and for hydrogenous PIB in benzene-D$_6$ (Θ solvent).

polymer and theta solvent overlap closely up to a Q of ~ 0.6 Å^{-1}, indicating that the molecular configurations are very similar for length scales ranging from R_g (~ 100 Å) down to $2\pi/Q_{max}$ (~ 10 Å; Q_{max} is maximum momentum transfer). Similar comparisons in the intermediate Q range have also been made with polycarbonate (59) and amorphous (molten) polyethylene (52, 60), and in each case, reasonable agreement was achieved with the RIS model.

Measurements in the IANS range are particularly sensitive to the incoherent background, which can be of the same order of magnitude as the coherent signal. The coherent intensity of singly scattered neutrons, $I(Q)$, is proportional to the thickness (t), transmission (T), and sample area (A). Thus the coherent differential scattering cross section (10) is given by equation 29:

$$\frac{d\Sigma}{d\Omega}(Q) = \frac{I(Q)}{tT}\frac{(\text{SDD})^2}{K_N A} \tag{29}$$

SDD is the sample–detector distance, and K_N is a calibration constant to convert the data to absolute units (61). Thus, measurements on samples with different dimensions (t and A) and transmission (T) may be normalized to the same volume to give a coherent cross section, which is an intensive (material) property independent of sample dimensions. This result is based on the assumption that neutrons are scattered only once before being detected, which has been shown to be a reasonable approximation for coherent SANS from polymers (62) and other materials (63). For incoherent scattering, however, this assumption is not valid, and 1–2-mm samples containing hydrogen (H_2O, hydrogenous polymers, etc.) give rise to appreciable multiple scattering, as illustrated in Figure 14, which shows the apparent cross section (56) produced by three hydrogenous PMMA blanks after normalization via equation 29. Because the data contain appreciable multiple scattering (which is not proportional to the thickness or transmission), the data cannot be normalized to a true cross section that is independent of the sample dimensions. However, the incoherent background is independent of Q, and empirical methods have been developed to subtract this background to a good approximation (24, 56, 58).

Other techniques that have been used to study the amorphous state include SAXS, light scattering (LS), and Rayleigh–Brioullin (RB) scattering. Whereas SAXS is sensitive only to density fluctuations, LS is sensitive also to the local anisotropy via the H_v (the depolarized Rayleigh ratio) component of the scattering, the magnitude of which is in reasonable agreement with the random coil model (64). RB scattering also reveals only short-range liquidlike correlations, with no evidence of long-range orientational order (65). SAXS arises from density fluctuations and is observed for all amorphous polymers (64), although it is often strongly sample dependent because of the presence of heterogeneities (catalyst residues, stabilizers, dirt, voids, etc.). Figure 15 shows absolute SAXS data for polycarbonate (66), which can be accurately modeled as the sum of (frozen-in) thermal density fluctuations plus small quantities (<< 1 vol %) of large (500–4000-Å) heterogeneities. When such impurities are removed, amorphous polymers are characterized by a very small, constant scattering

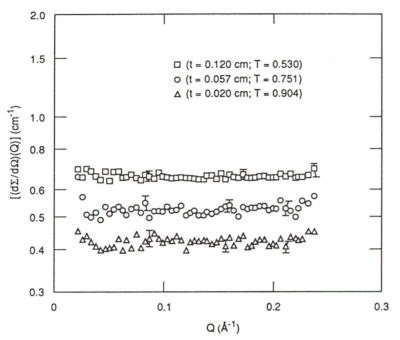

FIGURE 14. $d\Sigma / d\Omega(Q)$ versus Q for three PMMA-H blanks.

due to thermal density fluctuations (*64*). Thus SANS, SAXS, LS, RB scattering, and RDF techniques all provide complementary information for amorphous polymers, which is consistent with the random (Gaussian) coil model to a good approximation.

Diffusion and Deformation

The mode of diffusion of polymer chains in the bulk polymer has aroused much interest from both the theoretical and experimental viewpoints. According to deGennes, diffusion on a microscopic scale may be envisioned as the reptation of a chain along a tube formed by the entanglements of neighboring molecules. There has been considerable debate concerning the applicability of this concept over the distance scale ranging from a chain segment to the size of the overall radius of gyration. In the range of the overall radius of gyration, measurements have been performed previously by microdensitometry on samples of deuterated and hydrogenous polymers that were allowed to interdiffuse at an interface, which was then sectioned and examined by IR methods. Because of the limiting thickness of sections, this method was effectively limited to polymers with molecular weights of $<10^4$, which needed time scales of the order of a month for measurements (*67*). Bartels et al. (*68*) used SANS methods to extend the available range of diffusion coefficients that may be studied by preparing samples consisting of alternating layers of deuterated and hydrogenous polymers. As the temperature is raised, the spatial modulations in composition decay as the hydrogenous and deuterated molecules interdiffuse. The scattering grows progressively and finally

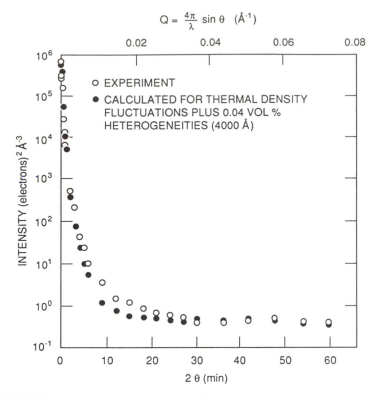

FIGURE 15. Absolute experimental SAXS from polycarbonate and calculations for frozen-in thermal density fluctuations plus various heterogeneities. (Reproduced with permission from reference 66. Copyright 1976 Marcel Dekker).

reaches the pattern corresponding to a uniform molecular mixture of deuterated and hydrogenous components. Analysis of the time dependence of the scattering yields the polymer diffusion coefficient (D), and because of the short distance scale probed by the SANS method, measurements may be extended by approximately 3 orders of magnitude beyond the limit of IR microdensitometry (68). Measurements on monodisperse fractions of hydrogenated polybutadiene (HPB) confirm the prediction that $D \sim M^{-2}$ (M is molecular weight) for long chains and give activation energies in good agreement with theoretical estimates (*see* Chapter 3 by W. W. Graessley). Similar SANS experiments have been undertaken to probe the diffusion behavior of branched polymers (69), for which diffusion coefficients (D) depart from the M^{-2} prediction (Figure 16); this technique should have important future applications for elucidating the mutual interdiffusion of different molecular species, one of which is labeled with deuterium.

An important application of the SANS technique is in the determination of the response of individual molecules in a bulk polymer to a macroscopic deformation or swelling of the sample. As shown in previous sections, SANS from partially deuterated samples can provide direct information about the deformation of the whole molecule down to the average statistical segment and offers the possibility of checking the various models that have been developed to relate the macroscopic and microscopic properties of polymers, for example, in theories of rubber elasticity (*see*

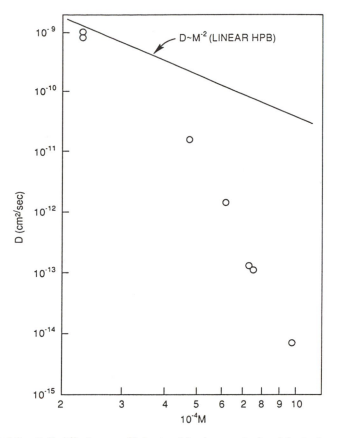

FIGURE 16. Self-diffusion coefficients of hydrogenated polybutadiene (HPB) three-arm stars (○) and for linear HPB (solid line) at 165 °C. (Reproduced from reference 68. Copyright 1984 American Chemical Society.)

Chapter 1 by J. E. Mark). A series of experiments has been undertaken to test these assumptions with respect to the deformation of individual chains in both chemically cross-linked polymers (networks) and linear polymers that contain only transitory (apparent) cross-links arising from physical entanglements.

Experiments on deformed samples are inherently more difficult than corresponding studies of isotropic materials, because deformed materials produce anisotropic scattering that must be analyzed via slices or sectors of the two-dimensional pattern, thus giving rise to inherently larger statistical errors. One way to avoid this problem is to use networks that are swollen by a solvent, because the scattering patterns from such samples are azimuthally symmetrical and thus may be radially averaged to give better statistical accuracy. Other difficulties arise from ill-defined networks with broad molecular weight distributions; mismatch between the lengths of the labeled and unlabeled chains; and appreciable numbers of dangling chains, loops, or entanglements trapped between adjacent chains. Ullman (70–72) summarized the differences between different types of networks and pointed out that randomly cross-linked chains deform to a greater extent than do end-linked chains. To minimize these effects and provide well-characterized model networks, many of the first experiments were carried out on polystyrene, although this polymer is a glass at room temperature and is unsuitable for some experiments, because elongation must be carried out at elevated temperatures. For linear polymers, any finite time interval between stretching and quenching to the glassy state can result in stress relaxation, which changes the deformation of the molecules from its initial state.

SANS experiments with deformed linear polymers and networks have been reviewed (10, 72–74), and Ullman (72) has compared data from networks with theoretical models that have been advanced for this type of system. For linear (uncross-linked) polymers, several experiments indicate that the radii of gyration perpendicular and parallel to the orientation direction transform largely in the same manner as the external macroscopic dimensions (affine deformation), provided that the molecular weight is sufficiently high and that the conformational relaxation processes are slow compared with the time in which the sample is deformed. Picot et al. (75) studied the uniaxial deformation of hot-stretched polystyrene and concluded that, for low elongations with macroscopic deformation ratios, λ_d, of < 1.7, affine deformation is the dominant mechanism. Deviations from affine behavior were found at high elongations ($\lambda_d > 3$), and an explanation was put forward in terms of a model for which affine behavior holds only for distances separating effective cross-links. Boué et al. (76) found that the transverse coil radius of gyration of hot-stretched linear polystyrene undergoes affine deformation at $\lambda_d < 3$ and studied its relaxation as a function of time and temperature. As expected, affine behavior is more likely for short times and deformation at temperatures close to the glass transition temperature, T_g, because the temporary entanglements have less time to relax. For longer times and higher temperatures, time–temperature superposition was used to compare the observed relaxation with reptation theory, although definite conclusions will probably require the application of these methods over a wider range of reduced time and Q. For extrusion-oriented, high-molecular-weight (M_w, ~500 000) polystyrene, the molecular deformation was nearly affine up to $\lambda_d = 10$, whereas for lower molecular weight material, the deformation was less than affine (77). For linear chains, the segmental orientation arises solely from the entanglements that interconnect all chains in a macroscopic network. Chains with molecular weights less than the critical entanglement molecular weight (M_c) and that are embedded in a matrix of longer chains do not enter into the entanglement network and thus should not be susceptible to orientation during stretching of the specimen, except for transitory flow effects.

For networks, chemical cross-linking does not appreciably change the molecular dimensions from those of the melt (78–80), and the change in dimensions introduced by swelling or stretching is much lower than that predicted by affine deformation of the whole chain (chain affine model). This result is not unexpected, because networks deform quite differently from linear polymers, and the chain affine model has never been seriously advanced for chemically cross-linked materials, except perhaps in cases when they are deformed at temperatures close to the glass transition temperature, T_g (70). The classical theory of rubber elasticity assumes that the junction points deform affinely and is often referred to as the "junction affine" model (81). The "phantom network" model, due to James and Guth (82), is based on the assumptions that the mean positions of the cross-link points deform affinely and that fluctuations in the cross-link points are

Gaussian and independent of the applied deformation. The junction affine model may be regarded as a special case of the phantom network model when fluctuations are suppressed (83). Other models, intermediate between the junction affine and the phantom network models, assume partial suppression of fluctuations at high elongations (83) or an end-to-end pulling mechanism, by which the junctions pull out the chain ends and leave the middle undeformed (78).

The emphasis of SANS experiments lies in the measurement of the radii of gyration parallel and perpendicular to the deformation axis and comparison with the R_g in the undeformed state. It should be noted, however, that the factor of 3, which normally appears in Guinier or Zimm plots (e.g., equation 27), applies only for isotropic systems and that other factors (e.g., 1 or 2) must be used in connection with various radii that may be defined for oriented materials. The different factors and systems of nomenclature that have been used in SANS orientation studies are summarized in reference 10. Formulae giving the variation of the ratio of SANS R_g values, α, parallel and perpendicular to the stretch direction have been summarized as a function of λ_d (72). For swelling experiments, the deformation is isotropic, the parallel and perpendicular deformation ratios are identical, and $\lambda_d = Q_d^{1/3}$, where Q_d is the ratio of volumes in the swollen and unswollen states. Experiments on uniaxially stretched, end-linked poly(dimethylsiloxane) networks (78) show that, for short chains, the molecular deformation may be described approximately by the end-to-end pulling model, but for longer chains, the observed deformation is lower than that predicted even by the phantom network model. The observed molecular R_g in tetrafunctional polystyrene networks (84) does not show any appreciable dependence on Q_d, and all the models considered overestimate the variation of R_g in the high swelling range. The observed molecular dimensions in polystyrene networks swollen in toluene (80) were intermediate between those predicted by the phantom network and chain affine models and even exhibited a maximum as a function of Q_d. To the author's knowledge, such a feature is not explicable by any current model, and thus, although much progress has been made in understanding deformation mechanisms, many questions remain to be answered. None of the models described here or envisioned in the classical theories of rubber elasticity or gel swelling can give a complete description of all aspects of deformation observed at the molecular level via SANS, and new mechanisms are needed to explain the experimental results.

The Crystalline State

Scattering techniques have been used since the beginnings of polymer science and actually predate the acceptance of polymers as molecules of very high molecular weight. Wide-angle X-ray diffraction was used to establish the size of the unit cell, which was similar in size to those of small molecules. The demonstration that unit cell dimensions of this order were consistent with polymer chains containing large numbers of repeat units was one of the major steps in establishing the macromolecular hypothesis (1). Similarly, SAXS data containing sharp peaks (e.g., Figure 8), reflecting large periodic structures, were one of the major sources of evidence for the existence of lamellae, and light scattering has been used to probe the morphology on longer length scales (*see* Chapter 4 by L. Mandelkern).

Molecular Arrangement in Lamellae

The arrangement of molecular chains within lamellae has been the subject of considerable debate, and SANS has provided direct information about this topic. It is generally agreed that semicrystalline polymers exhibit a lamellar morphology for material crystallized from both the melt and dilute solution. The thickness of lamellae is typically 100–500 Å, with amorphous polymer interspersed between the crystalline regions. The molecular chains are at an angle (0–45°) to the lamellar normals and are much longer than the lamellar thickness; thus, molecular chains traverse one or more lamellae several times. On the basis of considerations of density conservation at the crystalline–amorphous boundary, Flory (85) demonstrated that a considerable fraction (~0.5) of chains must return to the same crystal to accommodate the flux of chains (stems) emerging from the crystal. Discussion has centered on whether the molecule returns to the crystallite of origin at a nearest neighbor crystallographic site immediately adjacent to the point of exit or whether it reenters randomly over a distribution of lattice positions. Such considerations have important practical implications, for example, in considering the deformation mechanisms of crystalline polymers (*see Melting and Recrystallization*).

The information available from neutron scattering in different regions of Q-space is illustrated in Figure 17, which shows that SANS is sensitive to longer length scales and hence gives the overall size (R_g) of the molecule. The R_g remains virtually unchanged on crystallization from the melt (*10, 32, 60, 86–90*) and hence has an $M_w^{0.5}$ dependence in both the molten and crystalline states (Table III). Thus, the molecules crystallize with a distribution of mass elements similar to that in the melt (*60*) and hence are distributed over several lamellae in the crystalline state. For

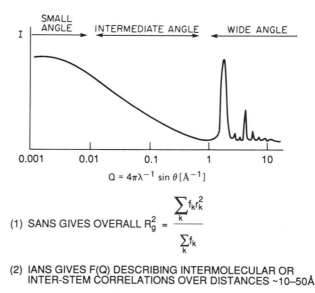

FIGURE 17. Information available from different regions of the neutron-scattering curve of mixtures of semicrystalline hydrogenous and deuterated polymers.

**Table III. Molecular Dimensions in Molten (Amorphous)
and Melt-Crystallized Polymers**

Polymer	Method of Crystallization	$R_g / M_w^{0.5}$ [Å/(g/mol)$^{0.5}$]	
		Melt	Crystalline
Polyethylene	Rapidly quenched from melt	0.46	0.46
Polypropylene	Rapidly quenched from melt	0.35	0.34
	Isothermally crystallized at 137 °C	0.35	0.38
	Rapidly quenched and annealed at 137 °C	0.35	0.36
Poly(ethylene oxide)	Slowly cooled	0.42	0.52
Isotactic polystyrene	Crystallized at 140 °C	0.26–0.28[a]	0.24–0.27
Poly(ethylene terephthalate)	Rapidly quenched	0.32–0.37[b]	0.36–0.40[b]

[a] Values quoted (88) are for amorphous samples (with atactic polystyrene matrix), annealed in the same way as the crystalline material (with isotactic polystyrene matrix).
[b] Dimensions in amorphous and crystalline states are similar, although absolute values are hard to determine because of transesterification effects that affect both R_g and M_w (89–90).

solution-crystallized material, however, the radius of gyration (R_g) is relatively independent of molecular weight and is generally markedly less than the dimensions in dilute solutions (91–92), indicating a much more compact configuration. The measurements of R_g were made at low Q values and contain no information about the mutual arrangement of stems, that is, straight sections of a chain traversing a crystalline lamella. Such information may be obtained from experiments in the intermediate-angle range (Figure 17), which is sensitive to the correlation of stems over distances of 10–50 Å. Measurements of this type were made for several systems (60, 92–95) and compared with various model calculations that simulate the chain trajectory (60, 92–102).

Figure 18 shows IANS data (96) for polyethylene that was quench-crystallized from the melt, along with the scattering function, $F_n(Q)$, defined by

$$F_n(Q) = (n + 1)Q^2 P(Q) = (n + 1)Q^2 \frac{\dfrac{d\Sigma}{d\Omega}(Q)}{\dfrac{d\Sigma}{d\Omega}(0)} \qquad (30)$$

In this equation, n is the number of bonds in the chain, and $P(Q)$ is the form factor of the molecule [$P(0) = 1$].

There is reasonable consistency between the data from several groups (60, 91–94), which are compared with model calculations based on Monte Carlo statistics as a function of the probability (p_{ar}), that the stem will fold adjacently along the (110) plane (96). Both the data and model lead to a Q^{-2} dependence for $d\Sigma(Q)/d\Omega$ and hence a plateau in the Kratky plot. The plateau levels differ by a factor of only ~2 for the extremes of random ($p_{ar} = 0$) and adjacent ($p_{ar} = 1$) reentry; hence, it is not surprising that different interpretations have been given to the IANS data. Thus it was

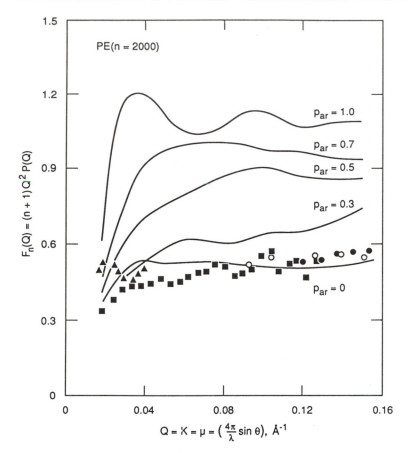

FIGURE 18. Calculation of $F_n(Q)$ as a function of the probability of regular (adjacent) folding (P_{ar}). Data are from references 60 (■), 93 (▲), 92 (○) and 94 (●).

concluded (*60, 97, 100*) that the IANS data were inconsistent with regular folding ($p_{ar} < 0.3$), although other model calculations, performed as a function of the number of stems folded adjacently in a central cluster (*99*), were interpreted in terms of much higher probabilities ($p_{ar} \approx 0.7$) of adjacent folding. The latter conclusion involved plotting the experimental data (*60*), measured by both SANS and chromatographic techniques, as a function of the molecular weight of the labeled chains. This procedure introduces further uncertainty ($\sim$30%, depending on which value of M_w is used) into the height of the plateau level in $F_n(Q)$ and hence makes the comparison less precise. Equation 30 shows that, for a high-molecular-weight material ($n > 1000$), $F_n(Q)$ is independent of molecular weight to a good approximation, because $d\Sigma(0)/d\Omega$ contains the molecular weight that is proportional to n. Cancellation of the parameter $d\Sigma(0)/d\Omega$, therefore, would make $F_n(Q)$ independent of molecular weight and lead to a lower estimate of the probability of adjacent folding ($p_{ar} < 0.3$) or the number of stems in a central cluster (*10*).

The initial comparisons of theory and experiment were made for quench-crystallized polyethylene to avoid a segregation artifact, which occurs in slowly cooled samples (*see Melting and Recrystallization*). This artifact does not occur for polypropylene (*87*) and hydrogenated polybutadiene (*32*), for which experiments show that the plateau level, and hence the conclusions drawn from it, is independent of the rate of cooling. A similar conclusion was reached for polyethylene (*103*), although the slowly cooled samples were cross-linked to suppress segregation.

Sadler (*102*) reviewed the data from both quenched and slowly cooled materials and estimated that the probability of adjacent folding within the rows is ~0.3–0.5 for melt-grown crystals. The average number of stems in adjacent sequences is ~$(1 - p_{ar})^{-1}$, and although the precise definition of p_{ar} varies among different models, probabilities of ≤ 0.5 are inconsistent with regular, uninterrupted folding of an appreciable number of stems ($> 2-3$) in one crystallographic plane. Longer sequences of adjacent stems would lead to an observable modulation of the wide-angle neutron-scattering pattern ($Q > 0.6$ Å^{-1}), and such features are not observed for melt-crystallized polyethylene (*104–105*).

Stem–Cluster Analysis

The fact that nearest neighbor reentry is improbable does not mean that reentry is completely random. Model calculations (*97, 99*) indicate that a large proportion of folds are relatively close and that the folds that are not adjacent are "near" and rarely involve stem separations greater than three nearest neighbors. This conclusion is also consistent with estimates of the distribution of distances of reentry made by using a method proposed for the evaluation of neutron-scattering data independent of detailed structural models (*106*). The only assumption of this approach is that the molecular structure can be described as consisting of "clusters" of stems belonging to the same molecule in each lamella.

The analysis leads to the average stems per cluster (N_c), the number of clusters per molecule, and the average number of tie molecules per chain. It was applied to melt-crystallized polyethylene (PE), polypropylene (PP), and poly(ethylene oxide) (PEO), and the results (Table IV) indicate that a typical molecule is distributed over 3–8 lamellae, depending on the molecular weight. The average distance between stems, $\langle a \rangle$, may be derived

Table IV. Evaluation of the Number of Stem Clusters per Molecule Applied to Various Polymers

	Polymer			
Item	PP[a]	PE[b]	PEO[c]	PET[d]
Number of stems per cluster, $\langle N_c \rangle$	12–13	5	4–8	6
Number of clusters per molecule	5	3	4–8	4
Distance between stems, $\langle a \rangle$ (Å)				
Random walk	38–44	26	19–26	21
Adjacent reentry	15–17	14	9–17	13

Source: Reproduced with permission from reference 106. Copyright 1985. Society of Polymer Science, Japan.
[a] Polypropylene, $M_w \approx 340 \times 10^3 - 415 \times 10^3$ (*106*).
[b] Polyethylene, $M_w \approx 40 \times 10^3$ (*106*).
[c] Poly(ethylene oxide), $M_w \approx 55 \times 10^3 - 125 \times 10^3$ (*106*).
[d] Poly(ethylene terephthalate), $M_w \approx 45 \times 10^3$ (K. P. McAlea, personal communication, 1985).

from the radius of gyration of stems belonging to one cluster $\langle R_{cc} \rangle$, on the assumption of a random walk or a linear arrangement of stems (regular folding). $\langle R_{cc} \rangle$ is typically 15–60 Å, and on the latter assumption, the values of $\langle a \rangle$ are 2–3 times greater than the distances involved in regular folding in one crystallographic plane ($\langle a \rangle \approx 5$ Å).

A similar conclusion was reached by Guenet (107), who modeled IANS data in terms of a "garland" model. The interstem spacings were ~8–10 Å (polyethylene) and ~13 Å (isotactic polystyrene), which involve second or "next-but-one" neighbor reentry. Regular next-but-one folding should produce a modulation of the WANS (wide-angle-neutron-scattering) pattern, although it is difficult to imagine why molecules should fold in such a fashion and it is not surprising that no such WANS effect is observed for isotactic polystyrene (108) or polyethylene (104–105). If a random walk is assumed, values of $\langle a \rangle$ are in the 20–40 Å range, which is consistent with the conclusion of near but not adjacent folding, after the model calculations described earlier.

The validity of the stem–cluster analysis (106) has been disputed by Gust and Mansfield (109) on the grounds that only the scattering from stems is calculated and that the amorphous material (folds, tie molecules, etc.) is ignored. As pointed out earlier, the first SANS and IANS experiments on melt-crystallized polymers (e.g., Figure 18) were undertaken on quench-crystallized samples (60) with crystallinity indices of ~65%; therefore, scattering from a substantial fraction of the sample is neglected in this case. Although it is not immediately obvious how this omission would affect the parameters derived from the analysis (Table IV), it is hard to see how the inclusion of the amorphous component could lead to results that are more consistent with regular folding. For slowly cooled materials with a minimum of supercooling (e.g., polypropylene), the amorphous fraction is much less so that its omission is less of a problem.

One potential objection (98, 99) to the type of model calculations indicated in Figure 18 is the possibility of anomalously high densities in the interfacial boundaries between amorphous and crystalline regions, which could result from space-filling considerations noted by Frank (110, 111) and Flory (85). If the stems are at an angle to the lamella normals or the transition between crystalline and amorphous regions is not abrupt and takes place over a distance of ~10 Å, space-filling anomalies can be avoided and do not invalidate the conclusions reached earlier (112, 113). Such considerations give strong support to the concept of an interfacial region between the crystalline and amorphous phases, which has been proposed on the basis of Raman and NMR data (see Chapter 4 by L. Mandelkern). Combined SANS and SAXS studies of polymer mixtures are also consistent with this idea (see Polymer Mixtures).

Superfolding

As mentioned earlier, the radius of gyration of polyethylene chains in solution-crystallized materials is much smaller than that in melt-crystallized samples (91–92) and is relatively independent of the molecular weight (Figure 19). The IANS data also show differences and exhibit a peak (91, 92, 102, 108) at $Q \approx 0.1$–0.2 Å^{-1} in the Kratky plot (Figure 20), which is not consistent with regular folding in one crystallographic plane (97). On the basis of these results, a "superfolding" model was proposed (91, 102). According to this model, the folding of a chain is not confined to

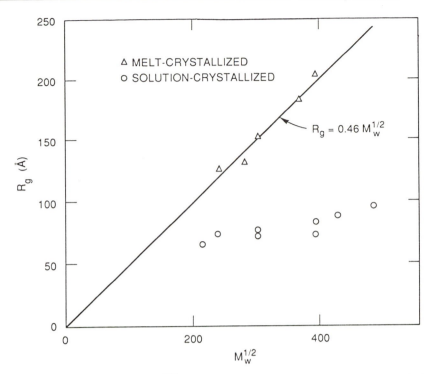

FIGURE 19. R_g versus $M_w^{1/2}$ for melt- and solution-crystallized polyethylene. (Reproduced with permission from reference 102. Copyright 1983 Elsevier Applied Science Publishers.)

a single layer, and after executing a number of folds in a given plane, the molecule continues to fold in an adjacent layer. This type of folding produces a peak in the Kratky plot, but comparisons with model calculations (*97, 102, 109*) indicate that the superfolding model overestimates the measured intensity. To produce agreement with experimental data, the adjacent stem arrangement must be "diluted" by a factor of ~2–3. One mode of "dilution" that has been put forward is an array of stems confined to several layers but not densely packed in any of them, where the stems are seldom adjacent in a given growth plane (*97*). An alternative proposal (*91, 92, 102, 108*) is that a molecule may be distributed over several (110) sheets, but ~75% of the folds are connected to adjacent sites.

These possibilities are illustrated in Figure 20, where $I_c(Q)$ is an equivalent intensity for infinitely thin stems (*108*) and $Q^2 I_c(Q)$ is proportional to $F_n(Q)$. The model calculations are represented schematically for a polyethylene lattice viewed along the chain direction (c axis). The solid circles indicate sites occupied by stems of a given molecule folding in the (110) plane. The model with approximately next-but-one folding fits the data well up to $Q \approx 0.6$ Å^{-1}, although if this type of folding is too regular, it generates a peak, as indicated in the upswing at $Q \approx 0.7$ Å^{-1}. This feature results from a spacing of ~10 Å ($Q \approx 2\pi/D$) and could probably be eliminated by introducing slightly more randomness on this length scale, although IANS is not sufficiently sensitive to serve as a unique fingerprint for a given stem sequence or mode of stem dilution. However, it seems that the neutron-scattering data rule out the possibility that a

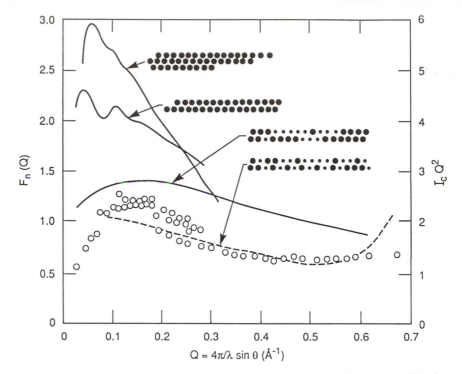

FIGURE 20. Kratky plots and model calculations for solution-crystallized polyethylene. ○, data; ●, sites occupied by stems of a given molecule in the (110) plane; —— and - - -, models for the stem arrangements shown.

typical molecule is regularly folded in one crystallographic plane over many stems without interruption. This model gained widespread support over the previous decades for both melt- and solution-crystallized materials. Similarly, the extremes of random configurations have been ruled out, with a large fraction of molecules usually folding in near reentry sites within a few nearest neighbors.

Melting and Recrystallization

The information provided by neutron scattering about the chain configuration in semicrystalline polymers has important implications for the mechanical properties of such materials. Because the radius of gyration is virtually unchanged on crystallization, the chains must have a similar distribution of mass elements in the solid and the molten states. Thus, the copious entanglements that exist in the melt must be duplicated in the solid state and concentrated in the amorphous regions (because of exclusion from the crystalline lamellae). The implications are that the neighborhood crystals are profusely interconnected and that these interconnections preclude the relative movements of crystals without their destruction.

Flory and Yoon (*114*) used these considerations to argue that large, irreversible deformations, such as those that occur in plastic flow or cold drawing, must result either in chain scission or destruction of preexisting crystalline regions. Because plastic flow is known not to be accompanied by an appreciable drop in molecular weight, it was concluded that the crystalline regions must melt and recrystallize during the deformation process. Similar conclusions were reached after analysis of the long-period shrinkage and orientation of drawn polymers (*115*). Such a mechanism contrasts with models (*116, 117*) proposing that the yield process involves homogeneous nucleation of screw dislocations within the crystallites. Deformation proceeds by motion of such dislocations, which zigzag between different slip planes to avoid obstacles on the crystallite surfaces, rather than by local melting and recrystallization.

Wignall and Wu (*118*) used SANS to investigate these different viewpoints by making use of a clustering or segregation artifact that has been encountered by several groups (*102, 119, 120*). The scattering theory developed to this point has been based on the assumption that the centers of gravity of the labeled molecules are randomly or statistically dispersed in the unlabeled molecules. It is known, however, that any departure from such a distribution, either by aggregation of the labeled molecules or by correlations in their trajectories or positions, can lead to excess scattering and hence anomalous values of M_w or R_g measured by SANS (*60, 119*).

Extensive studies of melt-crystallized polyethylene have shown that correlated aggregates of deuterated polyethylene (PED) molecules in hydrogenous polyethylene (PEH) are created in the melting region (125–135 °C), indicating that the nonrandom distribution is caused by the difference between the melting points (~6 °C) of PEH and PED homopolymers. Conversely, the anomalous molecular weights determined by SANS are unaffected by annealing outside the melting region and can be returned to their true values, as measured by chromatography, only by melting and quenching the sample (*119*), as illustrated in Figure 21a. This figure shows how the apparent molecular weight determined by SANS [calculated from the extrapolated ($Q = 0$) cross section after slow cooling from the melt] can exceed the true value by large factors. Conversely, the anomalous (or apparent) molecular weights determined by SANS are unaffected by annealing outside the melting region and can be returned to their true values, as measured by chromatography, only by melting and quenching the sample (Figure 21b).

The basic idea of the experiments was to take a mixed PEH–PED blend exhibiting an anomalous value of M_w and to apply solid-state deformation in temperature ranges where annealing alone is known not to affect the molecular weight determined by SANS. Large changes (> 5 times) in the value of M_w were observed and gave strong indirect evidence that melting and recrystallization occurs during solid-state deformation. These results are illustrated in Figure 22, which shows that the reduction in molecular weight produced by deformation is of the same order as that produced by melting and quenching the sample.

The initial measurements (*118*) were undertaken with a deformation ratio of ~13, and consequently, these results did not shed any light on that portion of the deformation process where partial melting begins. This may be expected to commence in the less perfect parts of the crystal and subsequently propagate to adjacent regions as the deformation proceeds. It is, therefore, reasonable to associate the onset of melting with the yield process, and subsequent experiments (*121*) were undertaken close to the yield point to investigate this suggestion. The undeformed blend had an extrapolated cross section, $d\Sigma/d\Omega$, of ~2500 cm^{-1} at $Q = 0$, compared with ~70 cm^{-1} for a homogeneous (statistical) distribution of PED in

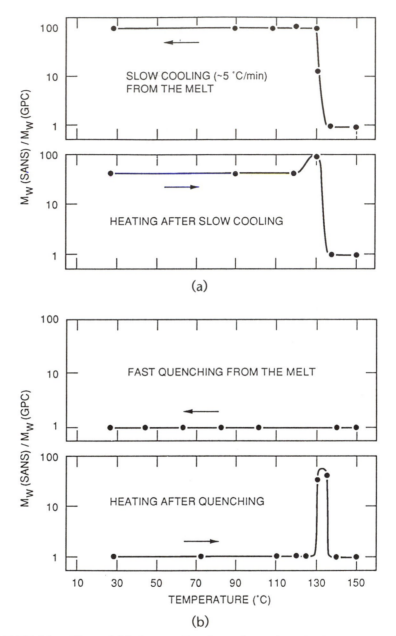

FIGURE 21. Effect of (a) slow cooling from the melt and reheating and (b) quenching from the melt and reheating on anomalous molecular weights of labeled PEH / PED samples as determined by SANS.

PEH. This result demonstrates that the blend is highly segregated, and Figure 23 shows how the cross section varies with the applied strain. The abrupt drop at a strain level of ~20% is very striking and corresponds to the onset of plastic flow. Below this strain, complete recovery took place after deformation. These findings are easily explained in terms of partial melting and recrystallization occurring from the onset of plastic deformation, although it is difficult to see how they can be explained by purely

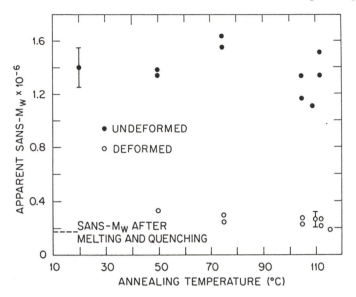

FIGURE 22. Effect of annealing and deformation on apparent M_w determined by SANS for 4.35 vol % PED molecules in PEH.

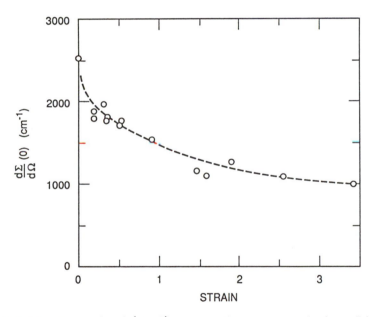

FIGURE 23. Extrapolated $(Q = 0)$ cross section versus strain for solid-state deformation of PEH–PED blends.

mechanical models. At the very least, these findings imply that large-scale reorganization takes place and that the level of molecular motion involved in the deformation process is of the same order of magnitude as that during the melting and recrystallization process.

The isotopic segregation signal was also used by Sadler and Barham (*122–124*) as an indicator of the melting and recrystallization mechanism in a study of drawing and necking of polyethylene fibers. Most of the samples were prepared with a statistical distribution of labeled molecules, and the appearance of excess scattering was taken to indicate that local melting had been produced by deformation, thus allowing molecules sufficient mobility to segregate on crystallization. Such effects were seen at drawing temperatures of ~70–90 °C and were interpreted in terms of melting and recrystallization produced by deformation. Of course, no excess scattering is expected if melting and recrystallization occur in such a fashion that the molecules are effectively quench-cooled and hence have insufficient time and/or mobility to segregate (Figure 21b), and this failure to segregate had been suggested (*121*) as an explanation for the absence of a segregation signal at temperatures below 70 °C.

Complementary SAXS studies by Chuah et al. (*125*) on high-pressure-crystallized polyethylene found no evidence of a melting and recrystalliza-tion mechanism when the polymer was deformed by solid-state extrusion. This type of system has been shown to consist of molecules in chain-ex-tended form (*126*). The melting and recrystallization mechanism was originally proposed (*114*) for systems in which the entanglements of the melt are conserved in the solid state and subjected to large-scale irre-versible deformation, and hence, this prediction would not necessarily apply to chain-extended systems. Similarly, systems subjected to small reversible deformation would not be expected to melt and recrystallize, and SAXS studies of oriented deformation of poly(ethylene terephthalate) fibers were interpreted in terms of a mechanical deformation (*127*).

Polymer Mixtures

One of the most significant applications of SANS has been in the area of polymer–polymer thermodynamics. Prior to this method, polymer–poly-mer phase behavior could be investigated only in the limit of phase separation by light scattering (cloud point) or by using various indirect thermal or mechanical spectroscopic methods. These methods could indi-cate macroscopic segregation but could not demonstrate fine-grained sepa-ration or intermixing at the molecular level. In principle, this information could be obtained from light scattering (LS) or SAXS (*128*), although only in very few cases is there sufficient electron density contrast between the components for these methods to be applicable. Deuterium labeling to-gether with SANS revolutionized this field and provided detailed thermo-dynamic information about the homogeneously mixed state, primarily via the strong SANS contrast, which can be produced by deuterating one of the components of the blend. In this and the following section, the discussion is restricted to mixtures containing two chemically different polymer repeat units polymerized either as individual homopolymers or as diblock copolymers. These examples illustrate the basic features of SANS for multicomponent systems and are readily extended to more complex mixtures.

SANS measurements were first used to investigate polymer compatibil-ity in the mid-1970s (*129, 130*). Most chemically dissimilar polymer pairs are incompatible, although the use of SANS demonstrated that polystyrene–poly(α-methylstyrene) and polystyrene–poly(phenylene ox-ide) formed truly compatible mixtures (*130–132*). The initial applications were based on an extension to polymer blends (*133*) of the Zimm analysis (*134, 135*), which was originally developed for LS and SAXS to give R_g, M_w, and the second virial coefficient (A_2). In principle, the Zimm analysis (equation 27) is limited to the regime where one of the species is dilute. The analysis has been extended (*134, 136*) to concentrated homogeneous mixtures to give the Flory–Huggins interaction parameter (χ), which is related to A_2 in dilute systems (*134*). For a blend of two polymer species A and S, with a fraction X_D of the A polymer chains labeled with deuterium and a protonated fraction $X_H = (1 - X_D)$, the coherent cross section (after subtracting the incoherent background) of the three-component mixture (H, D, and S) is given (*134*) by equation 31:

$$\frac{d\Sigma}{d\Omega}(Q) = X_H X_D (a_H - a_D)^2 NZ^2 P(Q) +$$

$$(X_D a_D + X_H a_H - a_S)^2 NZ^2 [P(Q) + NQ(Q)] \quad (31)$$

In this equation, a_D and a_H are the scattering lengths (equation 24) of the labeled and unlabeled repeat units (monomers) of species A, respectively; a_S is the scattering length of species S; and N and Z are the number of molecules per unit volume and polymerization index, respectively, for species A. $P(Q)$ and $Q(Q)$ are the intra- and interchain functions, respectively, for the species A; the equivalent functions do not appear for the species S because it is not labeled with deuterium. The terms in equation 31 can be separated by performing combined SAXS and SANS investigations on the same sample or, more usually, by performing two or more SANS investigations on the same sample or, more usually, by performing two or more SANS experiments in which X_D is varied for the same overall composition of species A and S (*134, 136, 137*). An important achievement is application (*138*) of the mean-field random phase approximation (RPA) to derive the structure factor, $S(Q)$, which contains information regarding both molecular architecture and thermodynamic interactions. $S(Q)$ is proportional to $d\Sigma/d\Omega$:

$$\frac{d\Sigma}{d\Omega}(Q) = V^{-1}(a_H - a_D)^2 S(Q) \quad (32)$$

In equation 32, V is the segment volume. On the basis of the assumption that the polymer constituents could be treated as ideal (Gaussian) coils (equation 28), unperturbed by the weak interactions between monomers, $S(Q)$ is given by equation 33:

$$S^{-1}(Q) = [\phi_A Z_A P_A(QR_{gA})]^{-1} + [(1 - \phi_A) Z_S P_S(QR_{gS})]^{-1} - 2\chi \quad (33)$$

In this equation, ϕ_A is the volume fraction of species A; and R_{gA} and R_{gS} are the radii of gyration of species A and S, respectively; Z_A and Z_S are the polymerization indices of species A and S, respectively; and $P_A(Q)$ and $P_S(Q)$ are the intrachain functions of species A and S, respectively. In the dilute limit ($\phi \rightarrow 0$), equation 33 reduces to the traditional Zimm equation used extensively in early SANS studies of polymer mixtures (*48, 134*).

Numerous research groups have successfully applied the RPA theory to evaluate $\chi(T, \phi, N)$ for various polymer–polymer mixtures over the past decade (*48, 134, 136, 137, 139–151*). An example is shown in Figure 24, which depicts the interaction parameter for poly(methyl methacrylate)

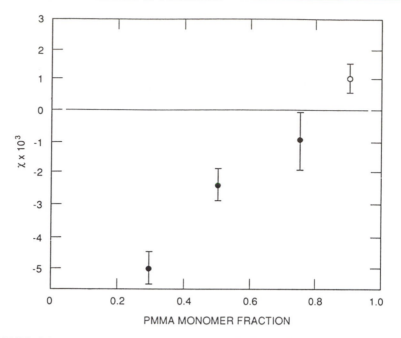

FIGURE 24. Interaction parameter between PEO and PMMA as a function of the monomer fractions of hydrogenous and deuterated PMMA. Data are from references 150 (●) and 151 (○).

(PMMA)–poly(ethylene oxide) (PEO) blends (*150, 151*). The values of χ obtained in two independent studies are in good agreement and confirm that the magnitude of χ is not only small but is also concentration dependent, like most other experimentally determined interaction parameters. For low PMMA concentrations, χ changes sign, indicating that the entropic, as opposed to enthalpic, contributions are dominant (*150, 151*).

Several studies have been directed at elucidating the limits of applicability of the mean-field approximation in polymer mixtures. At small scattering wave vectors, equations 32 and 33 reduce to the well-known Ornstein–Zernike form (*138*):

$$\frac{d\Sigma}{d\Omega}(Q) = \frac{\frac{d\Sigma}{d\Omega}(0)}{1 + Q^2\xi^2} \tag{34}$$

The composition fluctuation correlation length, ξ, is given by equation 35:

$$\xi \approx \left[\chi_s - \chi(T)\right]^{-\nu} \tag{35}$$

In equation 35, $\nu = 1/2$ in the mean-field limit, and $\nu = 0.63$ in the vicinity of the critical point (Ising regime); χ_s is the magnitude of the segment–segment interaction parameter at the stability temperature; and T is temperature. Schwahn et al. (*152*) were the first to report a transition from mean-field to non-mean-field behavior in polymer mixtures, on the basis of SANS results obtained from a polystyrene (PS)–poly(vinyl methyl ether) (PVME) mixture. Subsequently, Bates et al. (*153*) quantitatively verified these conclusions by using a model polyisoprene (PI)–poly(ethyl-ene–propylene) (PEP) mixture. The inset in Figure 25 shows an Ornstein–Zernike plot of representative SANS data taken at temperatures above the upper critical solution temperature ($T_c = 38$ °C) for a critical PI–PEP mixture, from which the extrapolated intensity $I(0)$ values at $Q = 0$ were obtained. Plotting of $I^{-1}(0)$ versus $T^{-1}(\chi \approx T^{-1})$ reveals a transition from $\gamma = 2\nu = 1$ (mean-field behavior) to $\gamma = 1.26$ (non-mean-field behavior) at ~30 °C above the critical temperature. These SANS crossover studies establish the limitations of mean-field theory, which has been used extensively in evaluating polymer–polymer thermodynamics.

The scattering from phase-separated systems has been treated by Koberstein (*154*), who derived an expression identical to, apart from

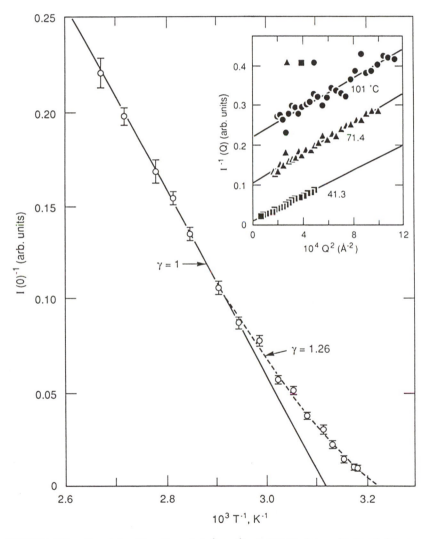

FIGURE 25. Ornstein–Zernike plot (inset) of SANS data obtained from a binary mixture of partially deuterated poly(ethylene–propylene) and polyiso-prene. The two branches of $I^{-1}(0)$ correspond to mean-field ($\gamma = 1$) and Ising ($\gamma = 1.26$) behavior. (Reproduced with permission from reference 153. Copyright 1990 American Physical Society.)

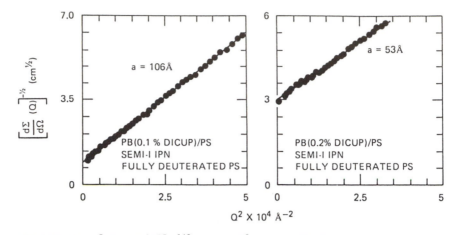

FIGURE 26. $[d\Sigma/d\Omega(Q)]^{-1/2}$ versus Q^2 for two PB–PS semi-interpenetrating polymer network systems with different amounts of cross-linker, DICUP (dicumyl peroxide).

nomenclature, equation 31. This expression may be used to separate the domain structure from the molecular configuration, and such experiments have been performed for polymer blends, interpenetrating polymer networks (IPNs), and block copolymers (*155–159*). Figure 26 shows a Debye–Bueche (D–B) plot (*158*) [$(d\Sigma/d\Omega)^{-0.5}$ versus Q^2] for phase-separated PS–polybutadiene IPNs. The PS molecules were fully deuterated so that the first term in equation 31 is zero ($X_H = 0$). After the incoherent background is subtracted, the remaining term may be analyzed to give the domain dimensions. The plots are linear, as expected if the D–B model is applicable, and the correlation lengths derived from the ratio of slope to intercept may be used to evaluate the domain dimensions (*156, 158*). SAXS can also provide similar information to SANS wherever there is sufficient electron density contrast, as demonstrated by Blundell et al. (*159*) for polyurethane–PMMA composites. If the electron density contrast is insufficient for SAXS (e.g., for polyolefins), strong SANS contrast may be created by deuterating one phase (*145, 155*).

For more complex systems, SAXS and SANS may be used in combination to provide complementary information to elucidate polymer morphology, as demonstrated by Russell et al. (*150*) for PMMA–PEO blends. In the amorphous (one-phase) state, the interaction parameter may be measured as a function of composition (Figure 24) and temperature. When the system is annealed, the PEO commences to crystallize, and the atactic PMMA is excluded from the lamellae. If the phase boundaries are sharp, the scattering should follow a simple Q^{-4} dependence at high Q values, as shown in Figure 26. Deviations from Q^{-4} (Porod's law or D–B model) dependence indicate diffuse interfaces and may be analyzed to give the thickness of the phase boundary (*150*), as shown in Figure 27 for semicrystalline blends of hydrogenous PEO and deuterated PMMA. Interestingly, the results suggest that the interface between crystalline and amorphous regions exhibits different characteristics depending on the radiation used. The interface appears to be quite sharp when studied by SANS, with a thickness (E_{SANS}) of ~5 Å, whereas when studied by SAXS, the interface is diffuse, with a thickness (E_{SAXS}) of ~20 Å. Figure 28

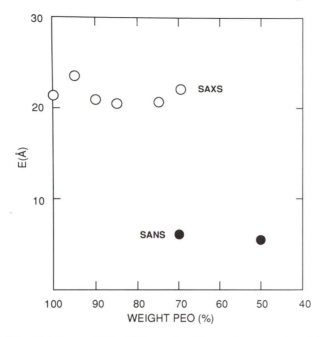

FIGURE 27. Thickness of the diffuse phase boundary as a function of composition, as measured by SAXS and SANS. The values remain constant as the composition is varied; however, there is a difference (~15Å) between the two measurements.

illustrates schematically how such results may be explained by using the concept of an interfacial region on the surface of polymer crystals, proposed by Mandelkern and co-workers (*see* Chapter 4). Proceeding along the normal to the lamellae, the electron density falls from the crystalline to the amorphous region as the order is dissipated. After sufficient order is lost (~15 Å), PMMA begins to mix with the amorphous PEO to form a mixture between the lamellae. In terms of the neutron-scattering length density (SLD), the situation is quite different, because PEO is hydrogenous (unlabeled) and PMMA is deuterated. Thus the SLD is small for PEO because of the cancellation between the scattering lengths of hydrogen and the other atoms (Table I). As the deuterated PMMA mixes with the amorphous PEO, the SLD increases dramatically over a distance of ~5 Å, and the slight differences between crystalline and amorphous PEO are negligible compared with the SLD of the hydrogenous PEO–deuterated PMMA mixture. Thus the effective interfaces measured by SAXS and SANS are different, as indicated in Figure 28, and this difference supports the idea that there is an interfacial region at the surface of the PEO crystals representing the distance over which the order is lost and from which PMMA is excluded.

Block Copolymers

Block copolymers chemically resemble binary polymer mixtures, and although they represent single-component systems, the local interactions between the chemically distinct repeat units can be described in the same way as previously indicated for binary mixtures. However, the chemical

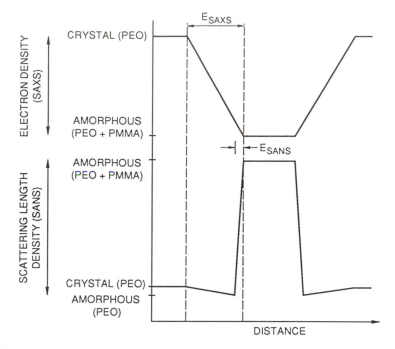

FIGURE 28. Schematic diagram of the electron density and the neutron-scattering-length density as a function of distance in a direction normal to the surface of the lamellae. The area between the dashed lines represents the transition zone between the crystalline lamellae and the homogeneous amorphous mixture. E_{SAXS} and E_{SANS} are the thicknesses of the interface as measured by SAXS and SANS, respectively.

bond between polymer blocks constrains intrachain block separation to a macromolecular scale, that is, on the order of the overall radius of gyration of the chain. The simplest molecular architecture is obtained by connecting a block of type A repeat units end-to-end with a block of type B repeat units to make a diblock copolymer. More complex molecules can be made by linking two or more diblock copolymers to make a starblock copolymer. The phase behavior is determined by the segment–segment interaction parameter, the overall degree of polymerization ($Z = Z_A + Z_B$, where Z_A and Z_B are the polymerization indices of the A and B blocks, respectively), and the composition ($f = Z_A/Z$). [In theoretical and experimental discussions of order–disorder transition (ODT), the degree of polymerization is often denoted by N. In this chapter, however, N has been used to denote the number of molecules per unit volume (*10*), and thus, the degree of polymerization is designated by Z.] At equilibrium, a block copolymer will be arranged such that the overall free energy is minimized. Decreasing temperature (i.e., increasing χ) favors a reduction in contacts between A and B segments. If the degree of polymerization (Z) is sufficiently large, reduced contact between segments A and B may be accomplished by local compositional ordering, as illustrated in Figure 29 for the symmetric case $f = 0.5$, where a lamellar morphology is observed. Alternatively, if either χ or the chain length (degree of polymerization) is decreased enough, entropic factors will dominate and lead to a compositionally disordered phase. The transition between ordered and disordered

ORDERED   DISORDERED

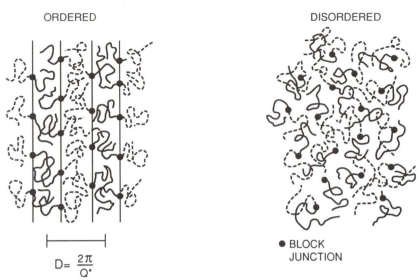

$$D= \frac{2\pi}{Q^*}$$

● BLOCK
JUNCTION

FIGURE 29. Schematic illustration of the speculated real-space morphology of a symmetric ($f = 0.5$) diblock copolymer melt.

phases was first treated by Leibler (*160*), who showed that an order–disorder transition (ODT) occurs when $\chi Z \approx 10$. Below this limit, the melt is disordered, although the connectivity of the two blocks leads to a correlation hole (*160*), which is manifested in scattering measurements (SANS or SAXS) as a peak corresponding to a fluctuation length, $D \approx R_g$. Above the limit, the system is ordered, and D corresponds to the lamellar spacing. The first (*160*) and subsequent ODT theories (*161*) were based on the assumption that the A–B interactions are sufficiently weak for the coils to remain Gaussian throughout the transition, and thus, the microdomain period scales as the square root of Z ($D \approx Z^{0.5}$). This limit is referred to as the weak-segregation limit, and the second limiting regime occurs for $\chi Z \gg 10$, where narrow interfaces separate the well-developed, nearly pure microdomains (*162*). Because the block joint is located at the domain interface and A blocks are excluded from the B phase (and vice-versa), the blocks are constrained to adopt configurations that are more extended than those for a Gaussian coil. Equilibrium is established by minimizing the total interfacial area under the entropic penalty of stretched configurations necessitated by the constraint of incompressibility. Thus the microdomain period will scale with a higher exponent than observed for Gaussian coils (*see* Table I) and is predicted to scale as $D \approx Z^{2/3}$ (*162, 163*).

The validity of the assumptions underlying ODT theories may be checked via small-angle-scattering measurements of the peak position ($Q^* = 2\pi/D$). The shape of the peak in the disordered region is a function of the interaction parameter, and Leibler (*160*) proposed that χ could be determined by fitting the theory to the scattering data. Similarly, the variation of Q^* with chain length could be used to test the predicted scaling behavior. Such experiments were performed by Bates and Hartney (*164*), as shown in Figure 30 for model poly(1,2-butadiene)-*block*-poly(1,4-butadiene) copolymers. In the disordered state, the theory (*160*) is shown for the case when $\chi = 0$ and also for the best fit value of χ, which

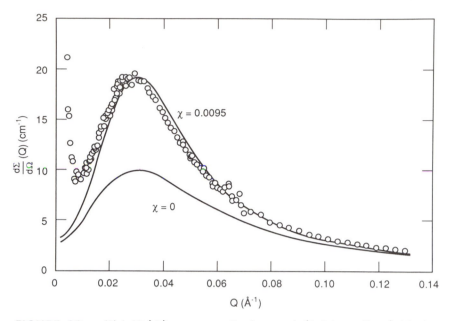

FIGURE 30. $d\Sigma/d\Omega(Q)$ versus Q for poly(1,2-butadiene)-*block*-poly(1,4-butadiene) copolymer to determine the Flory interaction parameter χ. (Reproduced from reference 164. Copyright 1985 American Chemical Society.)

leads to a reasonable description of the peak shape. When the phase transition is approached as the temperature is lowered ($\chi \approx T^{-1}$), the peak intensity diverges, and the peak position shifts (Figure 31). Such effects are not accounted for by theory, which assumes Gaussian statistics where the dimensions, and hence the peak position (Q^*), are determined by the degree of polymerization (Z). The peak shifts to lower Q values, indicating that the chain dimensions are stretching ($D \approx Q^{-1}$) as the ODT is approached, contrary to the assumptions of the theory.

Experiments to quantify the degree of stretching near the ODT have been performed by Almdal et al. (*165*) on model poly(ethylene–propylene)-*block*-poly(ethylethylene) (PEP–PEE) copolymers by measuring the peak position as a function of Z for both SAXS and SANS. Electron density contrast was sufficient for observation of the peak position by SAXS for only a limited number of samples, and most of the data were taken by SANS with deuterium-labeled material. However, for samples that can produce both signals, the close overlap of peak positions (Figure 32) serves as an internal cross-check of the results. Because $Q^* = 2\pi/D$, the scaling exponent (δ) may be measured via the variation of Q^* with Z ($Q^* \approx Z^{-\delta}$). The results indicate that departures from Gaussian statistics ($\delta = 0.5$) occur in both ordered and disordered phases. This observation is consistent with the shift in peak position as the ODT is approached (Figure 31) and shows that the coils undergo a Gaussian-stretched coil transition in the disordered state, where $\delta = 0.8 \pm 0.04$. Such data have indicated the deficiencies of the original theory (*160*) and have subsequently been used as the basis for new theoretical developments (*166, 167*). The study by Almdal et al. (*165*) indicates the close interaction between theory and

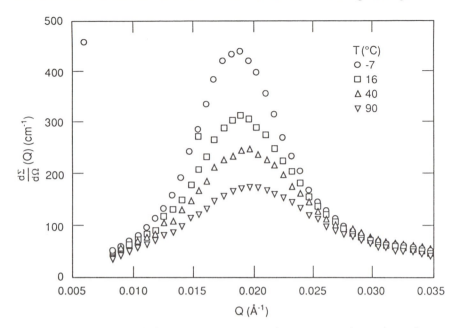

FIGURE 31. $d\Sigma/d\Omega(Q)$ versus Q for poly(1,2-butadiene)-*block*-poly(1,4-butadiene) copolymer as a function of temperature. (Reproduced from reference 164. Copyright 1985 American Chemical Society.)

experiment, which has often been stimulated by the new information provided by SANS over the past 2 decades. The possibility of measuring previously inaccessible parameters has stimulated theoretical developments (for example, *see* reference 160), which have led to new experiments (e.g., Figures 30–32), which have, in turn, stimulated improvements in theory (*166, 167*).

Latex Structure and Film Formation

Latexes constitute one of the most important forms of polymers; they account for about one-fifth of current polymer consumption and are widely used in coating and engineering applications. Many properties of latex polymers originate in the molecular conformation and structure of the polymer chains inside the latex particles, although there has been no general agreement on the actual structure of latexes. Because the latex interacts with its environment through its surfaces, understanding and control of surface properties are particularly important. Grancio and Williams (*168*) proposed a structure made up of a polymer-rich spherical core surrounded by a monomer-rich shell that serves as the major locus of polymerization, thus giving rise to a core–shell morphology. Latex structures based on this model, in which the first-formed polymer constitutes the core and the second-formed polymer makes up the shell, had earlier been described extensively in the patent literature and had also been advanced on the basis of dynamic mechanical data (*169*). However, the model has also been disputed, and the same evidence was often used to support conflicting viewpoints. To resolve such differences, characterization techniques are required that probe the internal structure of latex particles. With latex particle diameters of the order of 1000 Å, LS and SAXS may be used to measure intraparticle dimensions. Since the early

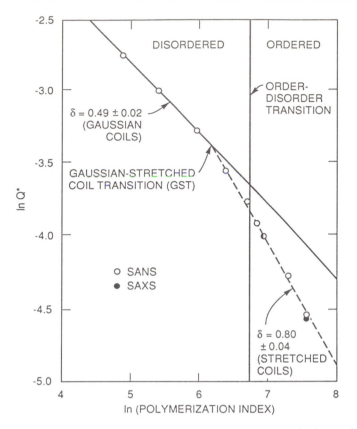

FIGURE 32. Gaussian-stretched coil transition for PEP–PEE block copolymers as a function of polymerization index.

1980s, neutron scattering has been used in combination with contrast-variation methods for isotopic labeling of particular chains generated at specific points in the polymerization process (*170–175*). The scattering contrast between normal (hydrogenous) and deuterated (D-labeled) molecules allows their locations and dimensions to be determined.

Figure 33 illustrates schematically how the core–shell hypothesis may be tested via SANS. The morphology of the latex core may be characterized by measurements in D_2O, which gives strong contrast with the hydrogenous polymer. For a homogeneous particle, the neutron-scattering cross section is given by equation 36:

$$\frac{d\Sigma}{d\Omega}(Q) = (\rho_m - \rho_p)^2 N_p V_p^2 P(Q) \tag{36}$$

In this equation ρ_m and ρ_p are the scattering-length densities of the medium and the particle, respectively; N_p is the number of particles per unit volume; V_p is the particle volume; and $P(Q)$ is the particle form factor [$P(0) = 1$]. For a solid sphere of uniform radius, R, $P(Q)$ is given by equation 37 (*176*):

$$P(Q) = 9(\sin QR - QR \cos QR)^2 / (QR)^6 \tag{37}$$

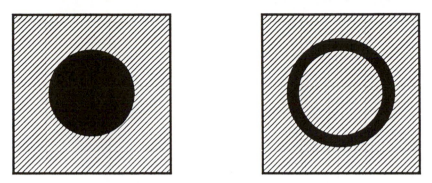

FIGURE 33.   SANS studies of polymer latex particles in D_2O-H_2O mixtures. (Left) SANS from hydrogenous latexes in D_2O gives the core morphology via the theoretical sphere-scattering (Bessel) function. (Right) SANS from latexes polymerized with a deuterated second monomer and examined in a D_2O-H_2O mixture chosen to match the core scattering-length density shows the hollow-shell-scattering function.

According to Grancio and Williams (*168*), polymerization takes place in a surface shell, and if the monomer feed is changed from hydrogenous to deuterated material, the result is a predominantly D-labeled shell. When examined by SANS in a H_2O-D_2O mixture that matches the scattering-length density of the protonated core (Figure 33), the scattering will arise from a hollow sphere with a particle form factor (*177*) given by equation 38:

$$P(Q) = 9(\sin QR - \sin QRl - QR \cos QR + QRl \cos QRl)^2 / Q^6 R^6 (1 - l)^6$$

$$(38)$$

In equation 38, $l = a/R$, and R and a are the outer and inner radii, respectively. At $l = 0$, this hollow-sphere-scattering function is reduced to the solid-sphere-scattering function. Thus the core–shell hypothesis can be tested by comparing SANS data for both cores and shells (Figure 33) with the model predictions (equations 36–38). (Experiments like those illustrated in Figure 33, where core scattering is eliminated, are often referred to in the literature as "contrast matching". This terminology is incorrect, because it is the scattering-length densities of the core and medium that are "matched", and thus the contrast between them is zero.)

Such experiments were undertaken by Fisher et al. (*171*), who studied PMMA latexes with deuterated shells of deuterated PMMA or deuterated PS polymerized on the surface. Similar experiments were undertaken by Wai et al. (*170*) on a partially deuterated PMMA shell polymerized on cores consisting of random PMMA–PS copolymers. The scattering from both cores and shells exhibits sharp maxima and minima for monodisperse particles, although in practice these sharp features are smeared by the finite experimental resolution (*178*). Desmearing procedures using indirect Fourier transform (IFT) methods, developed by Glatter (*179*) and Moore (*180*), were used to remove these instrumental effects and led to patterns showing the expected sharp minima with good agreement between the core radii determined independently by SANS and by light scattering (*170, 171, 181*).

In addition to instrumental resolution effects, the data can also be smeared by integrating over the finite range of particle radii, if the samples are not monodisperse. Such particle size distributions may be described by a zero-order logarithmic distribution (ZOLD) (*170–171*). For this distribution, the frequency of particles with radius R is a function of the average size and the standard deviation σ. Figure 34 shows desmeared SANS data for the copolymer core compared with the solid-sphere-scattering function (equation 37), by using the ZOLD with an average diameter, D, of 1010 Å and σ of 92 Å. In addition to the shape of the scattering envelope, the scattering intensity provides an independent check on the model if the concentration of particles and the scattering-length density are known. The absolute intensity at zero scattering angle is given by equation 36 with $P(0) = 1$. For the core, V_p is the volume of the latex particle in solution. For the shells, a core–shell structure was assumed for the calculations of absolute intensity, and V_p was taken to be the volume of the labeled polymer in the shell. The core and solvent scattering-length densities were matched (i.e., the core contrast was zero). The measured absolute cross sections and R_g values are shown in Figures 34 and 35, along with those calculated from the latex dimensions determined independently by light scattering (LS) and transmission electron microscopy (TEM). Both the R_g and intensity values agree with the model within the experimental error for both homopolymer (*171*) and copolymer (*170*) cores and shells and thus give strong support to the core–shell hypothesis for polymerization under monomer-starved conditions (*181*).

In both SANS studies, relatively high concentrations of latex particles (ϕ, ~0.25–10 vol %) were used to increase the signal-to-noise ratio of the experiment, and also because concentration studies (*171*) have indicated that $d\Sigma(0)/d\Omega$ is proportional to ϕ in this range. Such behavior is quite

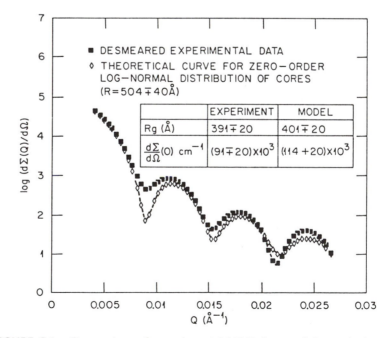

FIGURE 34. Comparison of experimental SANS data and theoretical scattering function for PS–PMMA core latexes in D_2O.

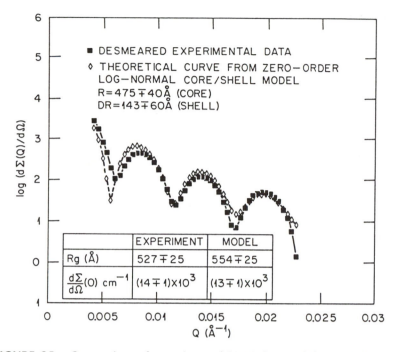

FIGURE 35.   Comparison of experimental SANS data and theoretical scattering function for PS–PMMA core latex with deuterated (D3) PMMA shell.

surprising in view of the known tendency of neutral particles to form correlated structures in concentrated solutions (*182, 183*). Calculations indicate that excluded-volume effects alone should perturb the radius of gyration and zero-angle cross section measured via SANS by ~10% for even a 1 vol % of particles (*182*). Two reasons have been suggested to account for the success of these studies with relatively large volume fractions of latexes (*181*). First, the above estimates of the perturbations to R_g and $d\Sigma/d\Omega$ refer to data analyzed only in the region of low Q ($Q < R_g^{-1}$) via Zimm or Guinier analysis. By contrast, the studies in question used IFT methods (*179, 180*), which make use of data in the whole measured Q range. The use of the whole Q range makes the analysis much less sensitive to concentration effects, which are exhibited predominantly at low Q values, and makes R_g and $d\Sigma(0)/d\Omega$ virtually independent of concentration in the range studied. The second reason for this behavior is illustrated in Figure 36, which shows estimates of the excluded-volume effect in concentrated (ϕ, ~10 vol %) latex solutions (*183*). In the absence of excluded-volume effects, the cross section is given by equation 36. The effect of excluded-volume interactions is to replace $P(Q)$ by the product $P(Q)S(Q)$, where $S(Q)$ is a structure factor [$S(0) = 1$ in the absence of interactions]. Figure 36a shows an estimate of $S(Q)$ calculated via the Percus–Yevick (PY) approximation (*182, 183*) for particles of different polydispersities (σ_s) in the Schultz distribution of diameters D ($\sigma_s = 0$ for monodisperse particles). By a fortuitous combination of circumstances, excluded-volume effects are manifested predominantly below the minimum Q for an average diameter, D, of ~1000 Å. Moreover these effects are further damped by particle polydispersity and are virtually absent when the whole curve, as opposed to the small Q region, is used in

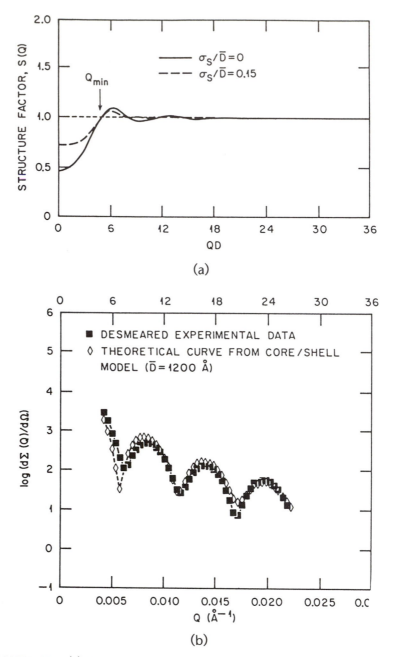

FIGURE 36. (a) Estimation of hard-sphere excluded-volume effect in concentrated ($\phi = 0.1$) suspensions via the Percus–Yevick approximation compared with (b) experimental data and core–shell model.

the analysis. Figure 36b shows experimental data for the hydrogenous poly(styrene-*co*-methyl methacrylate) latex with a deuterated (D3) PMMA shell compared with the core–shell model (*170*). Close examination of the curves for the range $3 \times 10^{-3} < Q < 7 \times 10^{-3}\,\text{Å}^{-1}$ shows that the experimental curve is slightly higher than the model, whereas for the range $7 \times 10^{-3} < Q < 9 \times 10^{-3}\,\text{Å}^{-1}$, the experimental data fall below the model. Comparison with the PY simulation indicates that such behavior is consistent with residual excluded-volume effects. However, because of the IFT methods used for desmearing, such effects do not perturb $d\Sigma(0)/d\Omega$, which agrees with the model prediction within experimental error.

Other studies of polymer latexes were undertaken by Goodwin et al. (*172*), who used SANS to measure the kinetics of swelling of polystyrene latexes in the monomer (styrene). The particle sizes resolved in this work ($\sim$4000 Å) were close to the upper limit of resolution of currently available SANS instrumentation. These experiments were also undertaken with low volume fractions of the dispersed latexes so that particle–particle interactions are minimized. At higher concentrations, the mutual arrangement of the particles is reflected in a structure factor, $S(Q)$, from which a latex–latex radial distribution function may be obtained via Fourier transformation. Experiments along these lines were performed by Alexander et al. (*173*) and Cebula et al. (*174*), and comparison with theoretical models (e.g., hard spheres) gives information about interparticle interactions. These techniques are relevant to a wide variety of polymeric, colloidal, and biochemical systems, and an excellent review of their application to the structure of micellar solutions has been given by Hayter and Penfold (*183*).

Information about the stabilization of nonaqueous dispersions of polymer particles by block copolymers may be obtained by SANS, and the possibility of deuterating one of the block lengths allows the determination of the state of dispersion of each block in the particle or on the surface. Such experiments were performed by Higgins et al. (*175*) for nonaqueous dispersions of PMMA and polystyrene particles stabilized by polystyrene-*block*-poly(dimethylsiloxane) copolymers.

The contrast variation and matching methods described earlier may also be used to provide information about the R_g of molecules constrained inside latex particles with diameters smaller than the unperturbed-coil dimensions. When deuterated and hydrogenous molecules are randomly (statistically) distributed in a latex that is suspended in a H_2O–D_2O medium that matches the average SLD of the particle, the component of the scattering due to the latex spherical form factor vanishes to leave only the scattering due to the D-labeled molecules within the latex (*181, 184–186*). A similar methodology was used to observe single-chain scattering within microphase-separated block copolymer domains (*157*). To the author's knowledge, the first application of such methods to study latexes was made by Sperling and co-workers (*184*), who measured end-to-end distances of molecules constrained from the relaxed state by factors of $\sim$2–4. When such latexes are molded together into a film at temperatures above the glass transition temperature, the molecules will expand to the relaxed configuration. The time dependence of the scattering from such systems may be analyzed to explore interdiffusion and relaxation processes, which take place as the latex particles coalesce to form a film. Table V shows how molecules originally constrained to $\sim$160 Å in the latex expand to the equilibrium R_g of $\sim$660 Å as the latexes are annealed in the solid state at $T = 160$ °C to form a polymer film (*185*). In addition to R_g, the M_w was monitored to ensure that the scattering resulted from single polymer chains. Aggregation effects resulting from a nonstatistical distribution of the hydrogenous and deuterated molecules produce excess intensity above the single-chain scattering (*184*). The M_w measured by SANS ($\sim$5 $\times$ 10^6) was close to that measured by other techniques, and the

Table V. Chain Relaxation of Polystyrene in Film Formation at 160 °C

Material	Diffusion Time	R_{gw} (Å)	$M_w \times 10^{-6}$
Latex		150	5.8
	0	171	4.4
	25 min	178	4.1
	50 min	208	4.6
	75 min	237	3.7
	20 h	645	6.1
Relaxed polymer		665	

Source: Data are from references 184 and 185.

relative constancy of this parameter (Table V) indicates that such effects were absent in this study. Similar methodology was used by Summerfield and Ullman (*187*), Anderson and Jou (*188*), and Hahn et al. (*189*) to explore interdiffusion and film formation.

Isotope Effects

In general, SANS studies of deuterium-labeled polymers are based on the assumption that the molecular configurations and interactions are independent of deuteration or, alternatively, that the interaction parameter between labeled and unlabeled segments of the same species, χ_{HD}, is zero. Nevertheless, several experimental observations have suggested that isotopic substitution does influence polymer thermodynamics. For example, deuterated and hydrogenous polyethylenes exhibit melting temperatures differing by ~6 °C, and their mixtures can segregate (e.g., Figure 21) in the solid state because of differential crystallization effects (*119, 190*). Also, the theta temperature of polystyrene solutions depends on the isotopic constitution of both polymer and solvent (*191*), and the critical temperature of polystyrene–poly(vinylmethyl ether) blends depends strongly on the isotopic labeling of the polystyrene (*192*). Thus, isotopic labeling may influence phase transitions, and Buckingham and Hentschell (*193*) suggested that this effect might arise from a finite interaction parameter between the hydrogenous and deuterated species ($\chi_{HD} \approx 10^{-4}$–$10^{-3}$). The first measurements of χ_{HD} were undertaken by Bates et al. (*194–199*), who delineated the circumstances under which demixing can occur. Figure 37 shows the scattering cross sections of mixtures of deuterated ($Z_D = 4600$) and hydrogenous ($Z_H = 960$) poly(1,4-butadiene) as a function of temperature. The extrapolated cross section for $Q = 0$ exceeds by large factors the value it would have (~100 cm^{-1}) if the interactions between hydrogenous and deuterated species were negligible. By regarding the hydrogenous and deuterated molecules as different species with volume fractions ϕ_H and ϕ_D, respectively, the random phase approximation (equation 33) may be fitted to the data with χ_{HD} as the only adjustable parameter (*194*). Similar experiments on polystyrene (*195*), poly(vinyl ethylene) (*198*), poly(ethylethylene) (*198*), and poly(dimethylsiloxane) (*199*) corroborate the existence of a universal isotope effect arising from the small differences in volume and polarizability of C–H and C–D bonds (*194, 195*). At room temperature, χ_{HD} is ~10^{-3}, and for sufficiently high

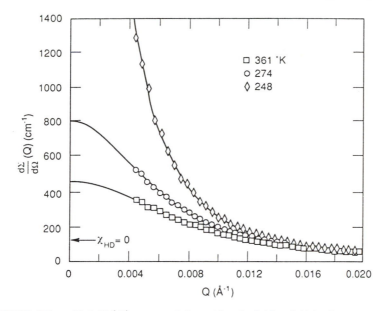

FIGURE 37. $d\Sigma / d\Omega(Q)$ versus Q for a blend of 69 vol % hydrogenous and 31% deuterated poly(1,4-butadiene) at the critical composition. The curves were obtained from the homogeneous mixture (RPA) scattering function by adjusting χ_{HD}.

polymerization indices ($> 2 \times 10^3$), the system will phase-separate (*200*), as illustrated in Figure 38, which shows data for two isotopic mixtures of polybutadienes (*176*). For $Q < 10^{-2}$ Å^{-1}, a strong excess intensity arises from the interfacial regions between the H- and D-rich domains. This component can exceed the single-chain scattering by several orders of magnitude.

Such results raise the important question of how prior SANS studies, as described in this chapter, were influenced by isotope effects. As explained earlier, initial SANS experiments on polymers relied on analogies with light scattering; that is, the limit of zero concentration was required to eliminate interchain scattering. Under such conditions, the isotope effect contributes almost insignificantly to the scattered intensity, as illustrated in Figure 39, which shows the SANS cross section for an isotopic polystyrene mixture with equal polymerization indices ($\sim 10^3$) and deuterated polymer at a concentration of 10%. This cross section is compared with the calculated signal assuming no isotope effect. The difference of $< 5\%$ indicates a worst case scenario for isotope effects in the early (1970s) SANS work on polymers. Upon recognizing that information about chain statistics could be obtained equally well from concentrated isotopic mixtures, many experiments were conducted under such conditions to enhance the intensity. It is under these conditions that isotope-induced segregation effects are manifested. In the bulk state, many of the systems studied are solids at room temperature and were exposed for only a limited time in the liquid state, for example, during melt pressing. The fact that the polybutadiene system, with a glass transition temperature below -90 °C, is a liquid at room temperature facilitates the attainment of equilibrium. Hence, isotope effects can be particularly dramatic in this system, although in practice, only a limited number of experiments en-

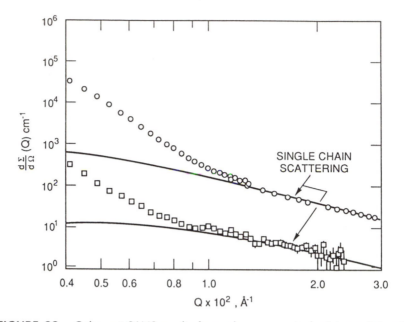

FIGURE 38. Coherent SANS results from phase-separated mixtures. Intensity in excess of the curves ($Q \leq 0.01$ Å^{-1}) derives from interfacial (Porod) scattering.

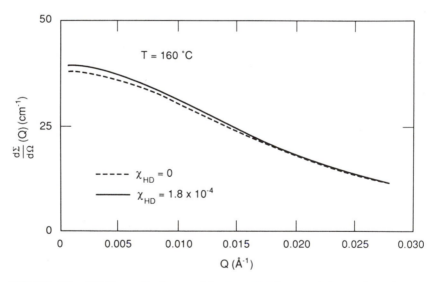

FIGURE 39. SANS results from a binary amorphous mixture of perdeuterated and normal atactic polystyrenes.

countered excess scattering due to this cause, for example, in contrast-variation studies of block copolymers (157) and SANS studies of interpenetrating polymer networks (201, 202). Thus, isotope effects occurred only infrequently in early SANS studies of polymers, although it is prudent to evaluate future experiments, on the basis of measured or calculated values of χ_{HD}, and to check for excess scattering by calibrating data on an absolute scale and comparing the measured and theoretical (calculated) intensities (*61*).

Conclusions and Future Directions

Scattering techniques have been one of the main sources of structural information since the beginnings of polymer science. Over the past 2 decades, SANS has been applied extensively to complement existing scattering methods (SAXS, WAXS, LS, etc.) and the examples in this chapter illustrate the new information that it has provided. In addition, inelastic neutron scattering has been used widely to probe polymer dynamics, although this methodology is beyond the scope of this chapter. Even within the SANS field, the applications are so numerous that a chapter of this length must inevitably omit significant research, and the reader is referred to other reviews (*10, 73, 74, 203*). A particularly clear introduction to the field has been given by Ullman (*204*), and the complementary aspects of neutron, light, and X-ray scattering, as applied to polymers and colloids, have been surveyed in a recent volume edited by Lindner and Zemb (*205*).

The use of SANS in polymer science has spread far beyond recognized experts in the technique, and most of the research surveyed in this chapter was undertaken by nonspecialists who applied the technique in areas of their own particular interests. This trend has been made possible by the development of national and international facilities that routinely provide assistance and access to instrumentation for a wide spectrum of external users and visiting scientists. In future, scattering measurements made with pulsed neutron (spallation) facilities should be of increasing importance, although to date, most of the SANS structural studies of polymers have been undertaken on reactor sources, which have been particularly suitable for this kind of research. Unfortunately, many of these reactors are at least 20 years old and cannot be expected to function indefinitely. This point is underscored by the extensive shutdowns of the three reactors with the highest core flux ($>10^{15}$ neutrons$\cdot$s$^{-1}\cdot$cm^{-2}) at the Institut Laue–Langevin in Grenoble, France, and the Oak Ridge and Brookhaven National Laboratories in the United States. The use of SANS has risen explosively since its beginning in Europe in the early 1970s, and industrial use in particular has been growing steadily as a result of the many practical applications of polymers. At present, over \$500 million of assured funding has been made available to construct or upgrade neutron facilities, with emphasis on SANS, outside the United States. Plans for a new world-class facility (*206*), when implemented, will give U.S. scientists access to state-of-the-art instrumentation in a field so important to materials science for the foreseeable future.

Acknowledgments

This research was supported by the Division of Materials Science, U.S. Department of Energy, under Contract No. DE-AC05-84OR21400 with Martin Marietta Energy Systems, Inc. The author wishes to thank his many co-workers for permission to include data and sections from joint publications, particularly F. S. Bates, J. B. Hayter, J. M. O'Reilly, T. P.

Russell, J. Schelten, L. H. Sperling, M. P. Wai, and W. Wu. Figures 6 and 9 were supplied by D. E. Engelman and A. H. Narten, respectively.

This chapter is dedicated to the memory of the late R. Ullman, who was one of the pioneers of SANS in this country. Bob was one of the first U.S. scientists to recognize the importance of the new techniques being developed in Europe in the early 1970s and to spend time there to learn how to apply them to polymers. He was a vigorous proponent of establishing these techniques in the United States, and he made extensive use of them when they were available. For many years a member of the American Chemical Society Polymer Division Speakers Bureau, he gave lectures on the subject "SANS in Polymers" to publicize the importance of this subject. He served for several years on the Oak Ridge Advisory Committee and was one of the first users to take advantage of the High-Flux Isotope Reactor after its restart in 1990. He was actively involved in SANS research and in supervising graduate students for as long as he was physically able. Those of us who worked with him will miss a very approachable colleague, who was always ready to debate his point of view.

References

1. Flory, P. J. *Principles of Polymer Chemistry*; Cornell University Press: Ithaca, NY, 1953; p 22.
2. Alexander, L. E. *X-Ray Diffraction Methods in Polymer Science*; Krieger: New York, 1969.
3. Wignall, G. D. In *Applied Fiber Science*; Happey, F. W., Ed.; Academic: London, 1978; Chapter 4.
4. Fischer, E. W., et al. *J. Macromol. Phys.* **1976,** *B12*(1), 41. Voigt-Martin, I.; Mijlhof, F. C. *J. Appl. Phys.* **1975,** *46*, 1165.
5. Lovell, R.; Windle, A. H. *Polymer* **1981,** *22*, 175. Mitchell, G. R.; Windle, A. H. *Colloid Polym. Sci.* **1985,** *263*, 280.
6. Cervinka, L. *J. Non-Cryst. Solids* **1985,** *75*, 63.
7. Havrilak, S.; Roman, N. *Polymer* **1966,** *7*, 387. Schneider, B., et al. *Macromolecules* **1971,** *4*, 715.
8. Inoue, Y.; Konno, T. *Makromol. Chem.* **1978,** *179*, 1311.
9. An introductory survey to diffraction methods has been given by J. M. Schultz (*Diffraction for Materials Scientists*; International Series; Prentice-Hall: Englewood Cliffs, NJ, 1982). The seminal treatise of A. Guinier and G. Fournet (*Small-Angle Scattering of X-Rays*; John Wiley: New York, 1955) is still an excellent introduction to the subject, though a more recent treatment has been given by O. Glatter and O. Kratky (*Small-Angle X-Ray Scattering*; Academic: London, 1982). The application of wide-angle X-ray diffraction methods to polymers is surveyed in reference 2, and a description of the information derived from light scattering has been given by M. G. Huglin (*Light Scattering from Polymer Solutions*; Academic: London, 1982). The application of a wide range of scattering techniques to amorphous polymers has been described in *The Physical Structure of the Amorphous State* (Allen, G., Petrie, S. E. B., Eds.; Dekker: New York, 1977).
10. Wignall, G. D. *Encyclopedia of Polymer Science and Engineering*; Grayson, M.; Kroschwitz, J., Eds.; Wiley: New York, 1987; Vol. 10, p 112.
11. Turchin, V. E. *Slow Neutrons*; Israel Program for Scientific Translations, Jerusalem, 1965.
12. Bacon, G. E. *Neutron Diffraction*; Clarendon: Oxford, England, 1971.
13. Maconnachie, A. *Polymer* **1984,** *25*, 1068.

14. Coyne, L. D.; Wu, W. *Polym. Commun.* **1989,** *30*, 312.
15. Hayter, J. B. In *Proceedings of Enrico Fermi School of Physics Course XC*; Degiorgio, V.; Corti, M., Eds.; North Holland: Amsterdam, Netherlands, 1985; p 59.
16. Schelten, J. *Kerntechnik* **1972,** *14*, 86.
17. Ibel, K. *J. Appl. Cryst.* **1976,** *9*, 196.
18. Schelten, J. In *Scattering Techniques Applied to Supramolecular and Nonequilibrium Systems*; Chen, S. H.; Chu, B.; Nossal, R., Eds.; NATO Advanced Study Series 73; Plenum: New York, 1981; pp 75–85.
19. Seeger, P.; Hjelm, R.; Nutter, M. *J. Mol. Cryst. Liq. Cryst.* **1990,** *180A*, 101.
20. Epperson, E.; Carpenter, J.; Crawford, K.; Thiagarajan, P., submitted to *J. Appl. Cryst.*
21. Maier-Leibnitz, H.; Springer, T. *J. Nucl. Energy* **1963,** *17*, 217. *Annu. Rev. Nucl. Sci.* **1966,** *16*, 207.
22. Koehler, W. C. *Physica* **1986,** *137B*, 320.
23. Abele, R. K.; Allin, G. W.; Clay, W. T.; Fowler, C. E.; Kopp, M. K. *IEEE Trans. Nucl. Sci.* **1981,** *NS–28*(1), 811.
24. Dubner, W. S.; Schultz, J. M.; Wignall, G. D. *J. Appl. Cryst.* **1990,** *23*, 469.
25. Akcasu, A. Z.; Summerfield, G. C.; Jahshan, S. N.; Han, C. C.; Kim, C. Y.; Yu, H. *J. Polym. Sci.* **1980,** *18*, 865.
26. Gawrisch, W.; Brereton, M. G.; Fisher, E. W. *Polym. Bull.* **1981,** *4*, 687.
27. Wendorff, J. H. *Polymer* **1982,** *23*, 543.
28. Fischer, E. W.; Wendorff, J. H.; Dettenmaier, M.; Lieser, G.; Voigt-Martin, I. *Physical Structure of the Amorphous State*; Allen, G.; Petrie, S. E. B., Eds.; Dekker: New York, 1977; p 41.
29. Ullman, D.; Renninger, A. L.; Kritchevsky, G.; Vander Sande, J. *Physical Structure of the Amorphous State*; Allen, G.; Petrie, S. E. B., Eds.; Dekker: New York, 1977; p 153.
30. Kirste, R. G.; Kruse, W. A.; Ibel, K. *Polymer* **1975,** *16*, 120.
31. Von Laue, M. *Ann. Phys.* **1918,** *56*, 497.
32. Crist, B.; Graessley, W.; Wignall, G. D. *Polymer* **1982,** *23*, 1561.
33. Guinier, A.; Fournet, G. *Small-Angle Scattering of X-Rays*; John Wiley: New York, 1955.
34. Hayashi, H.; Hamada, F.; Nakajima, A. *Macromolecules* **1976,** *9*, 543.
35. Narten, A. H. *J. Chem. Phys.* **1989,** *90*(1), 5857.
36. Fischer, E. W.; Stamm, M.; Dettenmaier, M.; Herschenroeder, P. *Polym. Prepr.* **1979,** *20*(1), 219.
37. Williams, C. E., et al. *J. Polym. Sci. Polym. Lett. Ed.* **1979,** *17*, 379.
38. Wignall, G. D.; Hendricks, R. W.; Koehler, W. C.; Lin, J. S.; Wai, M. P.; Thomas,
    E. L. T.; Stein, R. S. *Polymer* **1981,** *22*, 886.
39. (a) Tangari, C.; Summerfield, G. C.; King, J. S.; Berliner, R.; Mildner, D. F. R. *Macromolecules* **1980,** *13*, 1546; (b) Tangari, C.; King, J. S.; Summerfield, G. C. *Macromolecules* **1982,** *15*, 132.
40. Boué, F.; Nierlich, M.; Leibler, L. *Polymer* **1982,** *23*, 29.
41. Boothroyd, A. T. *Polymer* **1988,** *29*, 1555.
42. Wignall, G. D.; Ballard, D. G. H.; Schelten, J. *Eur. Polym. J.* **1974,** *10*, 861.
43. Kampf, A.; Hoffman, M.; Kramer, H. *Ber. Bunsen-Ges. Phys. Chem.* **1970,** *74*, 851.
44. Pechhold, W. R. *Kolloid Z* **1968,** *22*, 1.
45. (a) Yeh, G. S. Y. *Rev. Macromol. Sci.* **1972,** *1*, 173; (b) Yeh, G. S. Y.; Geil, P. H. *J. Macromol. Sci. Phys.* **1967,** *B1*(2), 235.
46. Kirste, R. G. *Jahresbericht 1969 des Sonderforschungsbereiches, Mainz* **1970,** *41*, 547.
47. Wignall, G. D. Imperial Chemical Industries (Runcorn) Memo PPR G19, 1970.
48. Cotton, J. P.; Farnoux, B.; Jannink, G.; Mons, J.; Picot, C. *C. R. Acad. Sci. (Paris)* **1972,** *275, 3C*, 175.

49. Kirste, R. G.; Kruse, W. A.; Schelten, J. *J. Makromol. Chem.* **1972**, *162*, 299; *Koll. Z. Z. Polym.* **1973**, *251*, 919.

50. Benoit, H.; Cotton, J. P.; Decker, D.; Farnoux, B.; Higgins, J. S.; Jannink, G.; Ober, R.; Picot, C. *Nature (London)* **1973**, *245*, 23.

51. Flory, P. J. *J. Chem. Phys.* **1949**, *17*, 303. Flory, P. J. *Principles of Polymer Chemistry*; Cornell University Press: Ithaca, NY, 1953; p 426.

52. Yoon, D.; Flory, P. J. *Macromolecules* **1976**, *9*, 294.

53. Debye, P. *J. Appl. Phys.* **1944**, *15*, 338. Flory, P. J. *Principles of Polymer Chemistry*; Cornell University Press: Ithaca, NY, 1969; p 295.

54. Coulton, G., et al. *Proceedings of the International Conference on Deformation, Yield and Fracture of Polymers;* Cambridge, United Kingdom, April 1–4, 1985.

55. Rawiso, M.; Duplessix, R.; Picot, C. *Macromolecules* **1987**, *20*, 630.

56. O'Reilly, J. M.; Teegarden, D. M.; Wignall, G. D. *Macromolecules* **1985**, *18*, 2747.

57. Yoon, D. Y.; Flory, P. J. *Polymer* **1975**, *16*, 645.

58. Hayashi, H.; Flory, P. J.; Wignall, G. D. *Macromolecules* **1983**, *16*, 1328.

59. Yoon, D. Y.; Flory, P. J. *Polym. Bull.* **1981**, *4*, 692. Flory, P. J. *Pure Appl. Chem.* **1984**, *56*, 305.

60. Schelten, J.; Ballard, D. G. H.; Wignall, G. D.; Longman, G.; Schmatz, W. *Polymer* **1976**, *27*, 751.

61. Wignall, G. D.; Bates, F. S. *J. Appl. Cryst.* **1987**, *20*, 28.

62. Goyal, P. S.; King, J. S.; Summerfield, G. C. *Polymer* **1983**, *24*, 131.

63. Schelten, J.; Schmatz, W. *J. Appl. Cryst.* **1980**, *13*, 385.

64. Fischer, E. W.; Wendorff, J. H.; Dettenmaier, M.; Lieser, G.; Voigt-Martin, I. *J. Macromol. Sci. Phys.* **1976**, *B12*(1), 41.

65. Patterson, G. D. *J. Macromol. Sci. Phys.* **1976**, *B12*(1), 61.

66. Uhlmann, D. R.; Renninger, A. L.; Kritvchevsky, G.; Vander Sande, J. *J. Macromol. Sci. Phys.* **1976**, *B12*(1), 153.

67. Klein, J.; Briscoe, B. J. *Proc. R. Soc. London A* **1979**, *365*, 53.

68. Bartels, C. R.; Crist, B.; Graessley, W. W. *J. Polym. Sci. Polym. Lett. Ed.* **1983**, *21*, 495; *Macromolecules* **1984**, *17*, 2702.

69. Bartels, C.; Crist, B.; Fetters, L. J.; Graessley, W. W. *Macromolecules* **1986**, *19*, 785.

70. Ullman, R. *J. Chem. Phys.* **1979**, *71*(1), 436.

71. (a) Ullman, R. *Macromolecules* **1982**, *15*, 582; (b) *Macromolecules* **1982**, *15*, 1395.

72. Ullman, R. In *Elastomers and Rubber Elasticity*; Mark, J. E.; Lal, J., Eds.; ACS Symposium Series 193; American Chemical Society: Washington, DC, 1982; Chapter 13.

73. Richards, R. W. *J. Macromol. Sci.* **1989**, *A26*, 787.

74. Higgins, J. S.; Maconnachie, A. In *Neutron Scattering*; Skold, K.; Price, D. L., Eds.; Academic: New York, 1987; Chapter 22, p 287.

75. Picot, C., et al. *Macromolecules* **1977**, *10*, 436.

76. Boué, F.; Nierlich, M.; Jannink, G.; Ball, R. *J. Physique* **1982**, *43*, 137. Boué, F. *Advances in Polymer Science*; Springer-Verlag: Berlin, 1987; Vol. 82, p 48.

77. Hadziioannou, G.; Wang, L.; Stein, R. S.; Porter, R. S.; *Macromolecules* **1982**, *15*, 800.

78. Beltzung, M.; Picot, C.; Rempp, P.; Herz, J. *Macromolecules* **1982**, *15*, 1594.

79. Fernandez, A. M.; Sperling, L. H.; Wignall, G. D. *Macromolecules* **1986**, *19*, 2572.

80. Davidson, N. S.; Richards, R. W. *Macromolecules* **1986**, *19*, 2576.

81. Wall, T. F. *J. Chem. Phys.* **1943**, *11*, 153.

82. James, H. M.; Guth, E. *J. Chem. Phys.* **1947**, *15*, 669; **1953**, *21*, 1039.

83. Flory, P. J. *Macromolecules* **1979**, *12*, 119. Flory, P. J.; Erman, B. *Macromolecules* **1982**, *15*, 800, 806; *J. Polym. Sci. Polym. Phys. Ed.* **1984**, *22*, 49.

84. Bastide, J.; Duplessix, R.; Picot, C.; Candeau, S. *Macromolecules* **1984**, *17*, 83.

85. Flory, P. J. *J. Am. Chem. Soc.* **1962,** *84,* 2857.
86. Lieser, G.; Fischer, E. W.; Ibel, K. *J. Polym. Sci.* **1975,** *13,* 29.
87. Ballard, D. G. H.; Cheshire, P.; Longman, G. W.; Schelten, J. *Polymer* **1978,** *19,* 379.
88. Guenet, J. M. *Polymer* **1981,** *22,* 313.
89. Gilmer, J. W.; Wiswe, D.; Zachmann, H. G.; Kugler, J.; Fischer, E. W. *Polymer* **1986,** *27,* 1391.
90. McAlea, K. P.; Schultz, J. M.; Gardner, K. H.; Wignall, G. D. *Macromolecules* **1985,** *18,* 447.
91. Sadler, D. M.; Keller, A. *Science (Washington, D.C.)* **1979,** *19,* 265.
92. Sadler, D. M.; Keller, A. *Macromolecules* **1977,** *10,* 1128.
93. Summerfield, G. C.; King, J. S.; Ullman, R. *J. Appl. Cryst.* **1978,** *11,* 548.
94. Stamm, M.; Fischer, E. W.; Dettenmaier, M.; Convert, P. *Discuss. Faraday Soc.* **1979,** *68,* 263.
95. Stamm, M.; Schelten, J.; Ballard, D. G. H. *Colloid Polym. Sci.* **1981,** *259,* 286.
96. Yoon, D. Y.; Flory, P. J. *Polym. Bull.* **1981,** *4,* 692. Flory, P. J. *Pure Appl. Chem.* **1984,** *56,* 305.
97. Yoon, D. Y.; Flory, P. J. *Discuss. Faraday Soc.* **1980,** *68,* 288.
98. Hoffman, J. D.; Guttman, C. M.; Dimarzio, E. A. *Discuss. Faraday Soc.* **1979,** *68,* 177.
99. Guttman, G. M.; Hoffman, J. D.; Dimarzio, E. A. *Discuss. Faraday Soc.* **1979,** *68,* 197; *Polymer* **1981,** *22,* 597.
100. Flory, P. J.; Yoon, D. Y. *Nature (London)* **1977,** *272,* 226.
101. Sadler, D. M.; Harris, R. *J. Polym. Sci. Polym. Phys. Ed.* **1982,** *20,* 561.
102. Sadler, D. M. In *The Structure of Crystalline Polymers*; Hall, I., Ed.; Elsevier: London, 1983; p 125.
103. Sadler, D. M.; Spells, S. J. *Polymer* **1984,** *25,* 1219.
104. Stamm, M. *J. Polym. Sci. Polym. Phys. Ed.* **1982,** *20,* 235.
105. Wignall, G. D.; Mandelkern, L.; Edwards, C.; Glotin, M. *J. Polym. Sci. Polym. Phys. Ed.* **1982,** *20,* 245.
106. Fischer, E. W.; Hahn, K.; Kugler, J.; Bom, R. *J. Polym. Sci. Polym. Phys. Ed.* **1984,** *22,* 1491. Fischer, E. W. *Polym. J.* **1985,** *17,* 307.
107. Guenet, J. M. *Polymer* **1980,** *21,* 1385. Sadler, D. M.; Spells, S. J.; Keller, A.; Guenet, J. M. *Polym. Commun.* **1984,** *25,* 290.
108. Spells, S. J.; Sadler, D. M. *Polymer* **1984,** *25,* 739.
109. Gust, M. W.; Mansfield, M. L. *Macromolecules* **1991,** *24,* 276.
110. Frank, F. C. In *Growth and Perfection of Crystals*; Doremus, H. R.; Roberts, B. W.; Tumbeill, D., Eds.; Wiley: New York, 1958; p 529.
111. Frank, F. C. *Discuss. Faraday Soc.* **1979,** *79,* 8.
112. Flory, P. J.; Yoon, D. Y.; Dill, K. A. *Macromolecules* **1984,** *17,* 862.
113. Yoon, D. Y.; Flory, P. J. *Macromolecules* **1984,** *17,* 868.
114. Flory, P. J.; Yoon, D. Y. *Nature (London)* **1977,** *272,* 226.
115. Juska, T.; Harrison, I. R. *Polym. Eng. Rev.* **1982,** *2,* 13; *Polym. Eng. Sci.* **1982,** *22,* 766.
116. Peterlin, A. In *Small-Angle X-ray Scattering*; Brumberger, H., Ed.; Gordon and Breach: New York, 1967; p 145.
117. Meinel, G.; Peterlin, A. *J. Polym. Sci. A-2* **1971,** *9,* 67.
118. Wignall, G. D.; Wu, W. *Polym. Commun.* **1983,** *24,* 354. Wu, W.; Wignall, G. D. *Polymer* **1985,** *26,* 661.
119. Schelten, J.; Wignall, G. D.; Ballard, D. G. H.; Longman, G. W. *Polymer* **1977,** *18,* 1111.
120. Summerfield, G. C.; King, J. S.; Ullman, R. *Macromolecules* **1978,** *11,* 218.
121. Wu, W.; Wignall, G. D.; Mandelkern, L. *Polymer* **1992,** *33,* 4137.
122. Sadler, D. M.; Barham, P. J. *Polymer* **1990,** *31,* 36.
123. Sadler, D. M.; Barham, P. J. *Polymer* **1990,** *31,* 43.
124. Sadler, D. M.; Barham, P. J. *Polymer* **1990,** *31,* 46.

125. Chuah, H. H.; Porter, R. S.; Lin, J. S. *Macromolecules* **1986,** *19,* 2732.
126. Ballard, D. G. H.; Cunningham, A.; Schelten, J. *Polymer* **1977,** *18,* 250.
127. Wu, W.; Zachmann, H. G.; Rickel, C. *Polym. Commun.* **1984,** *25,* 76.
128. Koch, T.; Strobl, G. R. *J. Polym. Sci. Polym. Phys. Ed.* **1990,** *28,* 343.
129. Kirste, R. G.; Lehnen, B. R. *Makromol. Chem.* **1976,** *177,* 1137. Kruse, W. A.; Kirste, R. G.; Haas, J.; Schmidt, J. B.; Stein, D. J. *J. Makromol. Chem.* **1976,** *177,* 1145.
130. Ballard, D. G. H.; Rayner, M.; Schelten, J. *Polymer* **1976,** *17,* 349.
131. Kambour, R. P.; Bopp, R. C.; Maconnachie, A.; MacKnight, W. J. *Polymer* **1980,** *21,* 133.
132. Wignall, G. D.; Child, H. R.; Li-Aravena, F. *Polymer* **1980,** *21,* 131.
133. Schmidt, B. J.; Kirste, R. G.; Jelenic, J. *Macromol. Chem.* **1980,** *181,* 1665.
134. Hadziioannou, G.; Stein, R. S. *Macromolecules* **1984,** *17,* 567. Stein, R. S.; Hadziioannou, G. *Macromolecules* **1984,** *17,* 1059.
135. Zimm, B. H. *J. Chem. Phys.* **1948,** *16,* 157.
136. Warner, M.; Higgins, J. S.; Carter, A. J. *Macromolecules* **1983,** *16,* 1931.
137. Hadziioannou, G.; Gilmer, J.; Stein, R. S. *Polym. Bull.* **1983,** *9,* 563.
138. DeGennes, P. G. In *Scaling Concepts in Polymer Physics;* Cornell University Press: Ithaca, NY, 1979; Chapter 5.
139. Herkt-Maetzky, C.; Schelten, J. *Phys. Rev. Lett.* **1983,** *51,* 896.
140. Higgins, J. S.; Walsh, D. J. *Polym. Eng. Sci.* **1984,** *24,* 555. Walsh, D. J.; Higgins, J. S.; Zhikuan, C. *Polym. Commun.* **1982,** *23,* 336.
141. Brereton, M. G.; Fischer, E. W.; Herkt-Maetzky, C. *J. Chem. Phys.* **1987,** *87,* 6078.
142. Schwahn, D.; Mortensen, K.; Springer, T.; Madeira, H. Yee; Thomas, R. *J. Chem. Phys.* **1987,** *87,* 6078.
143. Han, C. C.; Bauer, B. J.; Clark, J. C.; Muroga, Y.; Matsuchita, Y.; Okada, M.; Tran-cong, Q.; Chang, T.; Sanchez, I. *Polymer* **1990,** *29,* 2002.
144. Sakurai, S.; Hasegawa, H.; Hashimoto, T.; Hargis, I. G.; Aggarwal, S. L.; Han, C. C. *Macromolecules* **1990,** *23,* 451.
145. Lohse, D. J. *Polym. Eng. Sci.* **1986,** *26,* 1500.
146. Maconnachie, A.; Kambour, R. P.; White, D. M.; Rostani, S.; Walsh, D. J.; *Macromolecules* **1985,** *17,* 2645.
147. Jelenic, J.; Kirste, R. G.; Oberthur, R. C.; Schmitt-Strecker, S. *Makromol. Chem.* **1984,** *185,* 129.
148. Yang, H.; O'Reilly, J. M. In *Scattering, Deformation and Fracture in Polymers;* Wignall, G. D.; Crist, B.; Russell, T. P.; Thomas, E. L., Eds.; Materials Research Society: Pittsburgh, PA, 1987; Vol. 79, p 129.
149. Shibayama, M.; Yang, H.; Stein, R. S.; Han, C. C. *Macromolecules* **1985,** *18,* 2179.
150. Russell, T. P.; Ito, H.; Wignall, G. D. *Macromolecules* **1987,** *20,* 2213; **1988,** *21,* 1703.
151. Lefebvre, J. M. R.; Porter, S.; Wignall, G. D. *Polym. Eng. Sci.* **1987,** *27,* 433.
152. Schwahn, D.; Mortensen, K.; Yee-Madeira, H. *Phys. Rev. Lett.* **1987,** *58,* 1544.
153. Bates, F. S.; Rosedale, J. H.; Stepanek, P.; Lodge, T. P.; Wiltzius, P.; Fredrickson,G. H.; Hjelm, R. P. *Phys. Rev. Lett.* **1990,** *65,* 1893.
154. Koberstein, J. T. *J. Polym. Sci. Polym. Phys. Ed.* **1982,** *20,* 593.
155. Wignall, G. D.; Child, H. R.; Samuels, R. J. *Polymer* **1982,** *23,* 957.
156. Fernandez, A. M.; Sperling, L. H.; Wignall, G. D. *Multicomponent Polymer Materials;* Paul, D. R.; Sperling, L. H., Eds.; ACS Advances in Chemistry 211; American Chemical Society: Washington, DC, 1985.
157. Bates, F. S.; Berney, C. V.; Cohen, R. E.; Wignall, G. D. *Polymer* **1983,** *24,* 519.
158. Debye, P.; Bueche, A. M. *J. Appl. Phys.* **1949,** *20,* 518. Debye, P.; Anderson, H. R.; Brumberger, H. *J. Appl. Phys.* **1957,** *28,* 679.

159. Blundell, D. J.; Longman, G. W.; Wignall, G. D.; Bowden, M. *Polymer* **1974,** *15,* 33.

160. Leibler, L. *Macromolecules* **1980,** *13,* 1602.

161. Fredrickson, G. H.; Helfand, E. *J. Chem. Phys.* **1987,** *87,* 697.

162. Helfand, E. *Macromolecules* **1975,** *8,* 552. Helfand, E.; Wasserman, Z. R. *Macromolecules* **1976,** *9,* 879.

163. Semenov, A. N. *Zh. Eksp. Teor. Fiz.* **1985,** *88,* 1242 [*Sov. Phys. JETP* **1985,** *61,* 733]; *Macromolecules* **1989,** *22,* 2849.

164. Bates, F. S.; Hartney, M. A. *Macromolecules* **1985,** *18,* 2478.

165. Almdal, K.; Rosedale, J. H.; Bates, F. S.; Wignall, G. D.; Fredrickson, G. H. *Phys. Rev. Lett.* **1990,** *65,* 1112.

166. Barrat, J. L.; Fredrickson, G. H. *J. Chem. Phys.* **1991,** *95,* 1281. Malenkovitz, M.; Muthukumar, M. *Macromolecules* **1991,** *24,* 4199. Tang, H.; Freed, K. F. *Bull. Am. Phys. Soc.* **1992,** *37*(1), 367.

167. Olvera de la Cruz, M. *Phys. Rev. Lett.* **1991,** *67,* 85. Mayes, A. M.; Olvera de la Cruz, M.; Swift, B. W. *Macromolecules* **1992,** *25,* 944.

168. Grancio, M. P.; Williams, D. J. *J. Polym. Sci. A1* **1970,** *8,* 2617.

169. Hughes, L. J.; Brown, G. L. *J. Appl. Polym. Sci.* **1961,** *5,* 580. Dickie, R. A. *J. Appl. Polym. Sci.* **1973,** *17,* 65, 79.

170. Wai, M. P.; Gelman, R. A.; Fatica, M. G.; Hoerl, R. H.; Wignall, G. D. *Polymer* **1987,** *28,* 918.

171. Fisher, L. W.; Melpolder, S. M.; O'Reilly, J. M.; Ramakrishnan, V. R.; Wignall, G. D. *J. Colloid Interface Sci.* **1988,** *123,* 24.

172. Goodwin, J. W.; Ottewill, R. H.; Harris, N. M.; Tabony, J. *J. Colloid Polym. Sci.* **1980,** *78,* 253. Goodwin, J. W. *J. Colloid Interface Sci.* **1980,** *78,* 253.

173. Alexander, K.; Cebula, D. J.; Goodwin, J. W.; Ottewill, R. H.; Parentich, A. *Colloid Surf.* **1983,** *7,* 233.

174. Cebula, D. J.; Goodwin, J. W.; Ottewill, R. H.; Jenkin, G.; Tabony, J. *Colloid Polym. Sci.* **1983,** *261,* 555; *Discuss. Faraday Soc.* **1983,** *76,* 37.

175. Higgins, J. S.; Dawkins, J. V.; Taylor, G. *Polymer* **1980,** *21,* 627.

176. Lord Rayleigh. *Proc. R. Soc. London A* **1911,** *84,* 24.

177. Pecora, R.; Aragon, S. R. *Chem. Phys. Lipids* **1974,** *13,* 1.

178. Wignall, G. D.; Christen, D. K.; Ramakrishnan, V. R. *J. Appl. Cryst.* **1988,** *21,* 438.

179. Glatter, O. *Acta Phys. Austriaca* **1977,** *47,* 83. Glatter, O. In *Small-Angle X-ray Scattering;* Glatter, O.; Kratky, O., Eds.; Academic: London, 1982, Chapter 4.

180. Moore, P. B. *J. Appl. Cryst.* **1980,** *13,* 168.

181. Wignall, G. D.; Ramakrishnan, V. R.; Linne, M. A.; Klein, A.; Sperling, L. H.; Wai, M. P.; Gelman, R. A.; Fatica, M. G.; Hoerl, R. H.; Fisher, L. W.; Melpolder, S. M.; O'Reilly, J. M. *Mol. Cryst. Liq. Cryst.* **1990,** *180A,* 25.

182. Hayter, J. B.; Penfold, J. *Colloid Polym. Sci.* **1983,** *261,* 1022.

183. Hayter, J. B. In *Proceedings of Enrico Fermi School of Physics Course XC;* Degiorgio, V.; Corti, M., Eds.; North Holland: Amsterdam, 1985; Vol. 59.

184. Linne, M. A.; Klein, A.; Sperling, L. H.; Wignall, G. D. *J. Macromol. Sci. Phys.* **1988,** *B27,* 181.

185. Linne, M. A.; Klein, A.; Sperling, L. H.; Wignall, G. D. *J. Macromol. Sci. Phys.* **1988,** *B27,* 217.

186. Linne, M. A.; Klein, A.; Sperling, L. H. *ANTEC Plastics* **1985,** *85,* 303.

187. Summerfield, G. C.; Ullman, R. *Macromolecules* **1987,** *20,* 401. Summerfield, G. C.; Ullman, R. *Macromolecules* **1988,** *21,* 2643.

188. Anderson, J. E.; Jou, J. H. *Macromolecules* **1987,** *20,* 1544.

189. Hahn, K.; Ley, G.; Oberthur, R. *Colloid Polym. Sci.* **1988,** *266,* 631.

190. Wignall, G. D.; Schelten, J.; Ballard, D. G. H. *J. Macromol. Sci.* **1976,** *B12,* 75. Schelten, J.; Wignall, G. D.; Ballard, D. G. H.; Longman, G. W. *Colloid Polym. Sci.* **1974,** *252,* 749.

191. Strazielle, C.; Benoit, H. *Macromolecules* **1975,** *8,* 203.

192. Yang, H.; Hadziioannou, G.; Stein, R. S. *J. Polym. Sci. Polym. Phys. Ed.* **1983,** *21,* 159.

193. Buckingham, A. B.; Hentschel, H. G. E. *J. Polym. Sci. Polym. Phys. Ed.* **1984,** *18,* 853.

194. Bates, F. S.; Wignall, G. D.; Koehler, W. C. *Phys. Rev. Lett.* **1985,** *55,* 2425.

195. Bates, F. S.; Wignall, G. D. *Macromolecules* **1986,** *19,* 932.

196. Bates, F. S.; Wignall, G. D. *Phys. Rev. Lett.* **1986,** *57,* 1429.

197. Wignall, G. D.; Bates, F. S. *Makromol. Chem.* **1988,** *15,* 105.

198. Bates, F. S.; Muthukumar, M.; Wignall, G. D.; Fetters, L. J.; *Macromolecules* **1988,** *89,* 535.

199. Lapp, A.; Picot, C.; Benoit, H. *Macromolecules* **1985,** *18,* 2437.

200. Bates, F. S.; Dierker, S. B.; Wignall, G. D. *Macromolecules* **1986,** *19,* 1938.

201. Fernandez, A. M.; Widmaier, J. M.; Sperling, L. H.; Wignall, G. D. *Polymer* **1984,** *25,* 1718.

202. Weissman, J. G.; Sperling, L. H. *Macromolecules* **1985,** *18,* 1720.

203. Lohse, D. J. *Polym. News* **1986,** *12,* 8.

204. Ullman, R. *Polym. News* **1987,** *13,* 42.

205. *Neutron, X-Ray and Light Scattering*; Lindner, P.; Zemb, T., Eds.; North-Holland Delta Series; Elsevier: New York, 1991.

206. Moon, R. M.; West, C. D. *Physica* **1989,** *156B,* 522.

Index

Subject Index

A

Acetone, diffusion into poly(methyl methacrylate), 302

Acrylates, effect of side-chain length on glass transition temperature, 77t

Acrylonitrile–methyl methacrylate copolymer, NMR spectra, 275f

Adiabatic retraction, elastomers, 7

Adjacent folding, 344, 345

Adjacent reentry, 177, 178

Affine deformation, 9, 340

Affine model, 11–14

Aligned nematic phase, definition, 217

Alignment of liquid crystals by external fields, 225–226

n-Alkanes, crystallites, 148f

p-Alkylstyrenes, effect of side-chain length on glass transition temperature, 77t

Alpha transition, 189–191

Alternating copolymers, melting temperature, 151

Amorphous–crystalline interface, study by SAXS and SANS, 356, 357, 358f

Amorphous materials, study by radial distribution function, 314

Amorphous mixture, SANS results, 370f

Amorphous polymers
molecular dimensions, 334t
study by scattering techniques, 332–337
techniques for study, 336–337

Amorphous state, polymers, 146

Anisotropic properties, 220–223

Anisotropy
block copolymers, 241–242
macroscopic, 207, 241–242
side-chain polymer liquid crystal, 236
See also Macroscopic anisotropy

Annealing, effect on molecular weight determined by SANS, 349, 351f

Average molecular weights, types, 124

B

Band formation, 254

Barn, unit for cross section, 318

Basal plane interfacial energy, 178, 180

Beta relaxation, effect of branching, 190f

Beta transition
effect of counit composition, 193
polyethylene, 191, 192t
pseudo-glass transition temperature, 193
relation to interfacial content, 192

Beta transition temperature, factors, 192

Biased-chain trajectory, block copolymers, 242

Biaxial extension, 44–45, 46f

Bimodal networks, 32–33, 36–44

Bimodality, effect on strain-induced crystallization, 42

Bioelastomers, 47–48

Related Titles

Polymer Characterization: Physical Property, Spectroscopic, and Chromatographic Methods
Edited by Clara D. Craver and Theodore Provder
Advances in Chemistry Series 227; 511 pages; ISBN 0–8412–1651–7

Structure–Property Relations in Polymers
Edited by Marek W. Urban and Clara D. Craver
Advances in Chemistry Series 236; 832 pages; ISBN 0–8412–2525–7

Particle Size Distribution II: Assessment and Characterization
Edited by Theodore Provder
ACS Symposium Series 472; 407 pages; ISBN 0–8412–2117–0

Radiation Effects on Polymers
Edited by Roger L. Clough and Shalaby W. Shalaby
ACS Symposium Series 475; 633 pages; ISBN 0–8412–2165–0

Polymer Latexes: Preparation, Characterization, and Applications
Edited by Eric S. Daniels, E. David Sudol, and Mohamed S. El-Aasser
ACS Symposium Series 492; 462 pages; ISBN 0–8412–2305–X

Catalysis in Polymer Synthesis
Edited by Edwin J. Vandenberg and Joseph C. Salamone
ACS Symposium Series 496; 291 pages; ISBN 0–8412–2456–0

Colloid–Polymer Interactions: Particulate, Amphiphilic, and Biological Surfaces
Edited by Paul Dubin and Penger Tong
ACS Symposium Series 532; 290 pages; ISBN 0–8412–2696–2

Polymeric Delivery Systems: Properties and Applications
Edited by Magda A. El-Nokaly, David M. Piatt, and Bonnie A. Charpentier
ACS Symposium Series 520; ISBN 0–8412–2624–5

Irradiation of Polymeric Materials: Processes, Mechanisms, and Applications
Edited by Elsa Reichmanis, Curtis W. Frank, and James H. O'Donnell
ACS Symposium Series 527; 338 pages; ISBN 0–8412–2662–8

Polymeric Materials: Chemistry for the Future
By Joseph Alper and Gordon L. Nelson
110 pages; clothbound, ISBN 0–8412–1622–3;
paperback, ISBN 0–8412–1613–4

For further information and a free catalog of ACS books, contact
American Chemical Society
Distribution Office, Department 225
1155 16th Street, N.W., Washington, DC 20036
Telephone 800–227–5558